SOIL MICROBIOLOGY

(Fourth Edition of Soil Microorganisms and Plant Growth)

Prof. N.S. SUBBA RAO Ph.D., F.N.A., F.A.A.S.

Formerly Project Director and Head,
Division of Microbiology
Indian Agricultural Research Institute, New Delhi

Oxford & IBH Publishing Co. Pvt. Ltd.

New Delhi

(A Unit of CBS Publishers & Distributors Pvt Ltd)

CBSPD

CBS Publishers & Distributors Pvt Ltd

New Delhi • Bengaluru • Chennai • Kochi • Kolkata • Lucknow • Mumbai
Hyderabad • Jharkhand • Nagpur • Patna • Pune • Uttarakhand

Soil Microbiology

Fourth Edition

ISBN-13: 978-81-204-1383-2
ISBN-10: 81-204-1383-0

OXFORD & IBH
New Delhi
(A Unit of CBS Publishers & Distributors Pvt Ltd)

Published by **Satish Kumar Jain** and produced by **Varun Jain** for

CBS Publishers & Distributors Pvt Ltd
4819/XI Prahlad Street, 24 Ansari Road, Daryaganj, New Delhi 110 002, India.
Ph: 011-23289259, 23266861   Website: www.cbspd.com
                                    e-mail: delhi@cbspd.com

Corporate Office: 204 FIE, Industrial Area, Patparganj, Delhi 110 092
Ph: 011-4934 4934   Fax: 011-4934 4935
                          e-mail: publishing@cbspd.com; publicity@cbspd.com

Branches

- **Bengaluru:** Seema House 2975, 17th Cross, KR Road, Banasankari 2nd Stage, Bengaluru 560 070, Karnataka, India
  Ph: +91-80-26771678/79   Fax: +91-80-26771680   e-mail: bangalore@cbspd.com
- **Chennai:** 7, Subbaraya Street, Shenoy Nagar, Chennai 600 030, Tamil Nadu, India
  Ph: +91-44-26680620, 26681266   Fax: +91-44-42032115   e-mail: chennai@cbspd.com
- **Kochi:** 42/1325, 1326, Power House Road, Opp KSEB, Power House, Ernakulum Kochi 682 018, Kerala, India
  Ph: +91-484-4059061-65,67   Fax: +91-484-4059065   e-mail: kochi@cbspd.com
- **Kolkata:** 147, Hind Ceramics Compound, 1st Floor, Nilgunj Road, Belghoria, Kolkata-700056, West Bengal, India
  Ph: +033-25633055, 033-25633056   e-mail: kolkata@cbspd.com
- **Lucknow:** Basement, Khushnuma Complex, 7 Meerabai Marg (Behind Jawahar Bhawan), Lucknow-226001, UP, India
  Ph: +0522-4000032   e-mail: tiwari.lucknow@cbspd.com
- **Mumbai:** PWD Shed, Gala no 25/26, Ramchandra Bhatt Marg, Next to JJ Hospital Gate no. 2, Opp. Union Bank of India, Noorbaug, Mumbai-400009, Maharashtra, India
  Ph: 022-66661880/89   e-mail: mumbai@cbspd.com

Representatives

• Hyderabad	0-9885175004	• Jharkhand	0-9811541605	• Nagpur	0-8692091830
• Patna	0-9334159340	• Pune	0-9664372571	• Uttarakhand	0-9716462459

Printed at Chaman Enterprises, Daryaganj, New Delhi, India

Foreword

Among the basic assets of agriculture, soil is a vital component. Unfortunately however, integrated knowledge of different aspects of the soil-physics, chemistry, biology and topography is not very widespread. Consequently, we see in our country extensive abuse of the soil. While it may take a century to make one centimetre of top soil, all this work of nature can be destroyed in no time either because of erosion arising from improper management or other activities like removing good soil for making bricks. The care and maintenance of soil health is, therefore, essential for safeguarding the future of our agriculture.

Soil microorganisms play a pivotal role both in the evolution of agriculturally useful soil conditions and in stimulating plant growth. Dr. N.S. Subba Rao is to be congratulated on writing this exceedingly informative book on interaction between soil microorganisms and plant growth. In spite of the progress made in the production of chemical fertilizers, the following was the global situation in 1975 with reference to sources of nitrogen for crop production:

Origin	Millions of metric tonnes
Industrial fixation (fertilizer)	42
Biological fixation—total	175
Agriculutral soil	90
Crop legumes	40
Crop non-legumes	9
Meadows and graslands	45
Lightning	10
Combustion	20
Ozonisation	15

The importance of biological sources of nitrogen is hence obvious. In the tropics and sub-tropics, the potential for harnessing biological sources of nitrogen is particularly great. Although it is difficult to obtain definite values for the amount of nitrogen fixed by photosynthetic organisms in submerged rice fields because of many variables, estimates as high as 22–36 kg per hectare have been reported. In fact, in some recent experiments at the International Rice Research Institute, it has been indicated that free living bacteria may fix up to 50–60 kg per hectare. Factors like this may account for farmers obtaining a minimum yield of over one tonne per hectare even without any fertilization.

Dr. Subba Rao has dealt with in a masterly and authoritative way the different aspects of soil microorganisms, nitrogen fixation, phosphorus nutritions, trace element needs and the entire process of nitrification and de-nitrification. He has also dealt with interaction among pesticides, microorganisms and plants. This comprehensive treatment would enable students and others interested in this fascinating area of research to make an impact analysis' of the change introduced by the farming system on soil microorganisms.

In India, the only pathway of improved production available to us is increased productivity per unit of soil, water and time. To do this, we should understand all aspects of the building blocks of agriculture. This book is hence timely. Dr. Subba Rao deserves the gratitude of everyone interested in the improvement of biological productivity.

Dr. M.S. SWAMINATHAN, F.R.S.
Director-General,
Indian Council of Agricultural Research
and Secretary to the Government of India

Preface to the Fourth Edition

In this edition, the chapters on Pesticides and Biotechnology in Agriculture have been redone and made up-to-date. Similarly, additions have been incorporated in other chapters so as to render the contents in tune with recent developments. The title of the book has been changed to **Soil Microbiology** from the original title of **Soil Microorganisms and Plant Growth**. It is hoped that the new edition will prove useful to students and teachers dealing with Agriculture in general and Soil Science, Agricultural Microbiology and Environmental Science in particular because the book has been made comprehensive on all aspects dealing with issues relating to soil fertility, soil health and plant growth in relation to microbial activity.

1999

N.S. SUBBARAO
'SUMUKHA', 452, Eleventh
Main Road, R.M.V. Extension
BANGALORE - 560 080, INDIA

Contents

1. Introduction

The Origin of Life

The planet earth came into being about 4.5 billion years ago with a primeval atmosphere that was hostile to life. There were no continents as we see them today and the earth was one mass of archipelagoes of volcanic rock. Between the sun and the surface of the rocks, the ozone shield as we have in the present day atmosphere was lacking in the oxygen devoid primeval atmosphere and hence lethal U-V rays of the sun bathed the barren rocks and left them sterile. The composition of the primeval atmosphere has been debated but the primitive sea seems to have had plenty of dissolved ammonia, carbon dioxide, and abundant deposits of sulphates and iron.

Prior to the origin of life, it looks as if the precursors to living single celled microorganisms may have arisen either in a ball of ice or in a pond or in a cauldron (hot spring). It has been conjectured that ice caps of the ocean prevented the lethal U-V rays from entering the ocean and primitive organic chemicals arose in pockets of water. Interplanetary debris, hydrothermal vents and atmospheric reactions could have provided such compounds as formaldehyde, cyanide and ammonia. These then could have combined in water within a lattice of ice resulting in the formation of simple amino acids like lysine. A huge meteorite impact could have ultimately thawed the frozen world. Glaciers, volcanoes, geysers, individually or in combination with interplanetary debris may have been collected in a pond or small body of water. This elemental water may have concentrated on the internal surfaces of sheet-like minerals which attract certain molecules and serve as a catalysts in the subsequent reactions. Two aldehyde phosphate molecules may have combined to form a sugar phosphate and possibly ribonucleic acid (RNA). Before DNA and proteins evolved, life may have consisted solely of RNA molecules floating in water replicating, mutating and undergoing natural selection of their own. When the planet earth was still sizzling, gases and vital compounds were released from the molten magma. It is not unlikely that resting on a stabilising surface of pyrite, carbon monoxide and a methyl group may have combined towards the formation of activated acetic acid, a crucial chemical for synthesising other organic compounds. Asteroids, comets, meteorites and even specks of interplanetary dust may have helped in the formation of a 'primordial soup', the organic base of life. In fact, modern experiments under conditions simulating the primeval atmosphere have borne ample testimony to support the primodial soup hypothesis.

Since organic matter was virtually absent, it has been conjectured that chemosynthetic microorganisms were the pioneer colonizers of the primitive earth. Logically, sulphate reducing and iron oxidizing bacteria which utilized the sulphate and iron deposits of the primitive sea might have been the first organisms to get enriched and multiply. Fossil evidences seem to support such a contention. The earliest evidence can be traced as far back as precambrian period in what are known as 'stromatolites' which can still be seen in nutrient poor environments such as Lake Clifton along Australia's southwestern coast. These precambrian stones have a spongy outer layer made of oxygen loving cyanobacterial filaments and inner layer populated by bacteria that thrive in the absence of oxygen deriving energy by anaerobic metabolism of iron and sulphur. Rod-shaped bacteria-like organisms have been encountered as fossils in the border region between the Republic of South Africa and Swaziland which date back to 3.2 billion years. Similar bacterial deposits resembling iron bacteria have been encountered from the 'Gunflint cherts' from Ontario which are estimated to be two billion years old. The Gunflint flora also show fossils of algae which might have been both photosynthetic and nitrogen fixing. Thus, mineral decomposing organisms may have given rise to photosynthetic organisms (bacteria and algae), which in turn, provided the basic organic materials for diverse groups of heterotrophic bacteria and filamentous fungi to grow and evolve. At this juncture, the primitive atmosphere became increasingly oxygenic due to photosynthetic microorganisms and the earth's ozone shield started emerging gradually. The environment thus became somewhat hospitable to more microbial life.

At first the primitive microorganisms did not multiply systematically in the absence of mitosis or possess specialized structures such as mitochondria. Such prokaryotic cells gave rise to eukaryotic cells (about 1.2 to 1.4 billion years ago) which were capable of not only systematic division (mitosis) but also possessed enzyme activity by virtue of having well-developed mitochondria. Gradually, passive existence among microorganisms gave rise to symbiosis between microalgae and filamentous fungi resulting in the formation of lichens, which as we know have stood the test of time as pioneer colonizers of rocks in the process of soil formation.

The Origin of Land Plants

The oxygen content of the universe reached a threshold in the beginning of the Cambrian and Carboniferous periods. The first lowland plants may have appeared about 400 million years ago in the Upper Silurian rocks. They had rudimentary root and vascular systems. Upland plants appeared during the Carboniferous and the Permian periods. They were dominant throughout the Paleozoic era and were huge in size resulting in the accumulation of biomass. The formation of soil may have occurred in the Devonian period when microorganisms would have penetrated deeper into soil and come in contact with deep-seated root system in the late Carboniferous age. The upland plants began to increase in number in the Triassic age and angiosperms (flowering plants) appeared in the Cretaceous period (about 100 million years ago). The formation of organic soil which began in the Jurassic age, became extensive in the Cretaceous period together with the formation of soil profiles. With the extensive accumulation of dead plant biomass, saprophytic microorganisms flourished, the rhizosphere and mycorrhizal relations developed and nodule-bearing angiosperms became established side by side with the evolution of land plants. The origin of life and the biological events which took place on earth since its inception have been chronologically summarized in Fig. 1.

The Beginnings of Agriculture

The earliest hominids or man-like beings first appeared approximately one million years ago and the modern species of man, *Homo sapiens* evolved from his ancestors about half a million years ago. For a long time, man lived hunting, fishing and gathering food from nature (random harvest) until he learned the art of growing plants for food not so long ago (by about 8000 B.C.), which marked the beginnings of agriculture. When man wanted to produce his food intensively instead of merely collecting it from places where it occurred naturally, he began to clear jungles, sow seeds and raise a good harvest. Sooner or later, it was realized that the same soil cannot endlessly sustain plant growth and good productivity. Therefore, new areas were colonized and techniques for cultivation of plants were gradually introduced as evidenced by the writings of ancient civilizations of Mesopotamia (about 3000 B.C.), the Nile (about 2000 B.C.), the Indus Valley (about 1500–2000 B.C.), and of the Roman (around 700 B.C.) and the Chinese (around 2000 B.C.) empires.

The Biosphere

The actual estimates of time in the sequence of events described above are not important but the trend is clear enough to state that solar energy initiated photosynthetic life in the primordial earth as a consequence of which, earth's primitive atmosphere changed into an oxygenated condition. Since then, evolution of life has progressed on divergent lines and the elements, carbon, nitrogen, hydrogen and oxygen have undergone a series of recycling processes through assimilation and decomposition,

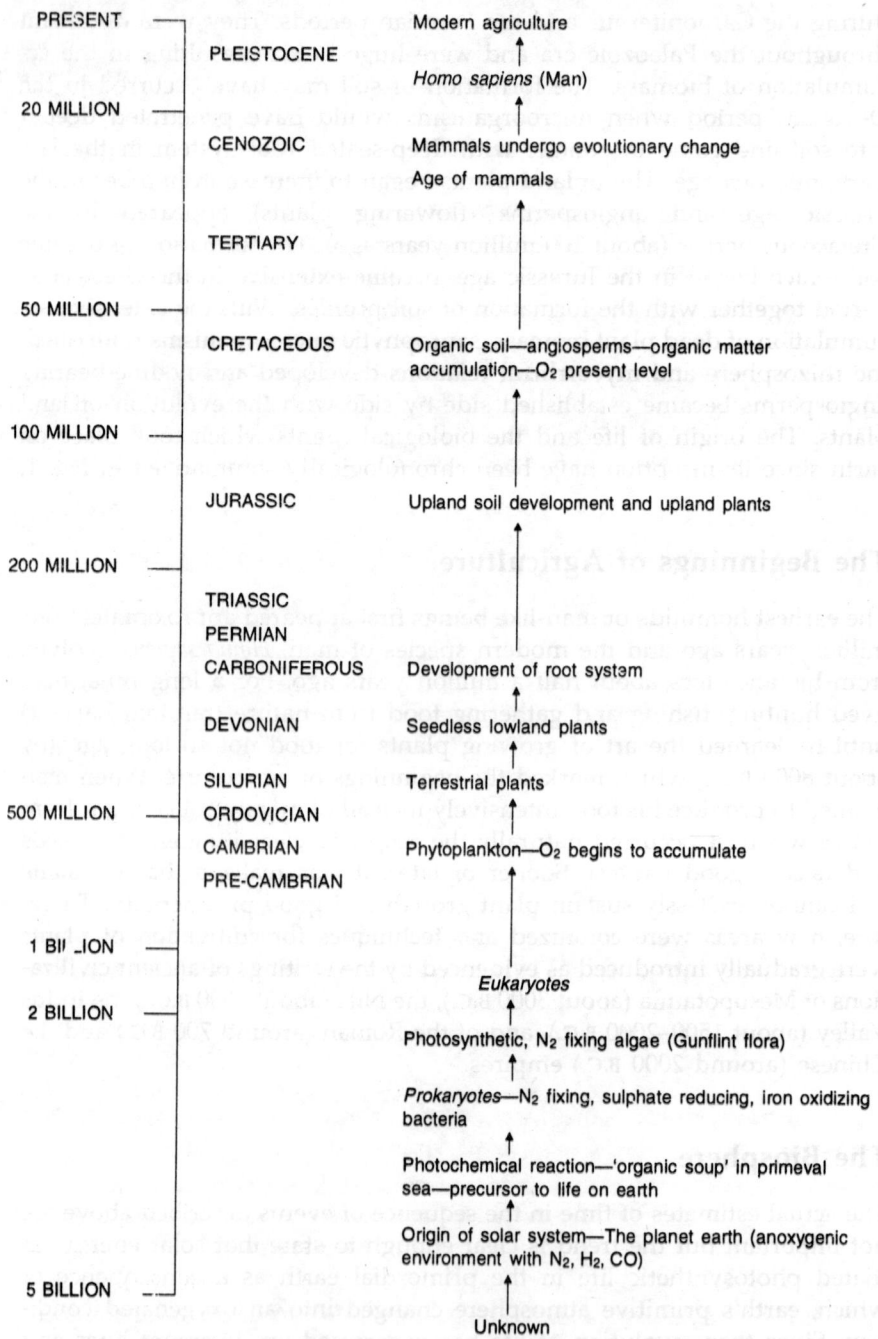

PRESENT	PLEISTOCENE	Modern agriculture
		Homo sapiens (Man)
20 MILLION	CENOZOIC	Mammals undergo evolutionary change
		Age of mammals
	TERTIARY	
50 MILLION	CRETACEOUS	Organic soil—angiosperms—organic matter accumulation—O_2 present level
100 MILLION	JURASSIC	Upland soil development and upland plants
200 MILLION	TRIASSIC	
	PERMIAN	
	CARBONIFEROUS	Development of root system
	DEVONIAN	Seedless lowland plants
	SILURIAN	Terrestrial plants
500 MILLION	ORDOVICIAN	
	CAMBRIAN	Phytoplankton—O_2 begins to accumulate
	PRE-CAMBRIAN	
1 BILLION		
2 BILLION		*Eukaryotes*
		Photosynthetic, N_2 fixing algae (Gunflint flora)
		Prokaryotes—N_2 fixing, sulphate reducing, iron oxidizing bacteria
		Photochemical reaction—'organic soup' in primeval sea—precursor to life on earth
		Origin of solar system—The planet earth (anoxygenic environment with N_2, H_2, CO)
5 BILLION		Unknown

Fig. 1   The chronology of events in the history of the earth with particular reference to origin of life and development of microorganisms and land plants.

thereby creating a 'biosphere' wherein microorganisms, plants, animals and man have lived in equilibrium for centuries. In recent years, attempts have been made to estimate the amounts of carbon and nitrogen transferred within the different ecosystems of the biosphere but the figures for the two elements in the biosphere may, however, be taken only as broad guidelines. Nonetheless, the pathways by which the elements get transferred within different ecosystems have been clearly understood and some of them fall in the purview of plant-soil microorganisms' interactions. Apart from the carbon and nitrogen transformations in the biosphere which are of great magnitude, other elements like oxygen, hydrogen, phosphorus, sulphur and iron undergo mutual transfers in the biosphere between ecosystems and some of these changes are also linked with plants and soil microorganisms.

The Turnover of Carbon

It has been estimated that the atmosphere has about 700 billion tonnes (*t*) of carbon and almost an equal quantity is locked up in the dead organic matter on land. The living plants on land possess about 450 billion *t* of carbon which naturally undergoes interchange with atmospheric carbon. Coal and oil resources, however, account for about 10,000 billion *t* of carbon.

Carbon fixation is the pivotal event in the biosphere and other cycles in nature depend on the extent and magnitude of this process. All organisms on land and sea which possess chlorophyll use solar energy to combine CO_2 and H_2O and form carbohydrates with the release of O_2 into the atmosphere. Part of the carbohydrate is directly consumed by plants for energy and in the process, CO_2 is liberated through roots and leaves. Some of the plant's carbohydrates are eaten by animals which give out CO_2 by respiration. Generations of plants and animals leave behind their remains which are decomposed by microorganisms in soil and in the process, the carbon in their tissues gets converted to CO_2 (Fig. 2).

The Turnover of Nitrogen

Just as carbon is the key element for the synthesis of carbohydrates, so is nitrogen for the synthesis of protein. The atmosphere is a vast reservoir of nitrogen but the N_2 of the atmosphere must necessarily combine with H_2 or O_2 before it can be assimilated by higher plants. The inert atmospheric nitrogen occupies approximately 75% by weight, and 78% by volume of the atmosphere and totals 3.8×10^{15} tonnes. However, the amount of nitrogen in the biosphere is dependent on the rates of transfer within the nitrogen cycle and can be estimated only within broad limits. Some authors have estimated that 400×10^6 *t* N/yr are transferred between the various nitrogen pools, with losses of 215×10^6 *t* N/yr by denitrification and 185×10^6 *t* N/yr by volatilization and combustion. Estimated

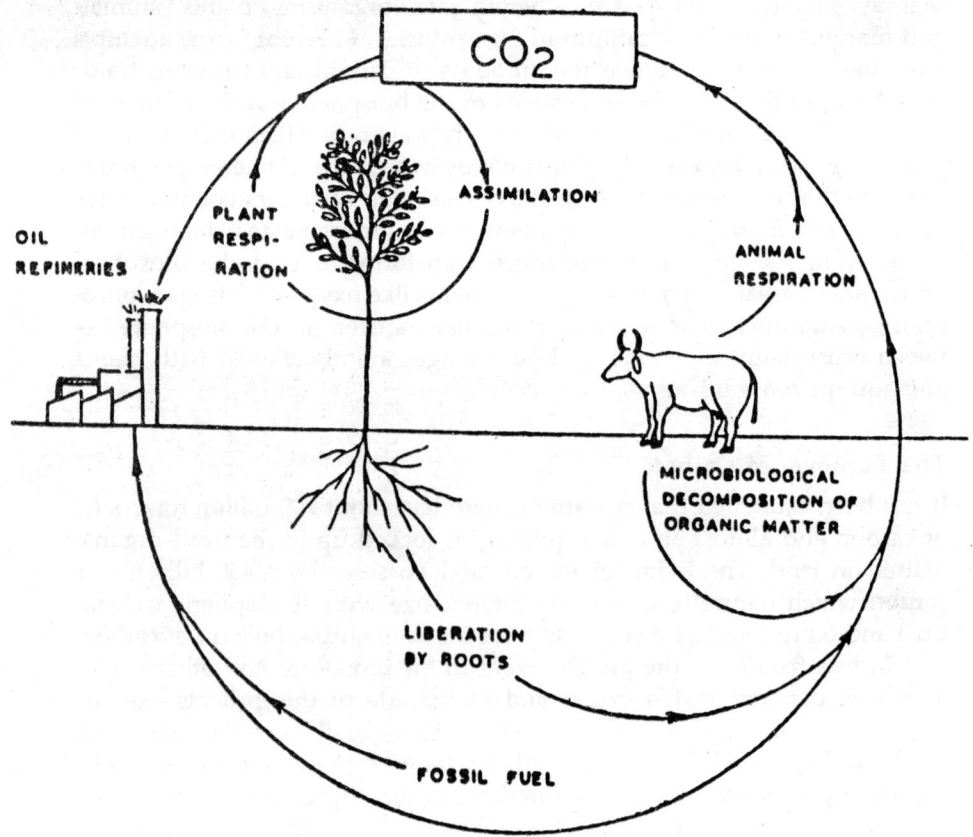

Fig. 2 The carbon cycle in the biosphere.

returns are $200 \times 106 \, t$ N/yr in precipitation (mainly volatilized ammonia), $170 \times 10^6 \, t$ N/yr from biological fixation and $30 \times 10^6 \, t$ N/yr from industrial fixation. These figures differ from those of others who estimate a biological nitrogen fixation rate of $92 \times 10^6 \, t$ N/yr. Nevertheless, these data provide a rough estimate and point out unmistakably that nitrogen fixed by biological means is more than three times that of industrial fixation.

Biological fixation of nitrogen takes place in root nodules of legumes and also on land and water by means of free-living bacteria and microalgae. Of minor importance is the fixation of atmospheric nitrogen through precipitation and lightning. Man has recently intervened in the nitrogen cycle by industrial fixation of N_2 in fertilizer factories where nitrogen and hydrogen are combined at high temperature and pressure with the help of an inorganic catalyst to produce ammonia. Denitrification processes, almost of equal magnitude, breakdown nitrates and return elemental nitrogen into the atmosphere making the cycle complete (Fig. 3).

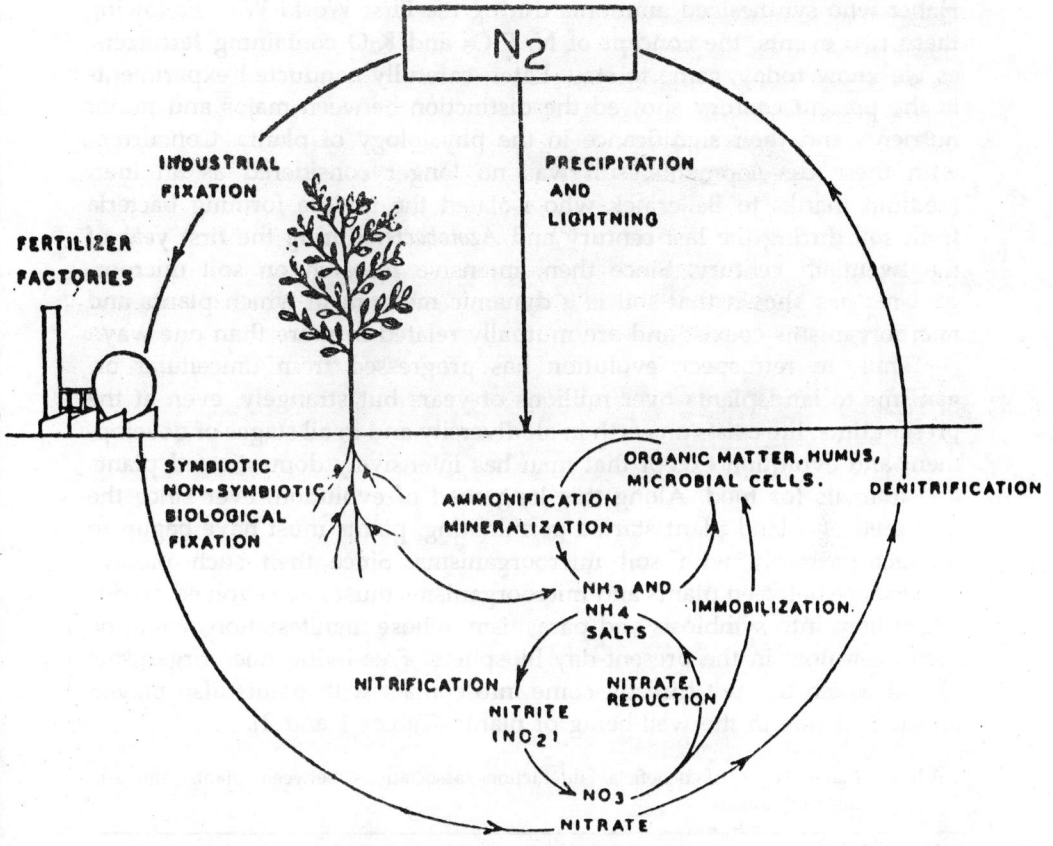

Fig. 3 The nitrogen cycle in the biosphere.

The Beginnings of Soil Science

Much of the earlier knowledge on soil science in relation to plant growth
was either passed on by word of mouth from one generation to the other
or was written in books with loose generalizations lacking experimental
evidences. Plant nutrients were vaguely called as 'principles' in rain water,
in soil and in plant and animal remains until the French scientists Antoine
Lavoisier in 1794 and J.B. Boussingau't in 1834 and the German chemist
Justus Von Liebig in 1840 attempted chemical analysis of plants and soils
and arrived at the conclusion that chemical elements in plants came from
soil and air. At Rothamsted in England, John Bennet Lawes in 1840 and
his associate J.H. Gilbert in 1842 produced superphosphate by chemical
treatment of crushed bones with sulphuric acid. The next major discovery
in the field of artificial fertilizers came from the German chemist Fritz

Haber who synthesized ammonia during the First World War. Following these two events, the concept of N, P_2O_5 and K_2O containing fertilizers, as we know today, came to stay. Later, carefully conducted experiments in the present century showed the distinction between major and minor nutrients and their significance in the physiology of plants. Concurrent with these developments, soil was no longer considered as an inert medium thanks to Beijerinck who isolated the nodule forming bacteria from soil during the last century and *Azotobacter* during the first year of the twentieth century. Since then, intensive research on soil microorganisms has shown that soil is a dynamic medium in which plants and microorganisms coexist and are mutually related in more than one way.

Thus, in retrospect, evolution has progressed from unicellular organisms to land plants over millions of years but strangely, even at the present time, life exists on earth in all diversity and in all stages of development and evolution except that man has intensively domesticated plants and animals for food. Along this long road of evolution, ever since the first seed of a land plant started germinating, plants must have begun to interact passively with soil microorganisms. Since then such passive coexistence between plants and microorganisms must have evolved on different lines into symbiosis and parasitism, whose manifestations could be seen even now in the present-day biosphere. Free-living microorganisms in soil which did not directly come into contact with plants also played an indirect role in the well-being of plants (Tables 1 and 2).

Table 1 Major types of beneficial interactions/associations between plants and soil microorganisms

Nature of interactions/associations	Examples of higher plants involved	Examples of microorganisms involved
Rhizosphere, Rhizoplane and phyllosphere microflora	All plants with roots and leaves	Bacteria, fungi and actinomycetes
Ectomycorrhizae	Forest trees—*Pinus*	Mostly basidiomycetous fungi—*Boletus, Lactarius, Armillaria*
Endomycorrhizae (AM fungal association)	Certain orchids, cereals, grasses and legumes	*Rhizoctonia, Endogone* and *Glomus*
Root nodules of nodulating legumes	Soybean, gram, lucerne etc.	*Rhizobium* spp.
Root nodules of plants other than legumes	*Alnus, Myrica, Casuarina*	Actinomycetous endophytes (*Frankia*)
Leaf nodules	*Psychotria, Pavetta*	*Klebsiella*
Algal associations with higher plants	*Cycas, Zamia, Heterozamia, Gunnera scabra, Azolla*	*Anabaena, Nostoc*
Associative symbiosis	Grasses, sorghum and millets	*Azospirillum*

Table 2 Major microbiological processes in soil by free-living microorganisms which indirectly influence plant growth

Nature of microbial processes	Examples of microorganisms involved
Aerobic decomposition of organic matter (cellulose, lignin, chitin etc.)	*Trichoderma, Fomes, Armillaria, Achromobacter, Nocardia, Streptomyces*
Anaerobic decomposition of organic matter	*Clostridium*, methane bacteria (*Methanobacter* and *Methanococcus*)
Non-symbiotic nitrogen fixation	*Anabaena, Azotobacter, Beijerinckia*
Nitrogen immobilization	Bacteria, fungi and actinomycetes
Nitrogen mineralization	*Pseudomonas, Bacillus, Serratia*
Nitrification	*Nitrosomonas, Nitrobacter*
Denitrification	*Pseudomonas, Achromobacter*
Phosphate solubilization	*Pseudomonas, Bacillus, Aspergillus*
Sulphur transformations	*Thiobacillus, Beggiatoa, Desulfovibrio*
Iron transformations	*Gallionella, Ferribacterium, Leptothrix*
Manganese transformations	*Aerobacter, Corynebacterium, Flavobacterium, Cladosporium*
Copper transformations	*Desulfovibrio, Clostridium, Escherichia*

Selected References

Alexander, M. 1977. *Introduction to Soil Microbiology* (Second edition). John Wiley and Sons, Inc., New York and London.

Delwiche, C.C. 1970. The nitrogen cycle. In *The Biosphere*. A Scientific American Book, pp. 69–80, W.H. Freeman and Co., San Francisco.

Fox, S.W. 1965. Simulated natural experiments in spontaneous organization of morphological units from protenoid. In *The Origin of Pre-biological Systems*, pp. 361–373, Ed. S.W. Fox, Academic Press, New York.

Hutchinson, G.E. 1970. The Biosphere. In *The Biosphere*. A Scientific American Book, pp. 1–11, W.H. Freeman and Co., San Francisco.

Krasilnikov, N.A. 1958. *Soil Microorganisms and Higher Plants*, Academy of Sciences, U.S.S.R., Moscow.

Oparin, A.I. 1953. *The Origin of Life*. Dover Publications, Inc., New York.

Oparin, A.I. 1961. *Life: Its Nature, Origin and Development*. Oliver and Boyd, Edinburgh.

Pirie, N.W. 1958. A discussion on anomalous aspects of biochemistry of possible significance in discussing the origin and distribution of life. *Proc. R. Soc., B*, **171**, 1–89.

Ponnamperuma, C. 1965. A biological synthesis of some nucleic acid constituents. In *The Origin of Pre-biological Systems*, pp. 221–236, Ed. S.W. Fox, Academic Press, New York, N.Y.

Russel, J.R. 1962. Soil Conditions and Plant Growth. Longmans, Green and Co. Ltd., London.

Rutten, M.G. 1971. *The Origin of Life by Natural Causes*. Elsevier Publishing Co., Amsterdam, London and New York.

Subba Rao, N.S. 1982. *Advances in Agricultural Microbiology*. Oxford and IBH Publishing Co., New Delhi.

Subba Rao, N.S. 1993. *Biofertilizers in Agriculture and Forestry*, 3rd Edition, Oxford and IBH Publishing Co., New Delhi.

U.S.D.A. 1957. *Soil*. The Year Book of Agriculture, The U.S. Govt. Printing Office, Washington, U.S.A.

Waksman, S.A. 1957. *Soil Microbiology*. John Wiley & Sons, Inc., New York.

Wright, E. and Stampp, K.M. 1964. *Illustrated World History*. McGraw-Hill, London.

2. Soil, the Natural Medium for Plant Growth

Soil in General

Soil is the outer covering of the earth which consists of loosely arranged layers of materials composed of inorganic and organic constituents in different stages of organization. It is the natural medium in which plants live, multiply and die and thus providing a perennial source of organic matter which could be recycled for plant nutrition. Soil provides the physical support needed for the anchorage of the root system and also serves as the reservoir of air, water and nutrients which are so essential for plant growth. The portion of the earth beneath the soil is known as the bed rock and does not contribute directly to the growth of plants.

The processes involved in the formation of soil are slow, gradual and continuous and are the sum total of environmental effects on rocks collectively known as the weathering of rocks. Rocks are exposed to sun, wind and rain. The changes brought about by fluctuations in temperature cause cracks on the surface of rocks in which the minerals released by the action of rain-water accumulate. The winds bring in particles of organic matter and sooner or later lichens establish on the rocks. The rocks gradually disintegrate resulting in a loose assemblage of rock materials which finally break down into smaller pieces of rocks. The process of weathering starts afresh in the smaller rocks and continues as a chain reaction. What has now been described in few words is a process which normally takes place in nature involving hundreds or thousands of years, depending on the nature of the parent rock material. In fact, weathering of rocks is a continuous phenomenon and will serve to add more and more soil to the surface of the earth.

There are different types of parent materials of rocks available for the formation of soil. Granite is slow to weather unlike limestone, sandstone or shale. The nature and chemical composition of the parent material have a direct bearing on the fertility of soil. For instance, soils derived from granite or sandstone are not as fertile as those derived from limestone.

Broadly, speaking, there are six soil zones in the world—the tundra soils having dwarf shrubs and mosses in very cold regions of the world, the podzolic soils of humid temperate forests, the chernozemic soils of subhumid, semi-arid temperate climates, the desertic (arid) soils of arid, temperate and tropical climates, the latsolic soils of humid and wet or dry tropical and sub-tropical climates and the stony mountain soils which may have any one of the above characteristics. Each soil zone or belt may consist of many soil types which can be clubbed into various soil groups depending on fertility, tilth, ability to hold water and resistance to erosion. In India, approximately 25 soil types are recognized. Among them, those which have been extensively studied are the red soils, the laterite soils, the black soils and the alluvial soils. Besides these types, nearly 14% of the total area of soil in India is occupied by forest trees and hills and certain parts of India are characterized by desert conditions and alkalinity which render them unfit for cultivation.

Each type of soil is characterized by the presence of different horizons which can be seen when a soil profile is obtained (Fig. 4). The formation of soil horizons (mostly distinguished as A, B, and C horizons) depends on climate, living organisms, parent rock material, topography and time, all of which control the weathering of rocks. All the three horizons and their subdivisions (often designated as A1, A2, A3, and so on) may not be present in all soils but a given soil is bound to possess at least some of them.

The A horizon is formed and differentiated by the accumulation of organic matter and its decomposition products through microbial activities. Transitional stages exist between individual horizons and, in practice, it is difficult to clearly demarcate such transitional areas. Nevertheless, the upper most horizon is rich in organic nutrients. The B horizon is characterized by the accumulation of silicate clay minerals and percolating humus particles while the horizon C consists of weathered parent material. The A, B, and C horizons are interlinked physically, biologically and chemically whereas the lowermost D horizon may consist of hard rock, layers of clay or sand which are not part of parent material but may have some relationship to the upper horizons of soil.

Physical Properties of Soil

Physical properties of a soil type depend on the size of particles in it. Particles above 2.0 mm are generally classified as gravel or stones and others as sand (between 0.05 and 2.0 mm), silt (0.002 to 0.05 mm) and clay (less than 0.002 mm). The texture of a soil depends on the percentage of sand, silt and clay in it (Fig. 5). Based on this yardstick, soils are designated as clay, sandy clay, clay loam, silty clay, silty clay loam, sandy loam, silty loam, loamy sand, sand and silt. Soil particles occupy roughly more than half the space in soil. The remaining space between the particles, called the pore space, is occupied by water and air. The

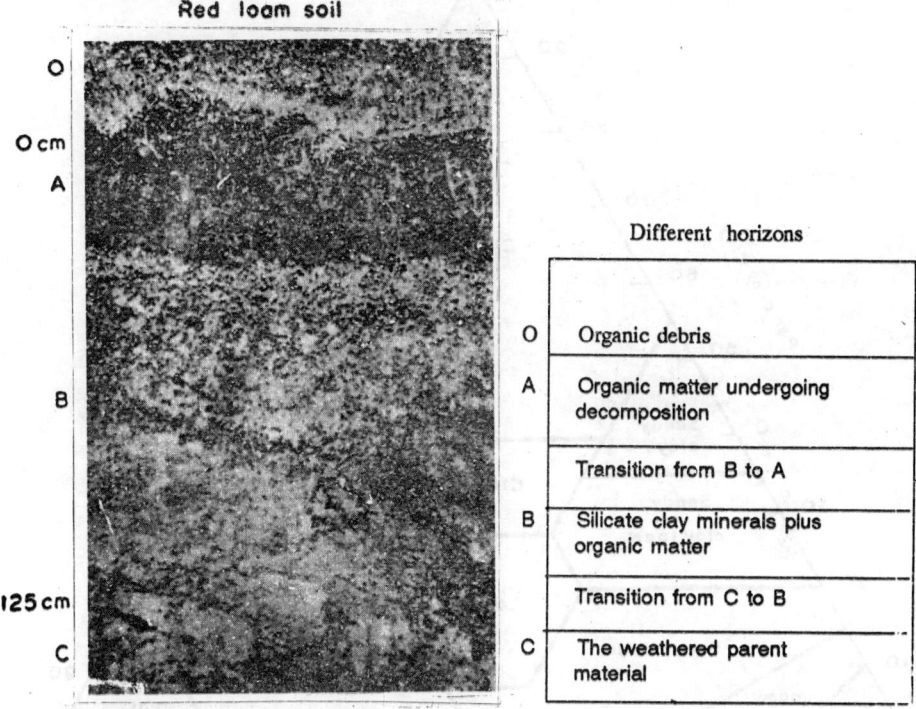

Red loam soil

O

O cm

A

B

125 cm

C

Different horizons

O	Organic debris
A	Organic matter undergoing decomposition
	Transition from B to A
B	Silicate clay minerals plus organic matter
	Transition from C to B
C	The weathered parent material

Fig. 4 The different horizons of a soil profile (Courtesy: Soil Survey Directorate, IARI, New Delhi).

bulk density of a soil is defined as the mass of a unit volume of dry soil which also reflects the total pore space of a soil. The pore size or porosity of soils together with bulk density determine the structure of soil. Sandy soils have single grained structure because of the general uniformity in particle size whereas in sandy clay or silty clay soils, the particles are grouped into aggregates. The stability of soil aggregates depends on the organic matter content of individual soils and the nature of the microbial products which bind the particles together. Soil aggregation is a vital factor in plant growth, since movement of air, water and transfer of energy are interlinked with the porosity of soil. Soil temperature is a very variable factor but nevertheless influences plant growth depending on the intensity of sunlight, day length, seasonal variations, rainfall and colour and texture of soil. Soil crusting is an important physical feature of soil which influences the emergence of seedlings. Soil crusts are formed by desiccation and may cause injury to the stems of seedlings resulting in crop failures.

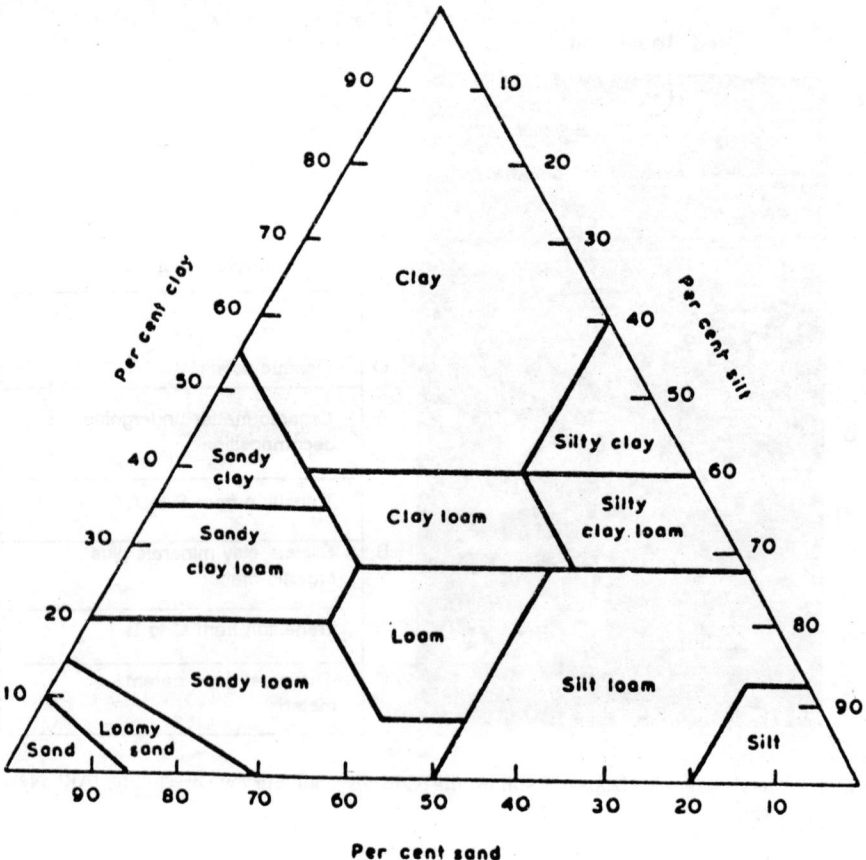

Fig. 5 The soil triangle showing the relationship between contents of clay, silt and sand in determining the different kinds of soil.

Other physical factors in soil affecting plant growth are pH and moisture holding capacity. Retention of water in soils is related to pore space and capillary action of soil particles while pH is dependent on the chemical condition of soil. Acid or alkaline soils are not generally as suitable for plant growth as neutral soils, since solubility and availability of plant nutrients are related to soil pH. Acid soils are usually characterized by excess availability of aluminium, iron, manganese, copper and zinc which may even prove toxic to plants. The reverse is generally true of alkaline soils and in such soils, plants show symptoms of deficiency to many of the elements listed above. Neutral soils, in particular, favour the growth of such microorganisms which are responsible for the conversion of organic forms of nitrogen, phosphorus and sulphur into inorganic forms which can be absorbed by plants.

Changes in temperature affect not only the physiological reaction of microbial cells but also the characteristics of soil environment. A sudden fall in growth rate of microorganisms can be the result of thermal denaturation of proteins and changes in the permeability of membrane at high as well as low temperatures. However, specific proteins of flagella and ribosomes of themophilic bacteria are more thermostable than those of mesophilic bacteria which grow at ambient temperatures. At low temperatures, microbes alter their cell structure especially the shape of proteins because cold hardiness of genes in ribosomes control the genetic coding of proteins and hence an organism's ability to function at low temperatures. Growth of most psychrophilic bacteria stops at zero degree temperature despite the fact that some species are capable of growing below the freezing point because of a mechanism to keep the cytoplasm unfrozen.

Chemical Properties of Soil

Soil is the medium from which plants normally derive their nutrients. The nutrients are carbon (C), hydrogen (H), oxygen (O), nitrogen (N), phosphorus (P), sulphur (S), potassium (K), calcium (Ca), magnesium (Mg), iron (Fe), manganese (Mn), zinc (Zn), copper (Cu), molybdenum (Mo), boron (B), and chlorine (Cl). Of these sixteen elements, Fe, Cu, Mo, B and Cl are considered as micronutrients since they are required in trace amounts for plant growth and the remaining ones are classed as macronutrients because they are required in large quantities.

The three main components of the soil which provide nutrients for plant growth are the organic matter, the derivatives of parent rock materials and the clay fraction. The nutrients are first released into the soil solution (soil water) before they get transferred into the root system of plants.

The organic matter in soils is a potential source of N, P and S for plant growth. Microbiological decomposition of organic matter is an essential step to release the bound nutrients in organic residues into an available form. The inorganic minerals come from rocks (sand and silt) which are made available to plants by mineral decomposition. On the other hand, the clay fraction of the soil provides secondary minerals and amorphous materials which are different from those derived from sand and silt. They are clay minerals and hydrous oxides. The clay minerals are composed of three mineralogical types—kaolinite, montmorillonite and illite. The hydrous oxides are compounds of iron and aluminium. Kaolinite has a crystalline structure composed of two-layer unit cells, one layer of which is made of aluminium and oxygen octahedra (eight-sided) and the other layer made of silicon and oxygen tetrahedra (four-sided) with silica and aluminium in a 1 : 1 ratio. The two layers are held together by common oxygen-hydroxyl linkage into a rigid structure that does not expand when wet and hence prevent the exchange of available cations. Montmorillonite has a unit comprising of one layer of silicon and oxygen tetrahedra, another layer of aluminium and oxygen tetrahedra and one more layer of silicon and oxygen tetrahedra. This unit is linked to another such unit by oxygen and hydroxyl bonds. Together, the silica and alumina are in a 2 : 1 ratio. Between the two units lies a space that expands when wet accommodating cations in an exchangeable form that are available for absorption of plants. Illite is similar in structure to montmorillonite and does not expand when wet but has the capacity to fix potassium (K^+) between non-expanding plates (Fig. 6).

The colloidal clay in soil acts as a large anion and absorbs cations. Having absorbed the cations, the clay particles act as reservoirs of exchangeable ions and release them for plant nutrition as and when required. Thus, the cation exchange or the base exchange properties of soil determine soil fertility and hence plant nutrition. The exchangeable cations are hydrogen, calcium, magnesium, potassium, sodium, ammonium, manganese, zinc, copper and aluminium. These ions possess different energies by which they are held to the solid phase of soil. The binding energy of ions also determines the ease with which the ions may be exchanged in soil. In fact, the property of base exchange in soils enables ions to be held in a readily available state for plant nutrition and at the same time prevents the loss of nutrients by leaching.

There are both positively and negatively charged ions in the soil solution. The positively charged ions are known as cations (K^+, Mg^{++}, Ca^{++}, Fe^{+++}, Mn^{++}, Zn^{++} and Cu^{++}) and the negatively charged ones are known as anions (NO_3^-, $H_2PO_4^-$, SO_4^{--}, Cl^-, $HB_4O_7^-$ and $HMoO_4^-$). The ions in the soil solution are in a state of continuous flux since absorption of ions by roots and the liberation of ions from soil particles are two independent processes whose common pool is the soil solution. Transfer of ions into roots involves the exchange of ions between the root and the soil solution. For instance, H^+ ions are released into soil solution from roots in exchange of cations. Likewise, anions are absorbed by roots in exchange of OH^- and HCO_3^- ions. Further, roots exhibit selective action in absorbing several ions from soil solution which is referred to as the internal regulation of absorption or selective absorption. External factors governing absorption by roots are the concentration of the soil solution and the relative preponderance of individual ions in the root region.

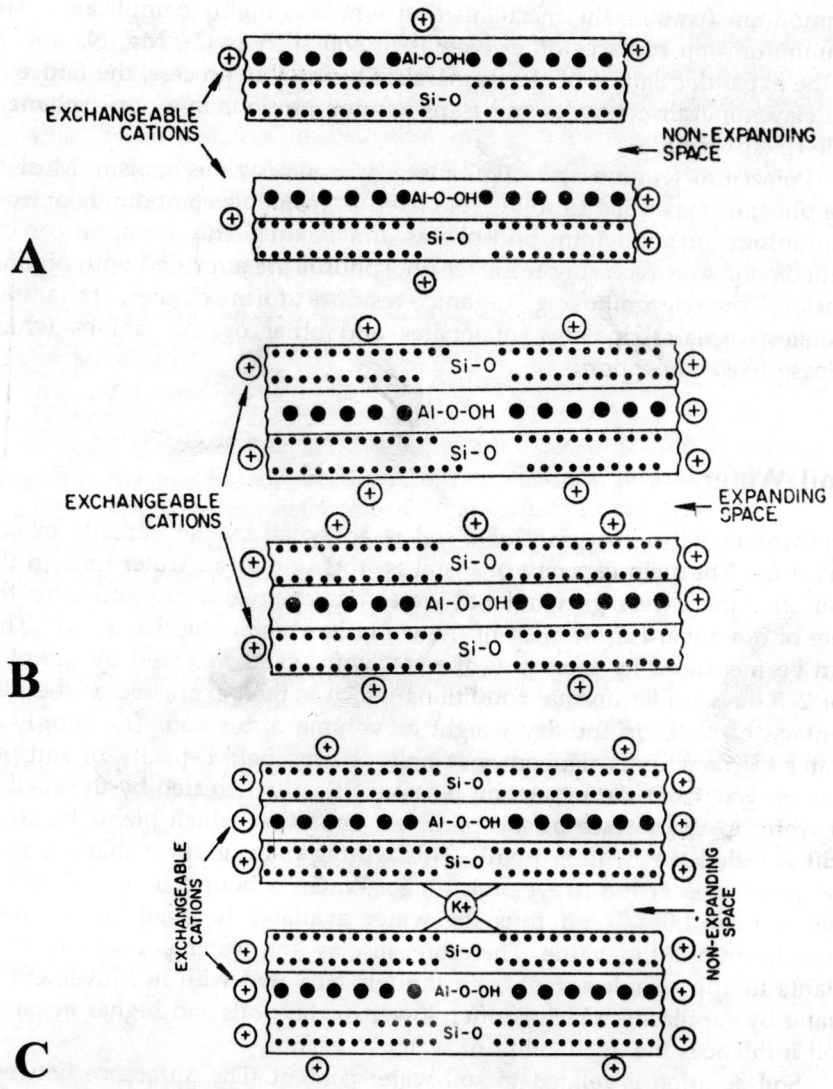

Fig. 6. Structure and composition of three mineralogical types of clay: A—Kaolinite; B—Montmorillonite; C—Illite (From Donehue *et al.*, 1971).

Fixation of Ions in Soil

It has been estimated that approximately 5% of the total nitrogen in surface soils and 60% of total nitrogen in sub-soils are held as non-exchangeable (fixed) ammonium. This property of soils to retain NH_4^+ ion is known as ammonium fixation, the mechanism of which is highly complicated. The ammonium ion replaces other cations in soil such as Ca, Mg, Na and H in the expanded lattice of clay minerals. During this process, the lattice of the clay minerals contracts and traps ammonium ions in a non-exchangeable (fixed) form.

Potassium fixation also takes place by a similar mechanism. Most of the phosphorus added to soil is also fixed as insoluble compounds of iron, aluminium, orthocalcium, phosphates and apatite. This situation can be remedied if soils receiving fertilizer phosphorus are amended with organic matter. The decomposing organic residues form citrates, tartarates, acetates, oxalates, malates, malonates and other organic anions which release fixed phosphorus.

Soil Water

The water holding capacity of a soil is governed by the porosity or soil structure. The field capacity of a soil is "the amount of water held in the soil after the excess gravitational water has drained away and after the rate of downward movement of the water has materially decreased." This can be measured by saturating the soil with water followed by draining for 2–3 days under normal conditions which is then expressed as the percentage of water in the dry weight or volume of the soil. The supply of water to plant roots depends not only on this field capacity of soil but also on water tension or suction tension (pF value) created by the suction of water to the surface of soil particles. The pF at which plants begin to wilt is called the 'wilting point'. This wilting point is reversible but later the plant dies at the irreversible stage. Water is bound if pF values are higher than 4.0–4.2 and thus the water available to plants is the field capacity minus this value. Therefore, in clay soil water is unavailable to plants to appreciable extent even though saturated with it. Movement of water by capillarity which is often lower in clay soils and higher in sandy soil influences the availability of water to plants.

Soil aeration is related to soil water content. The difference between the sum of pore space in a given sample of soil and the water content accounts for the air-filled space. Adequate aeration of soil can be achieved by a minimum air-filled pore space of 10 per cent by volume. Clay soil with 45 per cent of water has the lowest aeration among different soil types.

The occurrence of water in soil can be of three types—gravitational, capillary and hygroscopic. After heavy rain or irrigation, gravitational water is drawn through the soil. Upon drainage of excess water, water is retained at field capacity in soil pores and this water is known as capillary water. Hygroscopic soil water relates to that water absorbed by dry soil from an atmosphere of high relative humidity. Fungi are mostly tolerant to greater water stress than are bacteria. An example of microorganisms less tolerant to water stress is *Nitrosomonas*, followed by ammonifiers such as *Clostridium* and *Penicillium*.

Soil Atmosphere

Major gases in soil are N_2, O_2 and CO_2. The CO_2 content of soil air (0.3–1.0 per cent by volume) is higher than that of the atmospheric air (0.03 per cent by volume). This is due to the respiratory activity of soil microorganisms in which O_2 is consumed leading to lower O_2 content of soil atmosphere. In submerged paddy soils, at first anaerobiosis reduces O_2 but very soon rice roots excrete O_2 into the rhizosphere. However, reducing conditions prevail, depending on the depth. In other situations where waterlogging is a perennial feature (ex: marshy soils), plants adapt their metabolism. For instance, instead of accumulating toxic ethyl alcohol due to suppression of oxidative degradation of sugars, plants may end up accumulating malate. Under highly anaerobic conditions, toxic substances such as ethane, methane, hydrogen sulphide, cyanide, butyric acid and a number of other fatty acids may accumulate causing harmful effects to plant growth.

Methane Emissions from Lowland Rice Cultivation

Flooded rice fields are generally anoxic and thus provide excellent habitat for methanogenic microorganisms. Methane (CH_4) absorbs long-wave radiation and therefore is a green house gas that leads to the warming of the earth's surface. Together with CO_2 and nitrous oxide, methane traps the thermal radiation from the earth's surface and influences the atmospheric chemistry of ozone in the troposphere as well as the stratosphere through photochemical reactions. The present estimates of CH_4 emissions from wetland rice cultivation are placed around 25–170 Tg CH_4 per year. Recent studies however point out that Indian rice soils emit around 4 to 6 Tg methane per year, the upland rice cultivation registering lower values than lowland rice cultivation. Other estimates put the values around 20–150 Tg per year. It appears that the variability may be due to varietal differences of rice planted and to different agronomic practices followed besides the factor contributed by erratic rainfall.

One of the mitigation options found out by experiments is to grow rice cultivars that emit low levels of CH_4. A single midseason drainage may reduce seasonal methane emission rates by about 50 per cent but this practice may generate nitrous oxide, another green house gas. Application of gypsum that releases sulfide inhibits CH_4 formation and similarly application of rock phosphate or single superphosphate does the same trick.

Likewise, foliar application of technical grade organochlorine insecticide hexachlorocyclohexane (HCH) to rice plants retards CH_4 production. Most of the options to reduce methane emissions in rice cultivation are however location specific and have to be adopted cautiously so as to avoid lowering rice production.

Soil pH

The negative logarithm of the activity of H^+ in solution is denoted by pH which indicates the acidity/alkalinity of soil when measured in an aqueous or KCl extract. In the latter case, some of the H^+ ions absorbed to soil colloids are also replaced by K^+ and hence the pH indicated would be lower than water extract. The buffering capacity of soil, depending on the type of soil, also determines soil pH. Soils with high organic or inorganic colloids exert great buffering capacity because of saturation with cation species. The pH value of soils vary from 3–10. The acid sulphate and podzolic soils are low in pH whereas calcareous and alkali soils are very high in pH due to strong bases such as Na^+ or K^+ and weak acids such as HCO_3^-. The pH of soil determines the mineral contents as well as microbial composition. High pH releases K^+, Mg^{2+}, Ca^{2+}, Mn^{2+}, Cu^{2+}, and Al^{3+} by weathering processes of soil whereas low pH favours solubility of salts including carbonates, phosphates and sulphates. Generally, fungi are predominant in the rhizosphere under low pH conditions (< 5.5) and beneficial nitrogen fixing microorganisms are favoured by neutral pH. Plants adapt to soil pH and this depends upon species. Plants like rye, oats, potatoes and wheat adjust to pH range of 4.1–7.4 whereas, lucerne and sugarbeet favour pH optima in the range of 6.4–7.4. On the other hand, lupines have pH requirements in the limited range of 4.1 to 5.5. Liming soil is a well-known practice to correct pH of soils for plant growth. Generally, soils of temperate regions tend to become acidic mainly due to the production of H^+ ions by organic matter decomposition.

Saline soils (Solonchak) contain an excess of neutral salts such as chlorides and sulphates of Na^+ and Mg^{2+}. When ground water rises followed by evaporation in semi-arid conditions, salts surface as white patches. Alkali soils or solonetz (pH 7.5–10.0) are characterized by the presence of Na_2CO_3 and $NaHCO_3$. Sometimes alkali soils are black due to mixture of humic substances. The total ion concentrations of these soils are not conducive to plant growth. The degree of salinity is usually measured in the water extracts of soil in terms of electrical conductivity. This is expressed as mm hos/cm which is the reciprocal of electrical resistance. As the salt content increases the electrical conductivity also increases. Salt affected soils are common in many countries with conductivity values in the range of 2–20 mm hos/cm in saturated extracts.

Soil Structure

Movement of air and water in soil is governed by the structure of soil. The structure of soil, in turn, is dependent on stable aggregates of soil particles. A soil aggregate is a naturally occurring cluster of soil particles with a strong binding force among them. Soil organic matter including humus, polysaccharides and polyuronides produced by soil microorganisms help to cement soil particles together while filamentous fungi provide additional mechanical support.

Several workers have found that the aggregating influence of different groups of soil microorganisms can be graded in the order, fungi > streptomyces > gum producing bacteria > yeasts. Rapidly growing *Rhizopus*, *Mucor*, *Chaetomium*, *Fusarium*, *Cladosporium*, *Aspergillus* and *Rhizoctonia* are good examples of fungi known to secrete gums and provide mechanical support to bind soil particles. Examples of bacteria producing significant amounts of gum are *Azotobacter*, *Beijerinckia*, *Rhizobium*, *Xanthomonas* and *Bacillus*.

The lipopolysaccharides from gums of *Rhizobium* and *Agrobacterium* have been analysed and found to contain glucose, rhamnose, mannose, glucosamine and 4-0-methyl-glucoronic acid as major constituents. Other components of the gums are fructose, arabinose, galactose and xylose. It has been claimed that infection of root hairs of legumes is dependent on a water soluble polysaccharide of the rhizobial gum, although no conclusive evidence has been provided to support such a hypothesis.

Different soil components have been chemically extracted from soil and their effects on soil aggregation properties studied. Among the carbohydrates isolated from soil, dextran containing high amounts of uronic acid which is resistant to microbial degradation has been shown to have the best soil aggregating qualities. Several experiments have been carried out under laboratory conditions to demonstrate the effect of single or mixed cultures of microorganisms in soil amended with various carbon sources. The results of such experiments, however fascinating they may be, must be viewed with caution. It is also interesting to note that the involvement of polysaccharides in soil aggregation has been questioned by some workers based on results of experiments in which native polysaccharides of soil samples were destroyed by chemical treatments without any detriment to the property of soil aggregation.

The Role of Organic Matter

Soils receiving well decomposed organic manures have better soil aggregates than those receiving saw-dust and other types of not so easily decomposable organic wastes. The root hairs and exudates from roots together with sloughed off debris of the root cortex help in creating soil aggregates in the root zone which can be demonstrated by lifting a herbaceous plant carefully from soil. Since organic matter decomposition involves microorganisms, the products of microbial growth such as gums rather than the quantity of organic matter added is the deciding factor in the improvement of soil structure. The complex polysaccharides of microbial origin are resistant to microbial attack. This may be due to the ability of gums to form a complex with metals, other organic compounds and clay minerals in soil or due to physical protection offered by the aggregates preventing further microbiological degradation. Adequate evidence has not been forthcoming on these aspects to explain the persistence of microbially synthesised polysaccharides in soil and further research is needed on this aspect, under natural soil conditions.

Synthetic Soil Conditioners

These are synthetic compounds which act as stabilizers of naturally occurring soil aggregates. Their value lies in their inherent ability to resist microbial degradation and not in merely helping in the formation of new aggregates. They simulate microbial polysaccharides in binding soil aggregates and are known as Krillium-type conditioners—Polyacrylic acid (PAA), Polyacrylonitrile (PAN), hydrolysed Polyacrylonitrile (HPAN) and Vinyl acetate-maleic acid copolymer (VAMA). Such compounds are active in concentrations of about 0.1–0.2% (per unit dry weight of soil) in an aqueous phase. Soil conditioners have not been very popular because of the high cost and the necessity of frequent applications to obtain desired results.

In Relation to Root Growth

Although soil structure is a complex of several physical characteristics of soil, the growth pattern and extent of root system together with the attendant exudates and sloughed off debris of the root cortex contribute additional factors to the maintenance of soil structure, especially under intensive cultivation as in multiple cropping, mixed cropping and relay cropping. In other words, the root system of an actively growing plant or the remains of root system after harvest contribute to the organic matter status of soil and hence indirectly influence soil aggregation. In this context, leguminous crops are known to improve soil structure.

The growth of a plant depends not only on the capacity of soil to release nutrients, but equally on the capacity of the root system to absorb such nutrients in an efficient manner. An ideal root system is one which has the following characteristics: (1) capacity to spread quickly into a large volume of soil, (2) ability to spread spatially to hold the shoot in an erect position, (3) capacity to penetrate soil aggregates and soil layers of varied compaction and absorb water and nutrients from sub-soil zones, (4) capacity to solubilize and absorb maximum nutrients and moisture, especially under drought conditions, and (5) add large quantities of organic matter to different soil layers to improve physical properties of soil.

Varieties of one and the same crop show differences in their ability to add organic material through roots into the soil and thereby improve soil structure. For instance, in sub-soil depths, Kalyan Sona variety of wheat could add 170 kg/ha whereas Sharbati Sonora could add only 60 kg/ha. Root systems of different plants also differ in their abilities to absorb native soil nutrients and nutrients from inorganic fertilizers. This was verified from experiments in which ^{32}P tracer technique was used to test the feeding efficiency of roots of different varieties of wheat and rice. In wheat, relative per cent phosphorus feeding efficiency of Sonalika variety was 72.4 for soil phosphorus whereas it was 27.6 for fertilizer phosphorus. Similarly, the per cent feeding efficiency of soil phosphorus *versus* fertilizer phosphorus in NP 130 variety of rice were 76.3 and 23.7 respectively. Invariably, the feeding efficiencies of native nutrients were also uniformly better than nutrients from fertilizers, in several other varieties of wheat and rice.

Evaluation of Soil Structure

Soil structure is evaluated by determining one or more of the following parameters: (1) size of soil particles using dry or wet sieving methods, (2) pore-size and its distribution in soil, (3) saturated hydraulic conductivity of soil *in situ*, (4) bulk density of soil *in situ*, (5) infiltration rates of soil *in situ*, (6) hydraulic conductivity at compacted minimum bulk density, and (7) energy required to cause aggregate disintegration.

One of the methods followed in the Division of Agricultural Physics, Indian Agricultural Research Institute, for determining soil structure is the aggregate analysis by wet sieving method using a modification of the original Yoder-type sieve-shaker having four nests of sieves, each containing sieves with apertures ranging from 2.0, 1.0, 0.5, 0.25 and 0.10 mm. Air-dry soil samples (say 50 g each) having aggregates from 5 to 8 mm size are spread on the mesh of top sieve (2 mm) and are saturated with water by gradually increasing the level of water in the tank of the sieve-shaker. Samples are then sieved in water for 30 minutes by shaking at the rate of 33 cycles per minute. The water stable aggregates retained on each sieve are then quantitatively transferred to tared aluminium cans, oven-dried and weighed. Oven-dry soil aggregates of various sizes of each sample are transferred to a container having 10 ml of 5% solution of sodium hexametaphosphate with sufficient distilled water to disperse soil particles. After stirring the suspension for 15 minutes, the contents are washed in a stream of water on the same sets of sieves which were used earlier. The primary particles (sand) retained on each sieve are now oven-dried and weighed.

Moisture content of soil samples on oven-dry weight basis is also calculated separately to determine the percentage of aggregates. The calculation is as follows:

Percentage of aggregates > 0.25 mm =

$$\frac{\text{Weight of soil particles} > 0.25 \text{ mm} - \text{Weight of sand} > 0.25 \text{ mm}}{\text{Oven dry weight of soil sample}}$$

Based on the above data, aggregation index can be calculated by determining the area between the two curves obtained by plotting the accumulative percentages (fraction less than a particular size) of soil particles and sand against the sieve openings.

Selected References

Alexander, M. 1977. *Introduction to Soil Microbiology* (Second edition). John Wiley and Sons, Inc., New York and London.

Allison, F.E. 1952. Effect of synthetic polyelectrolytes on the structure of saline and alkaline soils. *Soil Sci*, **73**, 443–454.

Allison, F.E. 1973. *Soil Organic Matter and Its Role in Crop Production*. Elsevier Scientific Publishing Co., Amsterdam, London and New York.

Barooah, P.P. and Sen, A. 1964. Nitrogen fixation by *Beijerinckia* in relation to slime formation. *Arch. Mikrobiol.*, **48**, 381–385.

Bear, F.E. 1968. *Chemistry of the Soil*. Oxford and IBH Publishing Co., Calcutta, Bombay, New Delhi.

Biswas, T.D. 1964. Effect of organic manuring on the improvement of soil structure. *Pl. Fd. Rev.*, 4, 2.

Black, C.A. 1968. *Soil Plant Relationships*. John Wiley and Sons, Inc., New York, London.

Dakshinamurti, C. and Gupta, R.P. 1968. *Practicals in Soil Physics*. Division of Agricultural Physics, I.A.R.I., New Delhi.

Donehue, R.L., Shickluna, J.C. and Robertson, L.S. 1971. *Soils: Introduction to Soils and Plant Growth*. Prentice Hall, Inc., New Jersey, U.S.A.

Gaur, A.C., Subba Rao, R.V. and Sadasivan, K.V. 1972. Soil structure as influenced by organic matter and inorganic fertilizers. *Labdev. J. Sci. Technol.*, 10, 55–56.

Ghildyal, B.P. 1969. Influence of grass legume covers on organic matter accumulation and soil aggregation in a laterite sandy loam. *Indian J. Agric. Sci.*, **39**, 757–760.

Graham, P.H. and O'Brien, M. 1968. Composition of lipopolysaccharides from *Rhizobium* and *Agrobacterium*. *Antonie van Leeuwenhoek*, **34**, 326–330.

Gupta, K.G. and Sen, A. 1962. Aggregation of soil due to growth of *Rhizobium* spp. from some cultivated legumes. *Sci. Cult.*, **28**, 483–484.

Harris, R.F., Chesters, G. and Allen, O.N. 1966. Dynamics of soil aggregation, *Adv. Agron.*, **18**, 107–169.

Hedrick, R.M. and Mowry, D.T. 1952. Effect of synthetic polyelectrolytes on aggregation, aeration and water relationships of soil. *Soil Sci.*, **73**, 427–441.

Martin, W.P., Taylor, G.S., Engibous, J.C. and Brunett, E. 1952. Soil and crop responses from field applications of soil conditioners. *Soil Sci.*, **73**, 455–471.

McNeely, W.H. 1967., Biosynthetic polysaccharides. In *Microbial Technology*, pp. 387–402. Ed. H.J. Peppler, Reinhold Publishing Corporation, U.S.A.

Mehta, N.C., Streuli, H., Muller, M. and Deuel, M. 1960. Role of polysaccharides in soil aggregation. *J. Sci. Fd. Agric.*, **11**, 40–47.

Metting Jr., F.B. Ed. 1993. *Soil Microbial Ecology*, Marcel Dekker Inc., New York.

Mishustin, E.N. and Shilnikova, V.K. 1971. *Biological Fixation of Atmospheric Nitrogen.* Macmillan Co., New York.

Paul, P. and Clark, F.E. 1989. *Soil Microbiology and Biochemistry.* Academic Press, New York.

Prabhakara, J. 1970. Soil structure in relation to relay cropping. M.Sc. Thesis, I.A.R.I., New Delhi.

Russel, J.R. 1962. *Soil Conditions and Plant Growth.* Longmans, Green and Co. Ltd., London.

Subbaiah, B.V. and Oza, A.M. 1971. Root studies: Some practical basic considerations in relation to nutrient supply of high yielding varieties, pp. 616–624. In *International Symposium on Use of Isotopes and Radiation in Agriculture and Animal Husbandry Research*, New Delhi.

Waksman, S.A. 1957. *Soil Microbiology.* John Wiley & Sons, Inc., New York.

Yoder, R.E. 1936. A direct method of aggregate analysis of soils and the study of the physical nature of erosion losses. *J. Amer. Soc. Agron.*, **28**, 337–351.

3. Soil Microorganisms

Development of Soil Microbiology

The fertility of soil depends not only on its chemical composition, but also on the qualitative and quantitative nature of microorganisms inhabiting it (Fig. 7). The microorganisms inhabiting soil can be classified into bacteria, actinomycetes, fungi, algae and protozoa and the branch of science dealing with them and their activities in soil is known as soil microbiology. Unlike soil science whose origin can be traced back to Roman and Aryan times, soil microbiology emerged as a distinct branch of soil science only in 1838 after the French agricultural chemist and farmer, J.B. Boussingault showed that legumes can obtain nitrogen from air when grown in soil which was not heated. Fifty years later, a Dutch scientist, M.W. Beijerinck isolated bacteria from nodules on legume roots. Between these two major developments the science of medical bacteriology had, no doubt, been well established thanks to Robert Koch and Louis Pasteur who enunciated the fundamental principles of germ theory of disease. Strictly speaking, the development of microbiology as a branch of science can be dated back to the time of people who ground lenses from glass and saw microorganisms through them. Antony Van Leeuwenhoek, a linen draper from Holland (1632–1723) is credited with having made the first authentic drawings of microorganisms. The theory of spontaneous generation of microorganisms which had its roots in the age of Aristotle (384–322 B.C.) and which was revoked by the so-called experimental evidence of John Needham (1713–1781) was shattered by the conclusive experimental findings of Louis Pasteur (1822–1895). This was followed by Robert Koch's (1843–1910) famous Koch postulates concerning the authenticity of microorganisms as causative agents of disease. In 1878, Joseph Lister first obtained pure cultures of bacteria by serial dilutions in liquid media. Ten years later, in 1888, as mentioned earlier, the root nodule bacteria were obtained in pure culture by Beijerinck. The celebrated Russian microbiologist, S.N. Winogradsky (1856–1953) discovered the autotrophic mode of life among bacteria and established the microbiological transformation of nitrogen and sulphur. The enrichment culture technique involving the successive transfers of microorganisms growing in desired substrate for the isolation of sparsely occurring unusual types of microorganisms was the innovation of both Beijerinck and Winogradsky in their attempt to unravel more and more

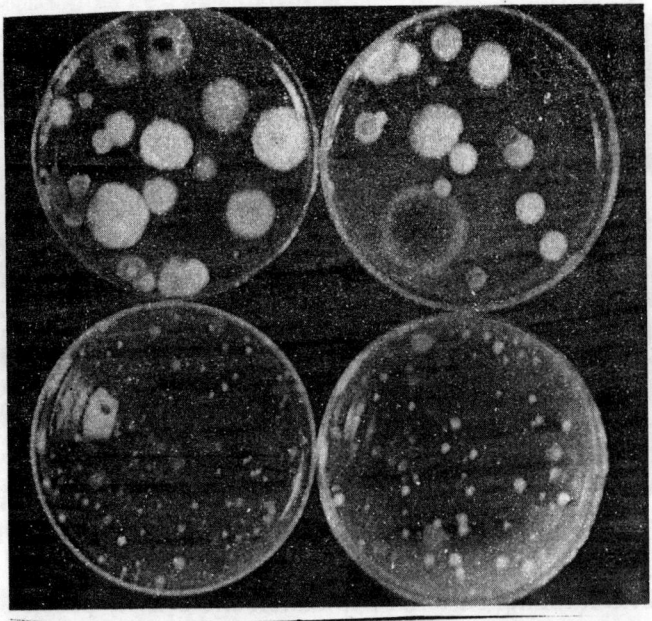

Fig. 7 Colonies of bacteria, actinomycetes and fungi appearing on agar plates, from aliquots of soil dilutions.

specialized types of microorganisms from soil. Thus, Beijerinck and Winogradsky (Fig. 8) may rightly be considered as pioneers in soil bacteriology, whose names can be equated with those of Robert Koch and Louis Pasteur in the field of medical and general microbiology.

Significant Developments in Soil Microbiology

There have been several notable contributions made by several scientists in soil microbiology since the beginning of the present century. Omeliansky in 1902 found out that anaerobic soil bacteria degrade cellulose. Lipman and Brown in 1903 in the U.S.A., studied ammonification of organic nitrogenous substances by soil microorganisms and developed the 'tumbler or beaker' method for studies on different types of transformations in soil. The work on rhizosphere started in 1904 by the German scientist Hiltner was later continued in the U.S.A. by Starkey in 1929, in Canada by Lochhead in 1940 and Katznelson in 1946, in Australia by Rovira in 1956, in Czechoslovakia by Matcura and associates in 1961 and in other countries including India by several workers. The importance of

Fig 8   Pioneers in the study of soil microorganisms: (A) Beijerinck and (B) Winogradsky.

protozoa in controlling bacterial population and activity in soil was enunciated in 1909 by Russel and Hutchinson at Rothamsted in England.

Side by side with these scientific developments, techniques in the
study of soil microorganisms such as the direct soil examination by Conn
in 1918 and the contact slide by Rossi in 1927 and Cholodny in 1930 were
innovated. In fact, modifications of some of these techniques are still in
vogue today.

The work of Rayner and Melin between 1921 and 1927 in Uppsala
marked the beginning of an intensive study of mycorrhiza. Among others
who subsequently got interested in mycorrhiza, mention may be made of
Harley in the U.K., Gerdemann in Germany, Marx, Trappe and Hacskaylo
in the U.S.A. and Bowen in Australia who worked on ectomycorrhizal
fungi. In 1936, Garrett established a school in the U.K. on soil fungi and
their ecological classification in relation to the nature of available native
substrate. Alexander Fleming in 1929 discovered penicillin which was followed by the discovery of streptomycin by Waksman in 1944. These findings led to an upsurge in the study of antibiotics in general (as evidenced
by large number of antibiotics discovered) and antibiosis in soil in particular.

In recent years, the occurrence and importance of arbuscular en-
domycorrhizal fungi (Ex. *Golmus, Acaulospora*) in roots of plants have been
repeatedly pointed out. Barbara Mosse at the Rothamsted Experimental
Station in England and several other workers in the U.S.A., India and New

Zealand have now realized the potential of AM fungi in phosphate mobilization into plants.

The work on microbiological transformations in soil was continued in the forties in Europe at the famous Pasteur Research Institute in Paris by Pochon and his associates. At the same time, Gerretsen and Mulder were busy in Wageningen on work concerning phosphate mobilization by soil microorganisms and the importance of molybdenum in the nitrogen metabolism of microorganisms. By about this time, Ruinen in 1956 developed the concept of phyllosphere (the microbiology of leaf surface) from her studies on the leaf surface microflora of Indonesian forest trees.

In the U.S.A., fundamental studies on soil bacteria were carried out by Van Niel in 1931 on chemo-autotrophic bacteria and bacterial photosynthesis, by Umbreit in 1947 on problems of autotrophy, by Barker on anaerobic fermentation by methane bacteria, by Starkey in 1945 on transformations of iron bacteria and by Allen and Allen in 1940 on soil bacteria in general and root nodule bacteria in particular. The contributions made by Fred and co-workers in 1932 in connection with nodule bacteria deserve special mention. Subsequently, Alexander in 1961 started a school of soil microbiology at the Cornell University, with particular reference to microbiological aspects of pesticide degradation.

The colourful microalgae fascinated many workers all over the world. Fritsch, Fogg and Stewart in the U.K. and Iyengar in India were the foremost investigators in the study of algae in general and microalgae in particular.

Biological Nitrogen Fixation

Biological fixation of atmospheric nitrogen was the topic for investigations by many workers. Fred, Baldwin and McCoy in 1932 from the U.S.A. put forth their exhaustive work on root nodule bacteria. At Rothamsted in England, Thornton in 1947 studied nodule bacteria from clovers. Jensen's method devised in 1942 for studying nodulation of plants on agar in test tubes helped in furthering investigations in this field. The German bacteriologist Bortels demonstrated in 1936 that an adequate supply of molybdenum was essential for accelerating nitrogen fixation by nodulating legumes. The Japanese scientist Kubo found out in 1939 that root nodules of legumes could be effective only in the presence of a red pigment in them. At Helsinki, Virtanen and his school in 1947 studied leghaemoglobin and the chemistry and mechanism of nitrogen fixation.

The discovery of consistent nitrogen fixation in cell-free extracts of Clostridium pasteurianum by Carnahan and others at the Du Pont Laboratory in the U.S.A. in 1960 was a landmark in the field of biological nitrogen fixation. Of equal importance was the development of the isotope method to quantify the amount of nitrogen fixed by the use of 15N and mass spectrometer developed by Burris and Wilson in 1957. The enzyme nitrogenase was isolated, characterized and most of its biochemical properties understood except the nature of intermediates between nitrogen and ammonia. Applying the knowledge obtained with asymbiotic nitrogen fixers, Bergersen in Australia elaborated the biochemistry of nitrogen fixation in legume root nodules in the sixties. Nutman in England from 1948 onwards gave us an insight into the hereditary mechanisms behind nodulation in legumes. At the same time, he proposed the theory of micro-invagination of root hairs as an explanation for the origin of infection threads in root hairs of clovers. The work on nitrogen fixation in nodulated plants other than legumes was taken up by Bond in the U.K. and Quispel, Silver and Becking in Europe who made considerable headway on this aspect between 1950 and 1970. During the same period, Fogg and Stewart in the U.K. intensified the work on nitrogen fixing blue-green algae.

One of the significant advances with regard to non-leguminous root nodulation has been the isolation of an actinomycetous endophyte (*Frankia* sp.) from root nodules of *Comptonia peregrina* in 1978 by Callaham, Del Tredici and Torrey. The isolated endophyte was slow growing and could re-infect and produce nodules on the roots of the same host plant.

Another important observation by Trinick from Australia in 1973 was the isolation of *Rhizobium* from the root nodules of the genus Trema (Parasponia) which has highlighted a unique association of Rhizobium with non-leguminous plants resulting in the formation of root nodules.

Bacteria containing fertilizers such as 'Azotobacterin' and 'Phosphobacterin' were extensively used in Russia to improve soil fertility. Krasilnikov and Mishustin in 1937 were busy in Russia on several problems relating to interactions between plants and soil microorganisms. The English translation of 'Mikrobiologiya' opened the door of Russian microbiology to the English-speaking world. The Australian group led by Vincent in 1954 was very active in the study of all aspects of nodulation, particularly in understanding the environmental factors in legume root nodulation as evidenced by the work of Gibson in 1965. While other Australian workers like Date, Brockwell and Roughley in 1962 were busy in the development of techniques involved in inoculant production and application to seed, Burton in 1950 in the U.S.A., was chiefly responsible for the establishment of legume inoculant research in North America with all its industrial implications.

The extensive use of the assay procedure utilizing the nitrogenase-catalyzed reduction of acetylene to ethylene coupled with the sensitive gas chromatographic analysis proposed by Hardy and his associates in 1968 of the Du Pont Laboratory in the U.S.A. may be regarded as a turning point in the field of biological nitrogen fixation. The ease and rapidity with which this technique could be used gave an impetus to many investigators to measure nitrogenase activity in situ in many natural ecosystems. The

Brazilian group of workers headed by Dobereiner showed how *Azotobacter paspali* could specifically inhabit the roots of the grass Paspalum notatum and fix abundant amounts of nitrogen, an observation which was later confirmed by the acetylene reduction reaction. Dobereiner also introduced the concept of 'associative symbiosis' or 'diazotrophic biocoenocis' after the rediscovery of *Azospirillum* within plant cells of many graminaceous plants. Blue-green algae and *Azolla* received greater attention in developing countries as a nitrogen supplement in rice cultivation.

Rapid advances have been made in the U.K. and the U.S.A. in our knowledge of the physiology, biochemistry and kinetics of the enzyme nitrogenase. Those active in this field are Eady and his associates in England and Burris and Orme-Johnson in Wisconsin. Pate and his associates in Australia have quantified the interrelationship between photosynthates (energy) and nitrogen fixation in legumes.

The rapid expansion in the field of bacterial genetics had its impact on the study of nitrogen-fixing microorganisms. Presently, our understanding of the genetic loci involved in the process of biological nitrogen fixation in *Azotobacter*, *Azospirillum*, *Rhizobium* and blue-green algae (cyanobacteria) has reached new heights due to the efforts of many groups of workers in Europe, Australia and the U.S.A. We are now aware that rhizobial genes control infection, nodulation and host range in legumes. Extrachromosomal genetic particles known as plasmids are known to carry nitrogen fixing traits. Other nodulation genes have been identified that are functionally and structurally conserved in rhizobia. In short, it has become clear that establishment of legume nodule symbiosis is the result of a multitude of communications and biochemical interactions between the legume host and the rhizobial symbiont. This area of plant-microorganisms relationships has been enlarged due to the efforts of several molecular biologists such as Postgate, Beringer, Dixon, Johnston, Kennedy, Brill, Ausubel, Helinksy, Fisher, Long, Denarie, Downie, Kondorosi, Pueppke, Rolfe, Gershoff, Broughton, Spanik, Stacey, Verma, Krishnan, Elmerich, Nuti and others all over the world.

Dommergues and associates from France and Senegal have discovered the occurrence of nodules on stems of Sesbania rostrata which fix nitrogen by virtue of which, the legume serves as an excellent green manure in low-land rice cultivation. Similarly, the group has also discovered nitrogen-fixing Stem nodules on *Casuarina* spp. caused by *Frankia*.

It is now known that rhizobia can fix nitrogen independently in cultures and this fact has been substantiated by 15N as well as acetylene reduction methods. The question whether rhizobia can establish endogenously within the cells of plant roots other than those of legumes and Parasponia, and induce the formation of nodules has been tackled by a team of workers headed by Cocking in England who have attempted to induce nodules on roots of rice, wheat and rape seedlings by inoculating the roots with the stem nodulating bacterial species, *Azorhizobium caulinodans*. Nodule-like structures capable of fixing nitrogen have no doubt been demonstrated in rice and rape seedlings under laboratory conditions but field applications of any of these laboratory findings are yet to be realised.

Studies on soil microorganisms in relation to soil fertility were enlarged in several agricultural universities in India through applied research programmes under the All India coordinated research projects funded by the Indian Council of Agricultural Research and the Department of Science and Technology. These projects were related to the field application of nitrogen fixing microorganisms and the role of soil microorganisms in the decomposition of organic matter. The main emphasis has been to minimize the use of chemical fertilizers in the cultivation of rice, legumes and millets.

Bacteria

Bacteria belong to the group of prokaryotes (with no defined nucleus) whose structures vary from those of eukaryotes (those possessing defined nucleus) in many ways (Table 3).

Table 3 Differences between eukaryotes and prokaryotes in cell structure

Structural Features	Eukaryotes (Fungi, Protozoa, Algae)	Prokaryotes (bacteria including blue greens)
1. Nuclear structure	Nucleus with chromosomes associated with histone Proteins	Single naked circular chromosome
2. Nuclear membrane	Present	Absent
3. Endoplasmic reticulum	Present	Absent
4. Mitochondria	Present	Absent
5. Ribosomes	80S type	70S type
6. Lysosomes	Present	Absent
7. Plasma membrane	Present, contains sterols	Present, no sterols except in mycoplasma
8. Cell wall	Absent or composed of cellulose or chitin	Complex structure with peptidoglycan layer, protein and lipids
9. Capsule	Absent	Frequently present

They are the most dominant group of microorganisms in soil and probably equal one half of the microbial biomass in soil. They are present in all types of soil but their population decreases as the depth of soil increases. In general, horizon A of a soil profile consists of more microorganisms than B and C horizons. Under anaerobic conditions (in the absence of oxygen), bacteria dominate the scene and carry on microbiological activities in soil since fungi and actinomycetes do not grow well in the absence of oxygen.

Bacteria live in soil (Fig. 9, 10) as cocci (spheres, 0.5 μ), bacilli (rods, 0.5 to 3.0 μ) or spirilli (spirals). The bacilli are common in soil whereas spirilli are very rare in natural environments. In 1925, Winogradsky classified soil microorganisms in general and bacteria in particular into two broad categories—the autochthonous and the zymogenous organisms. The autochthonous or indigenous population is always uniform and constant in soil since their nutrition is derived from native soil organic matter (example, *Arthrobacter* and *Nocardia*). On the other hand, zymogenous or fermentative organisms require an external source of energy and their normal population in soil is low (examples, *Pseudomonas* and *Bacillus*). When specific substrates are added to soil the number of zymogenous bacteria increases and gradually declines when the added substrate is exhausted. To this category belong the cellulose decomposers, nitrogen utilizing bacteria and those splitting ammonium into nitrate.

Growth in the presence or absence of oxygen is taken as the criterion to distinguish bacteria into anaerobic, aerobic and facultative anaerobic, that is, those capable of developing under oxygenated as well as non-oxygenated conditions. Under the Bergy's system of Determinative Bacteriology, bacteria are classified into taxonomic groups, orders, families, genera and species based on the classical Linnaen concept of binomial nomenclature. Ten orders are included in the class Schizomycetes. Of these, three orders—Pseudomonadales, Eubacteriales and Actinomycetales contain the species of bacteria which are predominantly encountered in soil.

The most common soil bacteria come under the genera *Pseudomonas, Arthrobacter, Clostridium, Achromobacter, Bacillus, Micrococcus, Flavobacterium, Corynebacterium, Sarcina* and *Mycobacterium*. *Escherichia* is encountered rarely in soils except as a contaminant from sewage whereas *Aerobacter* is frequently encountered and is probably a normal inhabitant of certain soils. Another group of bacteria common in soils is the myxobacteria belonging to the genera *Myxococcus, Chondrococcus, Archangium, Polyangium, Cytophaga* and *Sporocytophaga*. The latter two genera are cellulolytic and hence are dominant in cellulose-rich environments. Myxobacteria feed on other Gram-negative bacteria through lysis.

It is not easy to determine the total population of bacteria in any soil accurately. Apart from the inherent limitations of the soil dilution and plate methods, their numbers vary with the texture, water content and many other parameters especially the availability of organic substrates in soil. Bacteria can withstand extremes of climate although temperature and moisture influence their population. In Arctic zones where temperature is below the freezing point, bacteria can thrive as luxuriantly as they do in arid desertic soils where temperatures are very high. The inherent faculty of many bacteria to form spores possessing tough outer covering facilitates

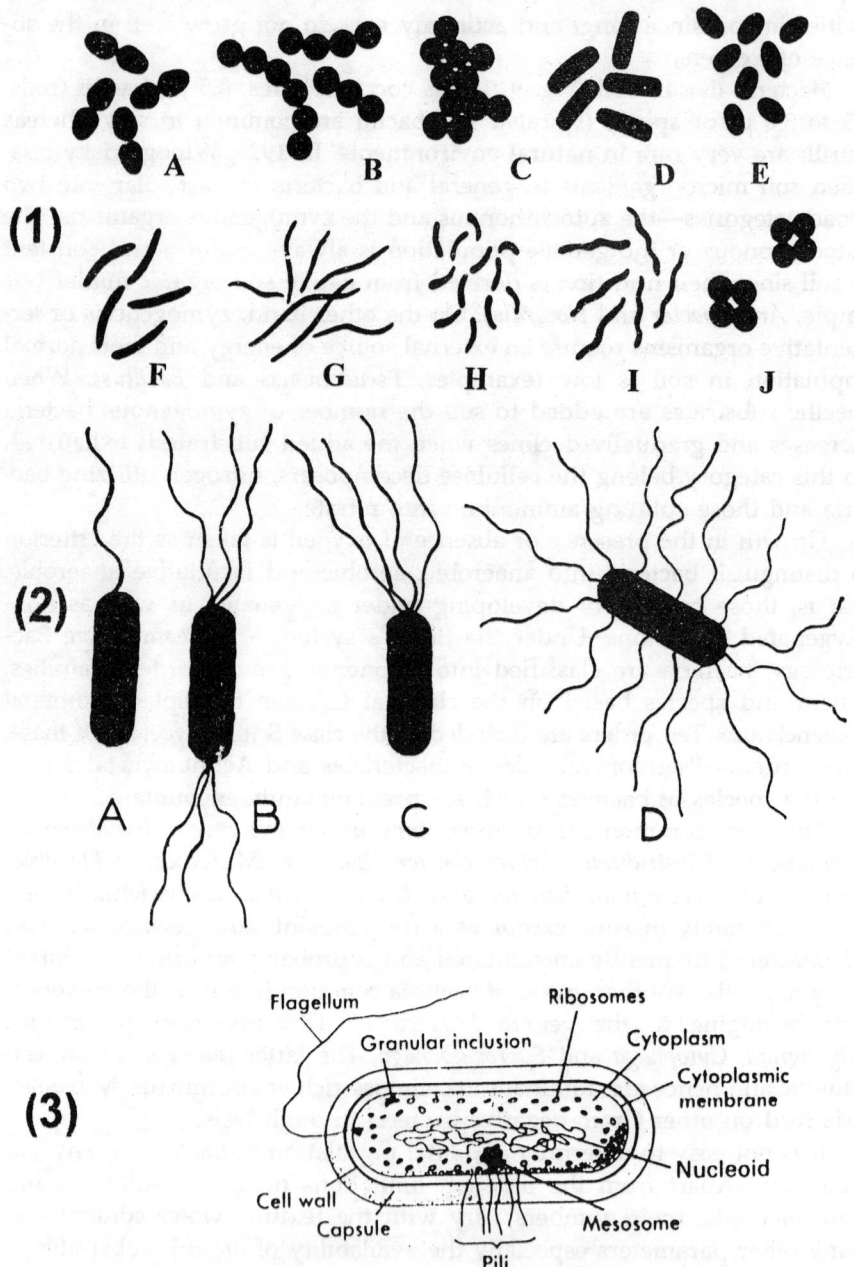

Fig. 9 Bacterial forms and structure: (1)—A, diplococci; B, streptococci; C, staphylococci; D, bacilli; E, coccobacilli; F, fusiform bacilli; G, filamentous bacillary forms; H, vibrios; I, spirilla; J, sarcinae; (2)—Position of flagella on the bacterial cell. *A.* monotrichous; *B.* amphitrichous; *C.* lophotrichous; *D.* peritrichous; (3)—structure of a typical bacterial cell.

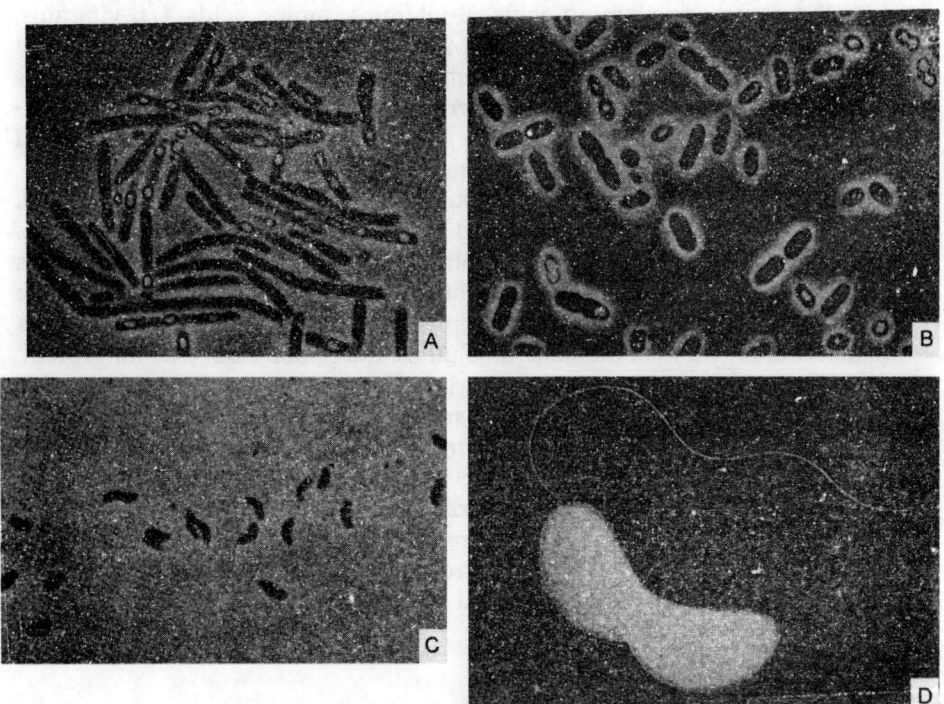

Fig. 10 Phase contrast photomicrograph showing (A) *Bacillus megaterium* containing intracellular endospores (from the textbook on *Microbiology—Molecules, Microbes and Man* by Nester, E.W., Roberts, C.E., McCarthy, B.J. and Pearsall, N.N. 1973; Holt, Rinehart and Winston Inc., New York); (B) *Azotobacter chroococcum* (from the textbook, *Fundamental Principles of Bacteriology* by Salle, A.J. 1967, Tata McGraw-Hill Publishing Co. Ltd., Bombay); (C) *Desulfovibrio* sp., vibriod cells of Hildenborough strain and (D) *Desulfovibrio* sp., a dividing cell with a single polar flagellum (from the textbook, *Microbiology—An Introduction to Protists* by J.S. Poindexter, 1971, The Macmillan Company, New York).

the survival of bacteria in all adverse environments. Survival by spore formation under extreme conditions ought to be differentiated from tolerance to different temperature ranges, which is one of the factors determining the population of bacteria in soil. Based on this criterion, bacteria are grouped as mesophiles (15 to 45°C), psychrophiles (below 20°C) and thermophiles (45 to 65°C). The mesophilic bacteria, however, constitute the bulk of soil bacteria. Other factors affecting bacterial population in soil are pH, farm practices, fertilizer and pesticide applications and organic matter amendments.

Bacteria are also classified on the basis of their nutritional requirements into those requiring amino acids, B-vitamins, amino acids + B-vitamins, unidentified growth factors in yeast extract or soil extract and soil extract + yeast extract. The source of B-vitamins and other growth factors in soils is difficult to explain and the occurrence of fastidious species in soil requiring growth substances can only be explained on the basis of mutual dependence of different bacterial strains on extracellular products.

Autotrophic as well as heterotrophic bacteria are present in soil. Autotrophs synthesize their own food whereas heterotrophs depend on preformed food for nutrition. Photoautotrophs are those whose food energy is derived through mediation of sunlight, as in the instance of photosynthetic bacteria as opposed to chemoautotrophs which oxidize inorganic materials to derive energy and at the same time utilize the carbon from CO_2 for growth. In the latter category, a group of bacteria known as obligate chemoautotrophs, are included which prefer specific substrates. Examples of this kind are *Nitrobacter* which utilizes nitrite, *Nitrosomonas* which utilizes ammonium, *Thiobacillus* which converts inorganic sulphur compounds to sulphate and *Ferrobacillus* capable of converting ferrous iron to ferric iron. Several of the reactions involved in nitrogen transformations in soil depend on the chemoautotrophic *Nitrobacter* and *Nitrosomonas* and hence chemoautotrophy of bacteria in soil is intimately related to crop production.

Cell Wall

The details of cell wall structure differ basically between Gram positive and Gram negative bacteria. Primarily, these two groups can be differentiated by Gram strain procedure (the name bearing the name of the microbiologist who devised the method). The cell wall of both Gram +ve and Gram –ve bacteria contain peptidoglycan and murein. Gram +ve bacteria have a thick (15 to 80 nm) peptidoglycan layer composed of chains of alternating subunits of N-acetylglucosamine and N-acetylmuramic acid. These chains are cross linked (bridged) *via* tetrapeptide units extending from the N-acetylmuramic acid which imparts rigidity to the cell wall. Species specific cell wall antigens are defined by the protein chains and bridges. Many Gram –ve and some Gram +ve bacteria have diaminopimelic acid (a precursor of L-lysine) in some tetrapeptide chains. All Gram +ve cell walls possess techoic acids bound to the peptidoglycan layer (Fig. 11).

In Gram –ve bacteria, the peptidoglycan layer is thinner (1–2 nm) than Gram +ve bacteria. Unlike Gram +ve bacteria, the negative ones have a phospholipid outer membrane outside the peptidoglycan layer. There is a periplasmic space between the outer membrane and cytoplasmic membrane. The outer membrane is protective in function against exposure to toxic substances, with porins (membrane proteins) serving to regulate the transport of materials through membrane pores. This regulatory mechanism is partly responsible for antibiotic resistance in Gram –ve bac-

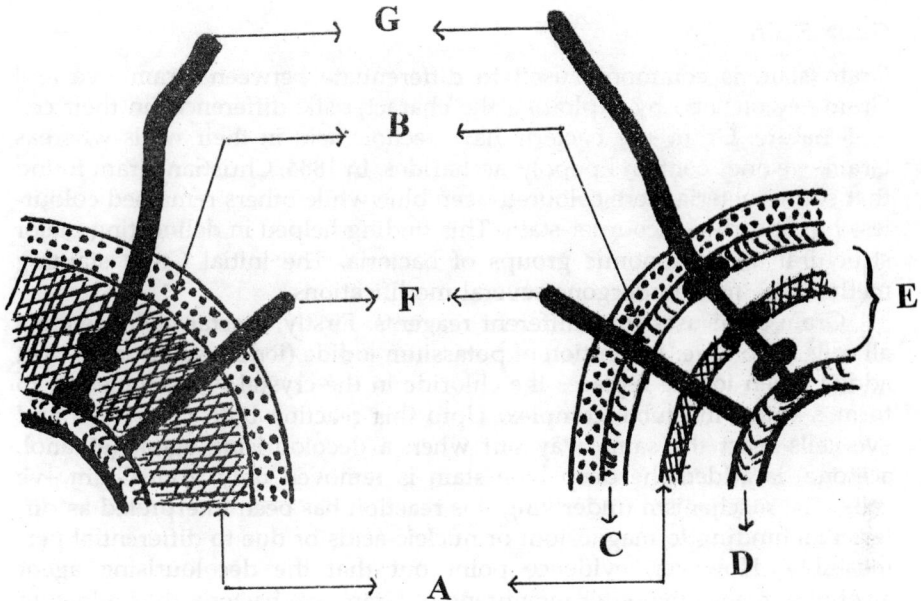

Fig. 11 Comparison between cell wall of Gram positive (left side of the figure) and Gram negative (right side of the figure) bacteria: Gram positive bacteria have a thick peptidoglycan layer (A) surrounded by a polysaccharide capsule (B) whereas Gram negative bacteria have a thin peptidoglycan layer (A) sandwiched between an outer membrane (C) and an inner cytoplasmic membrane (D); the space between the two membranes is known as the periplasmic space (E). The pilus (F) and the flagellum (G) are common to both types and are well embedded in the cell wall. Other differences are stated in the text.

teria. The presence of lipoprotein layer in the Gram –ve cell wall stabilises the outer membrane and cross links it to diaminopimelic acids in the peptidoglycan layer. Within the outside layer of the outer membrane, lipopolysaccharide (LPS) is embedded in Gram –ve bacteria which has a lipid component (endotoxin, lipid A) responsible for the toxic properties of this group of bacteria. The polysaccharide component has a common core to all Gram –ve bacteria but the variable terminal segment exposed on the outer surface serves as a major surface antigen. This variability of the terminal segment has been the key to differentiate bacterial isolates based on serological methods.

Many bacteria are enclosed within a polysaccharide slimy layer known as capsule which prevents destruction from outside agents and also helps in adhesion to substrates. There are also pili (fimbriae) which help in adherence by its hair like structure. Flagella made up of proteins help in the locomotion of bacteria. The number of flagella on a cell is of taxonomic value—monotrichous (single), lopotrichous (at both ends) and peritrichous (all over the surface). Some bacteria (*Bacillus* and *Clostridum*) often form spores to overcome dessication.

Gram Stain

Gram stain is commonly used to differentiate between Gram +ve and Gram –ve bacteria by exploiting the characteristic differences in their cell wall nature. Gram +ve bacteria have techoic acid in their walls whereas Gram –ve ones contain lipopolysaccharides. In 1884, Christian Gram found that some bacteria were coloured deep blue while others remained colourless but absorbed a counter-stain. This finding helped in delineating major structural and taxonomic groups of bacteria. The initial Gram staining method has now undergone several modifications.

Gram stain uses four different reagents. Firstly, crystal violet renders all cells deep blue. A solution of potassium iodide (iodine solution) is then added when iodide replaces the chloride in the crystal violet molecule to form a water insoluble complex. Upto this reaction both Gram +ve and –ve cells react the same way but when a decolourising agent (ethanol, acetone) is added the deep blue stain is removed only from Gram –ve cells. The mechanism underlying this reaction has been interpreted as differential binding to magnesium or nucleic acids or due to differential permeability. However, evidence point out that the decolourising agent alcohol damages the outer membrane of Gram –ve bacteria thus allowing the crystal violet iodine complex to leak out while the undamaged Gram +ve bacterial cells retain the stain. A second stain (counter-stain) safranin is taken up by the colourless Gram –ve cells and makes them visible.

Actinomycetes

These are soil organisms which have characteristics common to bacteria and fungi and yet possess sufficient distinctive features to delimit them into a distinct category. In the strict taxonomic sense, actinomycetes are clubbed with bacteria in the same class of Schizomycetes but confined to the order Actinomycetales. On agar plates, they can easily be distinguished from true bacteria. Unlike slimy distinct colonies of true bacteria which grow quickly, actinomycete colonies appear slowly, show powdery consistency and stick firmly to agar surface. A closer look at a colony under the compound microscope reveals slender unicellular branched mycelium (diameter of the hypha rarely exceeding one micron) forming asexual spores for propagation. They bear certain similarities to Fungi Imperfecti in the branching of the aerial mycelium which profusely sporulate and in the formation of distinct clumps or pellets in liquid cultures. Certain actinomycetes, on the other hand, resemble *Mycobacterium* and *Corynebacterium* in all respects both morphologically and physiologically including susceptibility to the attack by viruses. Viruses are known to attack bacteria and actinomycetes but not fungi. Actinomycetes differ from fungi in the composition of their cell wall. They do not have chitin and cellulose which are commonly found in the cell walls of fungi.

The number of actinomycetes increases in the presence of decomposing organic matter. As a rule, they are intolerant to acidity and their numbers decline at pH 5.0. The most conducive range of pH is between 6.5 and 8.0. Waterlogging of soil is unfavourable for the growth of actinomycetes whereas desertic soils of arid and semi-arid zones sustain sizeable population, probably due to the resistance of spores to desiccation. The percentage of actinomycetes in the total microbial population increases with the depth of soil and actinomycetes can be isolated in sufficient number even from soil samples obtained from the C horizon of a soil profile.

Delineation of species within different genera of actinomycetales has always been a difficult problem owing to many characteristics which are common with bacteria. For instance, *Mycobacterium* and *Mycococcus* of the family Mycobacteriaceae have many characteristics common to bacteria and in fact, are spoken of as bacteria in common parlance. Nevertheless, the order Actinomycetales has been classified into four families—Mycobacteriaceae, Actinomycetaceae, Streptomycetaceae and Actinoplanaceae. In the order of abundance in soils, the commonest genera of actinomycetes are *Streptomyces* (nearly 70%), *Nocardia* and *Micromonospora* (Figs. 12, 13) although Actinomyces, Actinoplanes and Streptosporangium have also been encountered occasionally.

Temperatures between 25 and 30°C are conducive for the growth of actinomycetes although thermophilic cultures growing at 55 and 65°C are common in compost heaps where they are numerically extensive and mostly belong to the genera *Thermoactinomyces* and *Streptomyces*.

Fungi

Next only to bacteria in abundance in soil, fungi dominate all soils and possess filamentous mycelium composed of individual hyphae. The hyphae may be uni-, bi- or multinucleate and non-septate (without cross walls) or septate. Asexual propagation by the production of spores or conidia takes place mitotically either with or without interception by a well-defined sexual cycle involving gametic fusion and subsequent production of spores through meiotic or reduction division. The size, shape and colour of conidia or spores and the physiological characteristics of cultures in artificial as well as natural substrates provide valuable taxonomic criteria in the classification of fungal isolates into well-defined genera and species.

All the environmental factors which influence the distribution of bacteria and actinomycetes also influence the fungal flora of soil. The quality and quantity of organic matter present in soil have a direct bearing on fungal numbers in soil since most fungi are heterotrophic in nutrition. Fungi are dominant in acid soils because acidic environment is not conducive for the existence of either bacteria or actinomyctes, resulting in the monopoly of fungi for utilization of native substrates in soil. They are also

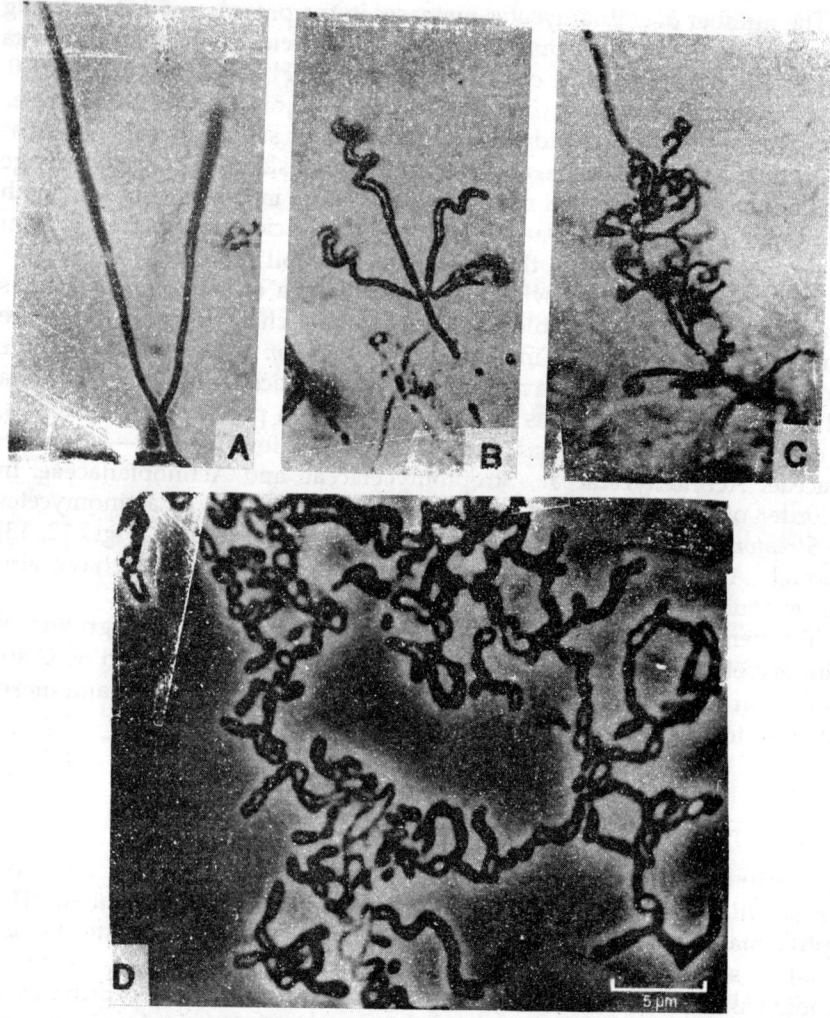

Fig. 12 Photomicrographs showing some soil actinomycetes—(A) *Streptomyces* sp. showing conidia on conidiophores and the vegetative mycelium is visible in the background; (B) *S. roseochromogenus*; (C) *S. coelicolor* (from the textbook on *Microbiology—An Introduction to Protists* by J.S. Poindexter, 1971, The Macmillan Co., New York) and (D) Phase contrast photomicrograph of *Nocardia asteroides* (from the textbook on *Microbiology—Molecules, Microbes and Man* By Nester, E.W., Roberts, C.E., McCarthy, B.J. and Pearsall, N.N. 1973; Holt Rinehart and Winston Inc., New York).

present in neutral or alkaline soils and some can tolerate pH beyond 9.0. Arable soils contain abundant fungi since they are strictly aerobic and excess of soil moisture decreases their numbers. Isolation of fungi from different horizons of soil profiles shows that these organisms exhibit selective preference for various depths of soil. Those fungi which are common in

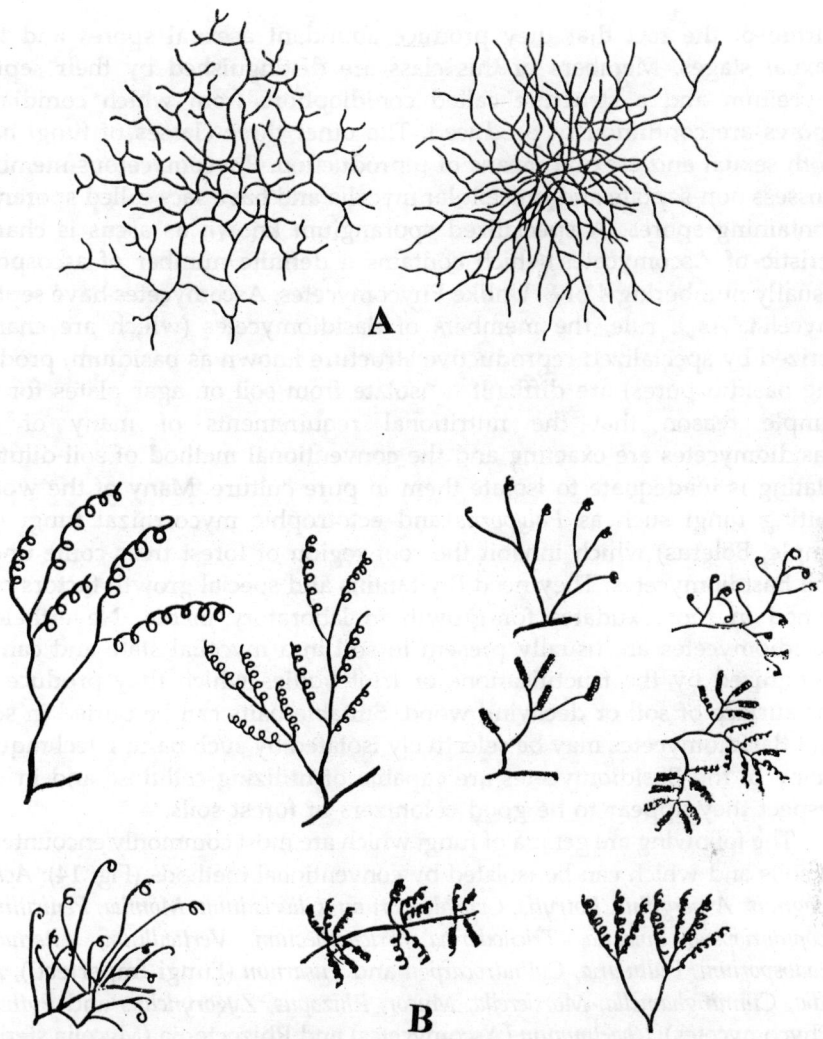

Fig. 13 General view of mycelium (A) and sporophores (B) of actinomycetes.

lower depths are rarely encountered on the surface of soils which may be explained on the basis of the availability of organic matter and the ratio between oxygen and carbon dioxide in the soil atmosphere at varying depths. Seasonal fluctuations in fungal numbers are not uncommon. Farm practices including crop rotation and fertilizer or pesticide applications influence the nature and dominance of fungal species.

Broadly speaking, fungi are classified into Phycomycetes, Ascomycetes, Basidiomycetes and Fungi Imperfecti. Many fungi which are commonly isolated from soil come under the class Fungi Imperfecti by virtue of the fact that they produce abundant asexual spores and lack sexual stages. Members of this class are distinguished by their septate mycelium and a structure called conidiophore from which conidia or spores are continuously produced. The other three classes of fungi have both sexual and asexual means of reproduction. Phycomycetous members possess non-septate and unicellular mycelia and have sacs called sporangia containing spores. A specialized sporangium known as ascus is characteristic of Ascomycetes which contains a definite number of ascospores usually numbering 4 or 8. Unlike Phycomycetes, Ascomycetes have septate mycelia. As a rule, the members of Basidiomycetes (which are characterized by specialized reproductive structure known as basidium, producing basidiospores) are difficult to isolate from soil on agar plates for the simple reason that the nutritional requirements of many of the basidiomycetes are exacting and the conventional method of soil-dilution plating is inadequate to isolate them in pure culture. Many of the wood-rotting fungi such as *Polyporus* and ectotrophic mycorrhizal fungi (example, Boletus) which inhabit the root region of forest trees come under soil Basidiomycetes. They need B-vitamins and special growth factors contained in root exudates for growth in laboratory media. Nevertheless, Basidiomycetes are usually present in soil in a mycelial state and can be recognized by the fructifications or fruit bodies which they produce on the surface of soil or decaying wood. Suitable baits can be buried in soils and Basidiomycetes may be selectively isolated by such baiting techniques. Many of the Basidiomycetes are capable of utilizing cellulose and in this respect they appear to be good colonizers of forest soils.

The following are genera of fungi which are most commonly encountered in soils and which can be isolated by conventional methods (Fig. 14): *Acrostalagmus, Aspergillus, Botrytis, Cephalosporium, Gliocladium, Monilia, Penicillium, Scopulariopsis, Spicaria, Trichoderma, Trichothecium, Verticillium, Alternaria, Cladosporium, Pullularia, Cylindrocarpon* and *Fusarium* (Fungi Imperfecti); *Absidia, Cunninghamella, Mortierella, Mucor, Rhizopus, Zygorynchus* and *Pythium* (Phycomycetes); *Chaetomium* (Ascomycetes) and Rhizoctonia (Mycelia sterilia, which fail to produce reproductive structures).

Many soil yeasts belonging to true Ascomycetes such as Saccharomyces (Fig. 15) and those belonging to Fungi Imperfecti such as Candida can also be isolated on acidified media (pH 4.0). Their numbers in soil are low and their significance in soil is poorly understood.

One of the primary functions of filamentous fungi in soil is to degrade organic matter and help in soil aggregation. Besides this property, certain species of *Alternaria, Aspergillus, Cladosporium, Dematium, Gliocladium, Helminthosporium, Humicola* and *Metarhizium* produce substances similar to humic substances in soil and hence may be important in the maintenance of soil organic matter. Some of the fungi capable of forming ectotrophic associations on the root system of forest trees such as pine belonging to

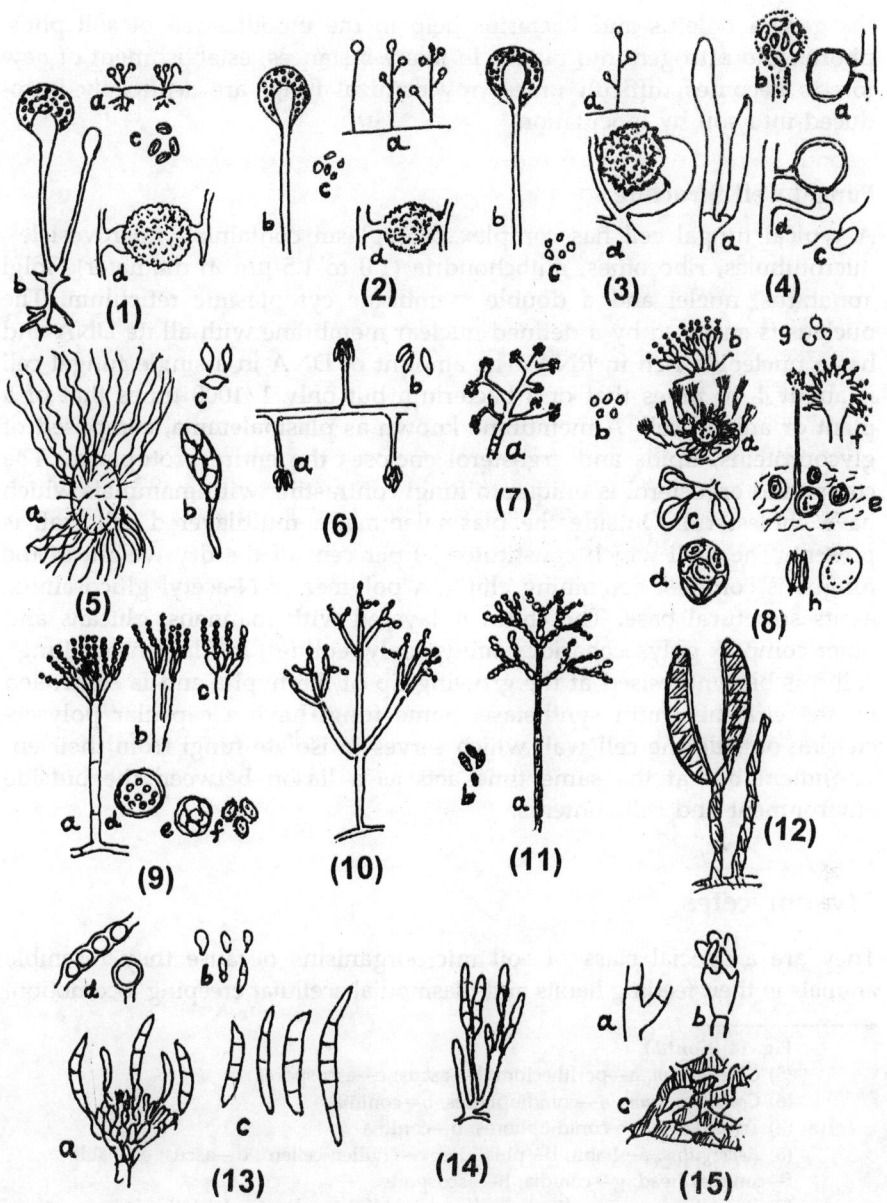

Fig. 14 Some examples of common soil fungi (genera) that come up on agar plate by soil dilution method (after Gilman, 1957).

(1) *Rhizopus.* a—habit; b—sporangiophores; c—sporangiospores; d—zygospore

(2) *Mucor.* a—habit; b—sporangiophore; c—sporangiospores; d—zygospore

(3) *Zygorhynchus.* a—habit; b—sporangiophore; c—sporangiospores; d—zygospore

(4) *Pythium.* a—sporangia; b—zoospore discharge; c—zoospore; d—oogonium

(Contd.)

the genera Boletus and Lactarius help in the mobilization of soil phosphorus and nitrogen into plants. In many instances, establishment of new forests becomes difficult unless mycorrhizal fungi are artificially introduced into soil by inoculation.

Fungal Cell Structure

A typical fungal cell has complex protoplasm containing microvesicles, microtubules, ribosomes, mitochondria (1.0 to 1.5 µm in diameter) Golgi apparatus, nuclei and a double membrane cytoplasmic reticulum. The nucleus is enclosed by a defined nuclear membrane with all its DNA and has a nucleolus rich in RNA. The amount of DNA in a single fungal cell is about 4–10 times that of a bacterium but only 1/1000 times that of a plant or animal cell. A membrane known as plasmalemma, composed of glycoproteins, lipids and ergosterol encloses the entire protoplasm. The compound ergosterol is unique to fungi contrasting with mammals which have cholesterol. Outside the plasmalemma, a multilayered cell wall is present. The wall which constitutes 90 per cent of the dry weight of the fungus is complex containing chitin, a polymer of N-acetyl glucosamine as its structural base. The chitin is layered with mannans, glucans and other complex polysaccharides amidst polypeptides. In filamentous fungi, chitin is biosynthesised at the growing tip of the hypha and is controlled by the enzyme chitin synthetase. Some fungi have a capsular polysaccharide outside the cell wall which serves to isolate fungi from their environment but at the same time acts as a liason between the outside environment and cell contents.

Myxomycetes

They are a special class of soil microorganisms because they resemble animals in their feeding habits and plasmodial acellular creeping locomotion.

Fig. 14 (Contd.)
(5) Chaetomium. a—perithecium; b—ascus; c—ascospores
(6) Cephalosporium. a—conidiophores; b—conidia
(7) Trichoderma. a—conidiophores; b—conidia
(8) Aspergillus. a—habit; b—phialides; c—"hullen-cellen"; d—ascus; e—habit; f—conidial head; g—conidia; h—ascospores.
(9) Penicillium. a—conidiophore; b, c—conidial heads; d—cleistothecium; e—ascus; f—ascospores
(10) Verticillium. conidiophore and conidia
(11) Cladosporium. a—conidiophore; b—conidia
(12) Helminthosporium. conidiophores and conidia
(13) Fusarium. a—conidial head; b—microconidia; c—macroconidia; d—chlamdospores
(14) Cylindrocarpon. conidiophores and conidia
(15) Rhizoctonia. a—hypha; b—basidium and spores; c—sclerotial hyphae.

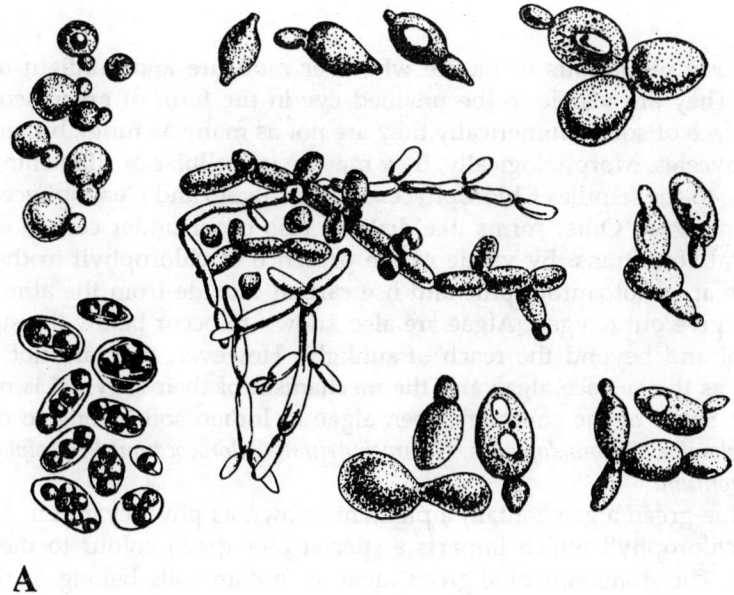

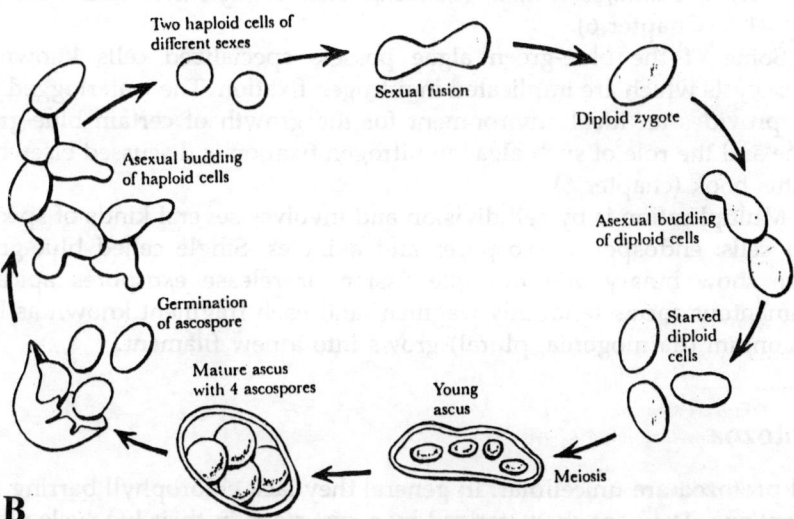

Two haploid cells of
different sexes

Sexual fusion

Diploid zygote

Asexual budding
of haploid cells

Asexual budding
of diploid cells

Germination
of ascospore

Starved
diploid
cells

Mature ascus
with 4 ascospores

Young
ascus

Meiosis

Fig. 15   A—Different forms of yeasts; B—life cyle of *Saccharomyces cerevisiae* (Bakers yeast)—
From Frobisher, 1968.

They also resemble fungi because reproductive structures and spores are formed. They are abundant in decaying vegetation or dung, often attaining 50 cm or more in size. The genus *Physarum* (100 species) is the largest genus of myxomycetes.

Algae

Soil algae are ubiquitous in nature wherever moisture and sunlight are available. They are visible to the unaided eye in the form of green scum on the surface of soils. Numerically they are not as many as fungi, bacteria or actinomycetes. Morphologically, they may be unicellular or filamentous and belong to the families Chlorophyceae (green algae) and Cyanophyceae (blue-green algae). Other forms like diatoms also occur under certain environmental conditions. By virtue of the presence of chlorophyll in their cells, algae are photoautotrophic and use carbon dioxide from the atmosphere and give out oxygen. Algae are also known to occur below the surface of soil and beyond the reach of sunlight. However, they are not as numerous as the surface algae and the mechanism of their survival is not very clear. Some of the common green algae in Indian soil belong to the genera: *Chlorella, Chalmydomonas, Chlorochytrium, Chlorococcum, Protosiphon* and *Oedogonium*.

The blue-green algae contain a pigment known as phycocyanin in addition to chlorophyll which imparts a special blue-green colour to these organisms. The dominant blue-green algae in Indian soils belong to the genera: *Chroococcus, Aphanocapsa, Lyngbya, Oscillatoria, Phormidium, Microcoleus, Cylindrospermum, Anabaena, Nostoc, Scytonema* and *Fischerella* (Fig. 41 in Chapter 6).

Some of the blue-green algae possess specialized cells known as heterocysts which are implicated in nitrogen fixation. The waterlogged rice soil provides an ideal environment for the growth of certain blue-green algae and the role of such algae in nitrogen fixation is discussed elsewhere in this book (chapter 6).

Multiplication is by cell division and involves several kinds of specialized cells: endospores, exospores and akinetes. Single celled blue-green algae show binary and multiple fission or release exospores apically. Filamentous forms randomly fragment and each fragment known as harmogonium (harmogonia, plural) grows into a new filament.

Protozoa

Soil protozoa are unicellular. In general they lack chlorophyll barring few exceptions. They are characterized by a cyst stage in their life cycle which can help the species to withstand adverse soil conditions. Barring a few genera which reproduce sexually by fusion of cells, the rest of them reproduce asexually by fission. The flagellated protozoans belonging to the class Mastigophora are predominant in soil. Important genera are *Allantion, Bodo, Cercobodo, Cercomonas, Entosiphon, Heteromita, Monas, Oikomonas, Sainouran, Spiromonas, Spongomonas* and *Tetramitus* (Fig. 16). Unlike the flagellates which move with the help of their flagella numbering

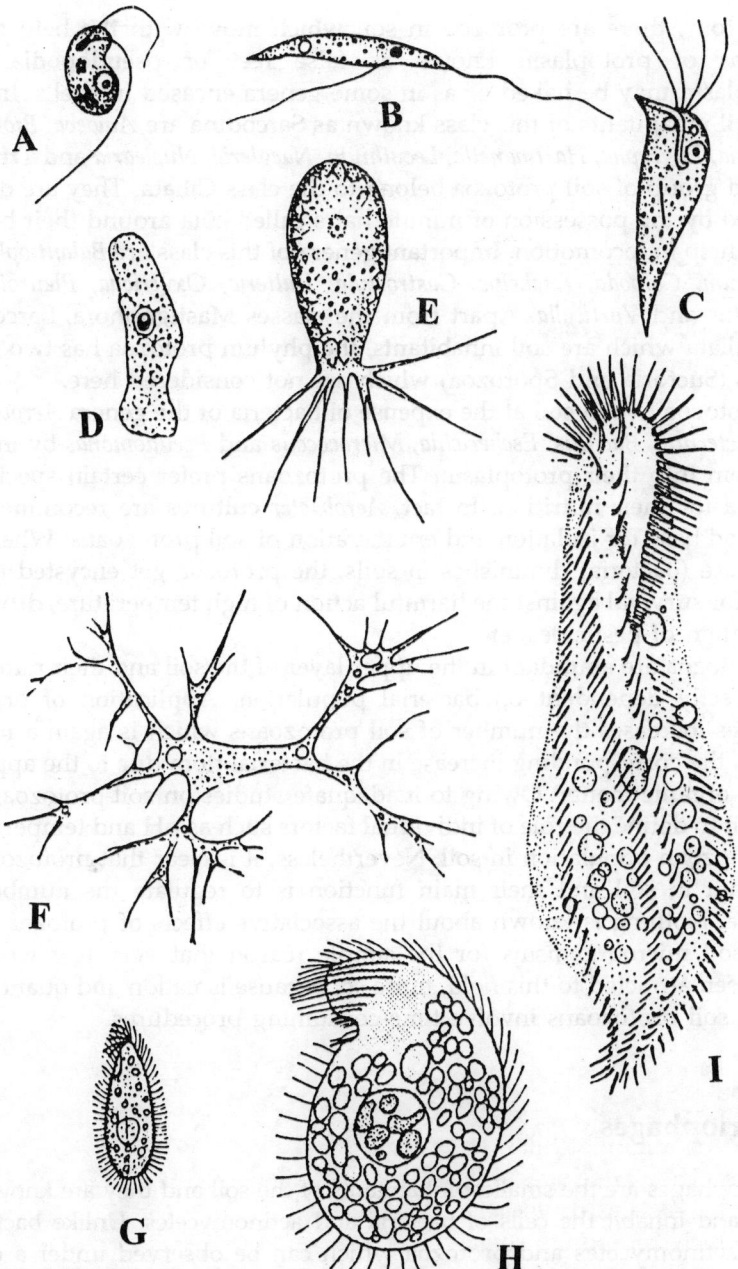

Fig. 16 Some examples of soil protozoa: A—*Bodo*; B—*Cercobodo*; C—*Tetramitus*; D—*Naegleria*; E—*Euglypha*; F—*Biomyxa*; G—*Balantiophorus*; H—*Colpoda*; I—*Uroleptus* (From Alexander, 1961).

up to four, there are protozoa in soil which move with the help of ex-
trusions of protoplasm known as false feet or pseudopodia. The
protoplasm may be naked or as in some genera encased in shells. Important soil inhabitants of this class known as Sarcodina are *Amoeba, Biomyxa,
Difflugia, Euglypha, Hartmanella, Lecythium, Naegleria, Nuclearia* and *Trinema.*
A third group of soil protozoa belong to the class Ciliata. They are distinguished by the possession of minute hairs called cilia around their bodies
which help in locomotion. Important genera of this class are *Balantiophorus,
Colpidium, Colpoda, Enchelys, Gastrostyla, Halteria, Oxytricha, Pleurotricha,
Uroleptus* and *Vorticella.* Apart from the classes Mastigophora, Sarcodina
and Ciliata which are soil inhabitants, the phylum protozoa has two more
classes (Suctoria and Sporozoa) which are not considered here.

Protozoa live in soil at the expense of bacteria of the genera *Aerobacter,
Agrobacterium, Bacillus, Escherichia, Micrococcus* and *Pseudomonas* by ingest-
ing them into their protoplasm. The protozoans prefer certain species of
bacteria for their nutrition. In fact, *Aerobacter* cultures are recommended
as a food base for isolation and enumeration of soil protozoans. When the
food base (bacteria) diminishes in soils, the protozoa get encysted (form
cysts) for survival against the harmful action of high temperature, drought,
application of pesticides, etc.

Protozoa are abundant in the upper layer of the soil and their numbers
are directly dependent on bacterial population. Application of organic
manures increases the number of soil protozoans which is again a reflection on the corresponding increase in the bacterial flora due to the application of organic matter. Owing to inadequate studies on soil protozoa, it is
difficult to define the role of individual factors such as pH and temperature
on protozoan population in soil. Nevertheless, it is clear that protozoa are
abundant in soil and their main function is to regulate the number of
bacteria. Nothing is known about the associative effects of protozoa with
other soil microorganisms for the simple reason that very few workers
have been attracted to this field, more so, because isolation and quantification of soil protozoans involve time-consuming procedures.

Bacteriophages

Bacteriophages are the smallest inhabitants of the soil and they are known to
attack and inhabit the cells of bacteria and actinomycetes. Unlike bacteria,
fungi, actinomycetes and protozoa which can be observed under a compound microscope, bacteriophages can be seen only under an electron
microscope because of their minute size. Some authors prefer to make a
sub-division known as actinophages while referring to phages attacking
actinomycetes. The phages attacking blue-green algae are known as
cyanophages. Although phages cannot be seen without the help of an

electron microscope, the lysis caused by the action of specific phages on their hosts can be seen as 'plaques' on agar plates. Bacteriophages can pass through bacterial filters since their size rarely exceeds 0.05 to 0.01 µ in diameter. The bacteriophage has a head-like and a tail-like structure (Figs. 17, 18). The tail attaches itself to the surface of the bacterium and gains entry into host's protoplasm. Lysis sets in when the bacteriophage multiplies resulting in the liberation of many more units to reinfect new bacterial cells. It is too early to assess the importance of bacteriophages in the overall influence of soil on agricultural productivity since sufficient information on this aspect is lacking to make any generalization.

Fungal Viruses

Although it was stated earlier that fungi are distinguishable from bacteria and actinomycetes in their freedom from virus attack, several workers have come to recognize the existence of fungal viruses in recent years. Prior to 1968 convincing evidence did not come forth regarding the presence or replication of fungal viruses and they were invariably referred to as virus-like bodies. Viruses have been reported to occur in over 60 species from some 50 genera of fungi and most reports include studies using an electron microscope. The most extensively studied system is the mycovirus of *Penicillium chrysogenum*. Examinations of virus particles for electrophoretic mobility, sedimentation coefficient, buoyant density and serological specificity reveal differences among those viruses infecting species of *Penicillium, Aspergillus, Periconia* and *Ustilago*. Accurate determinations of molecular weight and amino acid content of different viruses have been made and electron microscope studies suggest that young apical regions of hyphae of *Penicillium* are free of virus particles whereas the older regions of hyphae contain many particles. Viruses have also been observed in sections of fungal spores. In three species of *Penicillium*, viruses are sometimes associated with lytic plaque formation (Fig. 19) which are not often reproducible, probably due to altered sensitivity of the host. The relationship between fungal viruses and the metabolism or genetics of their hosts is poorly understood.

Microscopy

Since the discovery of lens and compound microscope in the 16th century and microorganisms by Antoine van Leeuwenhoek (1632–1723), several types of microscopes have been commercially made available. These instruments can provide structural details of microorganisms that are beyond the reach of the unaided human eye. Two features of microscopes related to magnification and resolution limit provided by the lenses

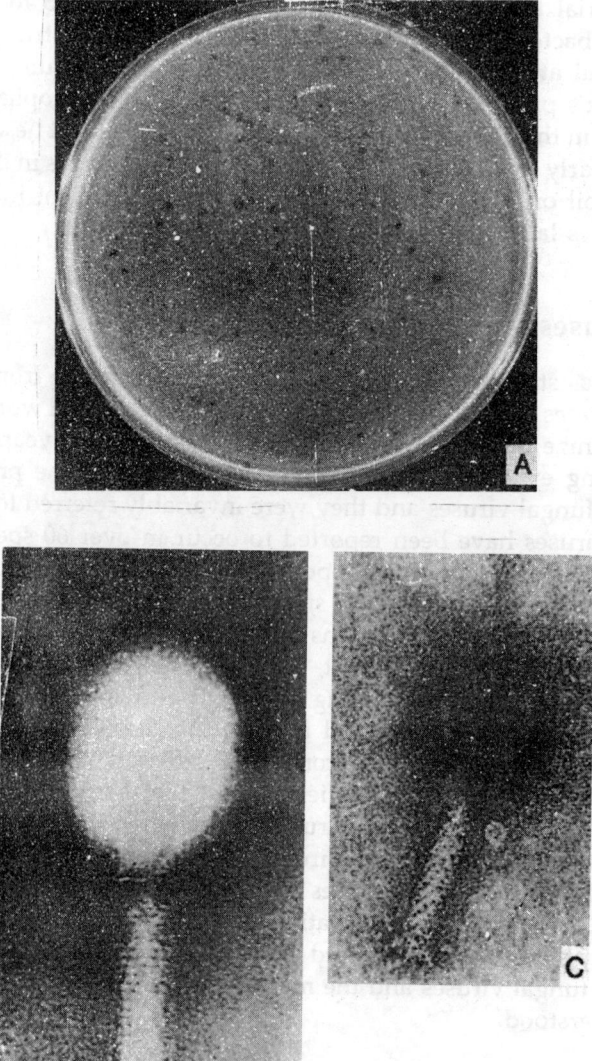

Fig. 17 A—the formation of plaques (black dots) due to bacteriophage specific to *Rhizobium trifolii* on agar plates seeded with a suspension of the bacterium and the phage; (B) Electron micrograph of *Escherichia coli*, phage T4 showing head and tail (from the textbook on *Microbiology—An Introduction of Protists* by J.S. Poindexter, 1971, The Macmillan Co., New York); and (C) Electron micrograph of negatively stained particle showing whole virion infecting the blue-green alga *Plectonema boryanum* (Courtesy: P.K. Singh, New Delhi).

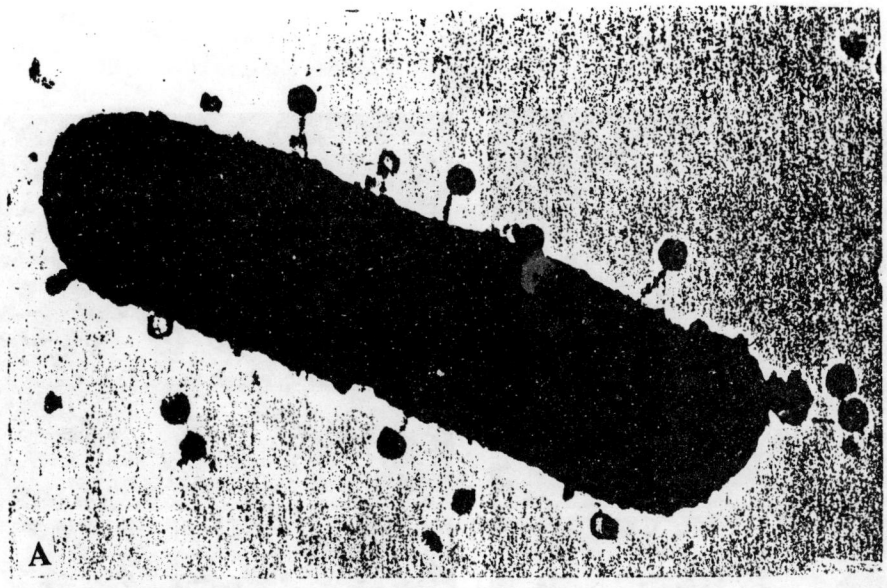

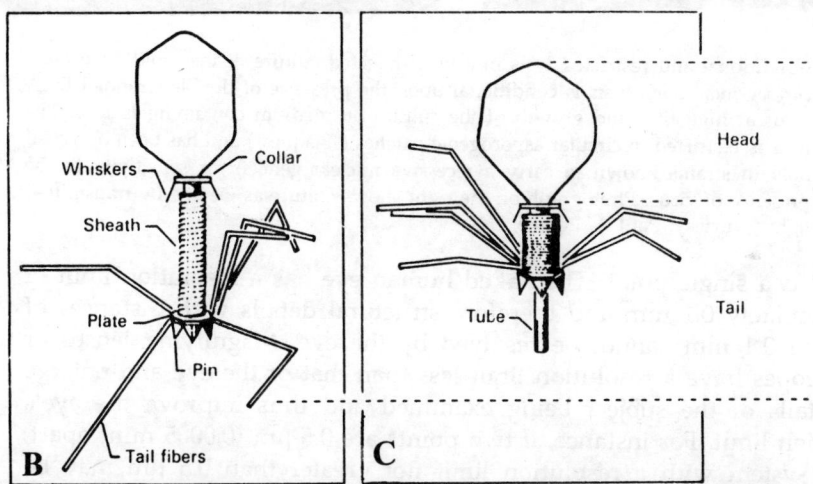

Fig. 18 A—Bacteriophages of *E. coli* adsorbed to the host cell by tips of their tails (from a highly enlarged electron micrograph); B—Diagram of a bacteriophage before injection into the host cell, and C—the same as in B after injection into the host cell (from Davis *et al.*, 1980).

or lens systems used. Magnification of an object results in the multiplication effect of the overall dimensions of the objects viewed. The resolution limit of a lens (or lens system) is the smallest distance by which two points can be separated and still be seen as distinct points; blurred image results when this resolution limit becomes less because the two points merge and

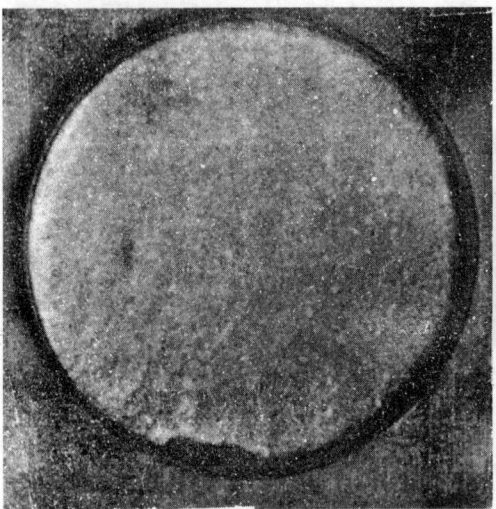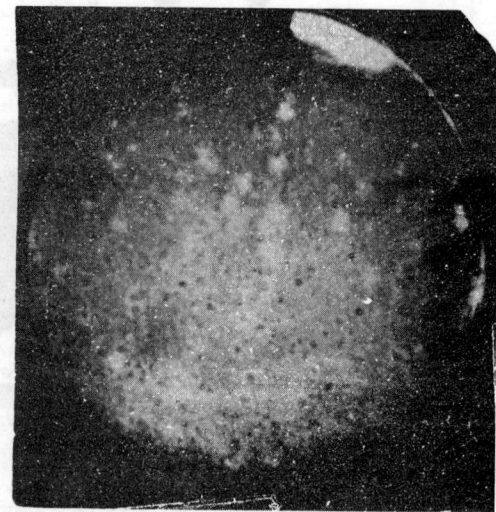

Fig. 19 Conditional and restricted lysis in a virus-infected culture of the mold *Penicilliur* *chrysogenum*. Such lysis is conditional upon the presence of double-stranded RNA virus at high titer and growth of the culture on medium containing lactose. The lysis is restricted to circular asporogenic patches ("Plaques") and has been observed only in strains known to carry a recessive nuclear gene(s) for sensitivity to the presence of virus. Photograph on the right shows culture as viewed by transmitted light (courtesy: Paul A. Lemke, U.S.A.).

appear as a single point. The naked human eye has a resolution limit of approximately 0.1 mm and therefore structural details with distances of less than 0.1 mm cannot be resolved by the eye. Magnifying lenses or microscopes have a resolution limit less than that of the eye and enlarge the details of the subject being examined and thus improve the eye's resolution limit. For instance, if two points are 0.5 μm (0.0005 mm) apart, a lens system with a resolution limit not greater than 0.5 μm may be needed to see the two points distinctly. To achieve this resolution limit, the lens system must magnify the image delivered to the eye atleast 200 times so that the two points are apart by 0.1 mm or higher (200 × 0.0005 mm = 0.1 mm). This magnification comes close to or above the natural resolution limit of the naked eye.

Compound Light Microscope

Two lens systems are used to enlarge the image in the compound light microscope. There is an 'objective' lens placed near the object on a slide whose image is further magnified by the 'ocular' or 'eye piece'. The combined magnification capacities of the objective and the eye piece results in the magnified image seen by the unaided eye. If the objective is 10× and the eye piece is 100× the image will have $10 \times 100 = 1000$ magnification. These microscopes carry a condensor lens placed under the slide that focusses light on the specimen making the microscopic field bright. The beam of light can be controlled by a centrifugally operating diagphram (Fig. 20).

The resolution limit is dependent on the 'numerical aperture' (NA) which reflects the light-gathering ability of the lens and the wavelength of visible light used. The mathematical equation for arriving at the resolution limit is as follows: Resolution limit = $\dfrac{\text{wave length}}{\text{NA}}$. The best resolution limit can be had by using the shortest wavelength of visible light and an

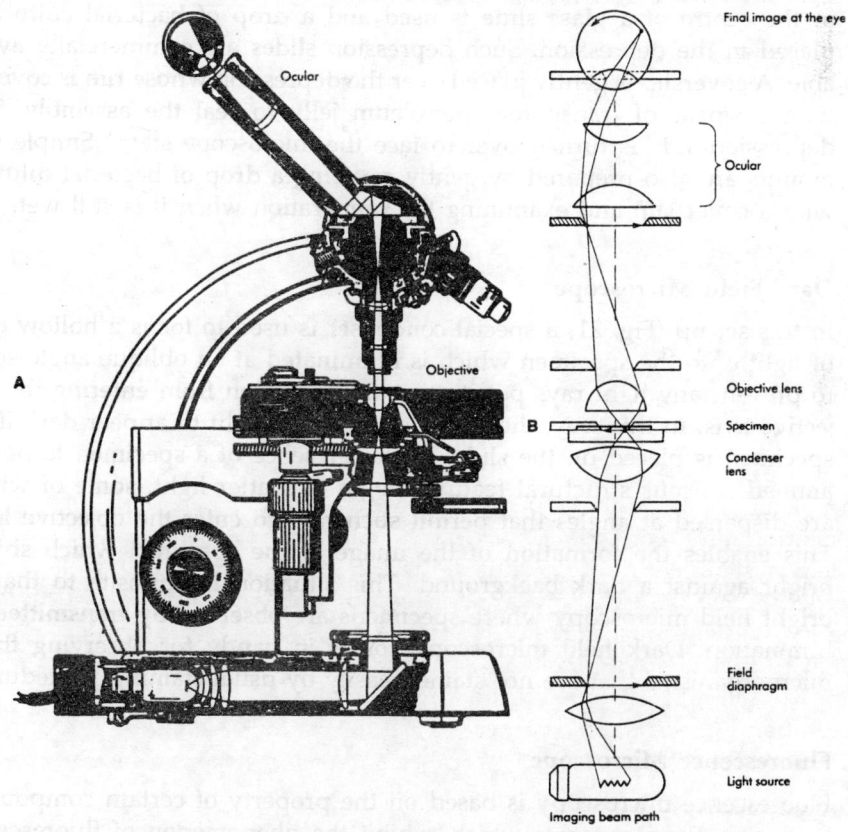

Fig. 20 **A,** typical compound microscope. **B,** diagrammatic representation of light paths established in compound microscrope. (*Courtesy of Carl Zeiss, Inc., Thornwood, N.Y., from Howard et al. 1994*).

objective lens of maximum numerical aperture. The oil immersion objective lens can provide about 1.4 numerical aperture which in combination with blue light of wavelength 0.5 μm, a resolution of about 0.2 μm can be obtained with a compound microscope.

Bright Field Microscope

When transmitted light instead of direct beam of light is used in a compound microscope, some structures in the specimen being examined appear darker than others due to differential absorption of light by different parts of the specimen, thereby providing contrast not seen when viewed under direct light. Both live specimens to understand motility of microorganisms as well as stained preparations can be viewed under a bright field microscope. Wet mounts or 'hanging drop' preparations come in handy for observing live specimens. To prepare a hanging drop slide a depression in the centre of a glass slide is used and a drop of bacterial culture is placed in the depression. Such depression slides are commercially available. A coverslip is gently placed over the depression whose rim is covered with a smear of transparent petroleum jelly to seal the assembly. The depression side is turned over to face the microscope stage. Simple wet mounts are also prepared by gently covering a drop of bacterial dilution with a cover slip and examining the preparation when it is still wet.

Dark Field Microscope

In this set up (Fig. 21) a special condenser is used to focus a hollow core of light onto the specimen which is illuminated at an oblique angle so as to prevent any light rays penetrating the specimen from entering the objective lens. In this way, the microscopic field ought to appear dark if no specimen is placed on the slide. In the presence of a specimen to be examined, various structural features begin to scatter light, some of which are dispersed at angles that permit such rays to enter the objective lens. This enables the formation of the image of the specimen which shines bright against a dark background. This situation is opposite to that of bright field microscopy where specimens are observed by transmitted illumination. Dark field microscopy comes in handy for observing those microorganisms that are not stained easily by usual staining procedures.

Fluorescence Microscope

Fluorescence microscopy is based on the property of certain compounds known as fluorochromes which exhibit the phenomenon of fluorescence under ultraviolet (UV) light. These compounds have the capacity to absorb light from the short visible to long UV portion of the spectrum depending upon the fluorochrome used and re-emit the absorbed energy in the form

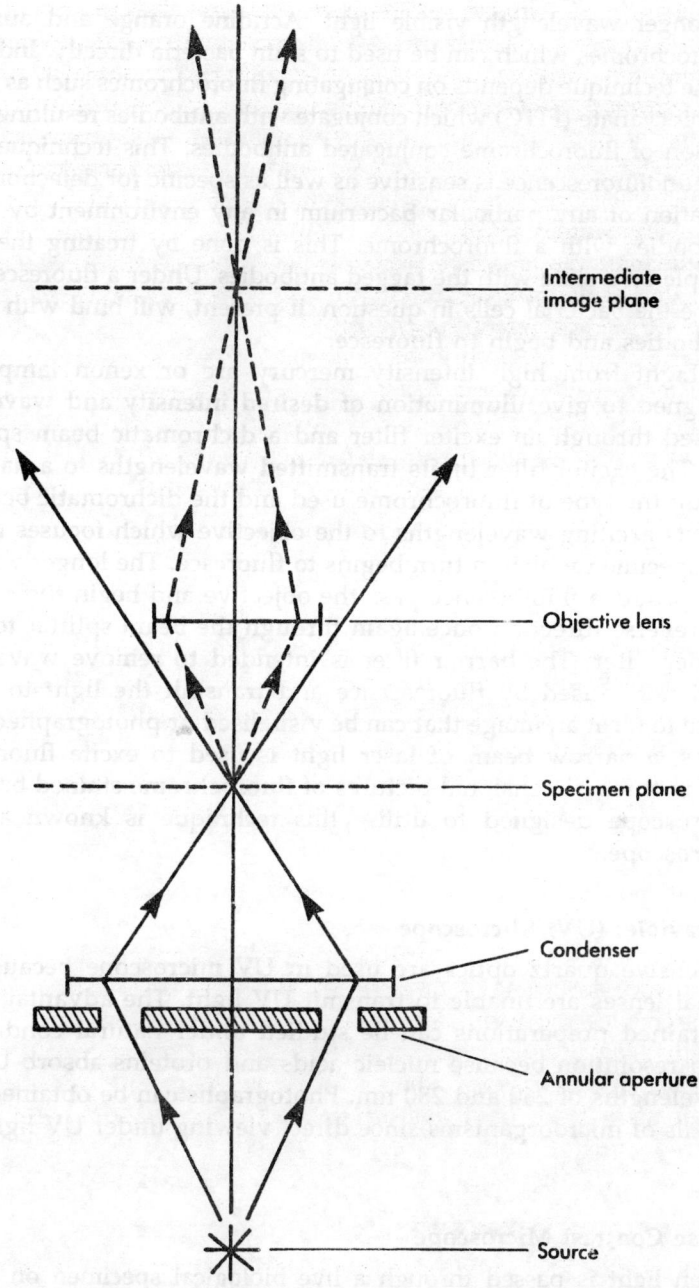

- - - - Intermediate
image plane

- - Objective lens

- - - - Specimen plane

Condenser

Annular aperture

Source

Fig. 21 Operation of dark field condenser to achieve dark field microscopy. The annular aperture in the condenser focuses a hollow cone of light onto the specimen. Only rays scattered by the specimen (*dashed lines*) enter the objective to form an image. (From Howard, *et al.* 1994).

of longer wavelength visible light. Acridine orange and auramine are fluorochromes which can be used to stain bacteria directly. Indirect usage of the technique depends on conjugating fluorochromes such as fluorescein isothiocyanate (FITC) which conjugate with antibodies resulting in the formation of fluorochrome conjugated antibodies. This technique known as immunofluorescence is sensitive as well as specific for detection and identification of any particular bacterium in any environment by tagging its antibodies with a fluorochrome. This is done by treating the suspected sample on a slide with the tagged antibodies. Under a fluorescence microscope the bacterial cells in question, if present, will bind with the tagged antibodies and begin to fluoresce.

Light from high intensity mercury arc or xenon lamps that are designed to give illumination of desired intensity and wavelengths is passed through an exciter filter and a dichromatic beam splitter (Fig. 22). The exciter filter limits transmitted wavelengths to a narrow band to suit the type of fluorochrome used and the dichromatic beam splitter directs exciting wavelengths to the objective which focuses them on to the specimen which in turn begins to fluoresce. The longer wavelengths that produce fluorescence pass the objective and begin their passage in the reverse direction once again through the beam splitter to reach the barrier filter. The barrier filter is intended to remove wavelengths of light not caused by fluorescence and transmit the light to the ocular piece to form an image that can be visualised or photographed. In recent times, a narrow beam of laser light is used to excite fluorescence to obtain three dimensional pictures of fluorochrome stained bacteria. The microscope designed to utilise this technique is known as confocal microscope.

Ultraviolet (UV) Microscope

Expensive quartz optics are used in UV microscope because conventional lenses are unable to transmit UV light. The advantages are that unstained preparations can be studied under natural conditions with high resolution because nucleic acids and proteins absorb UV light at wavelengths of 260 and 280 nm. Photographs can be obtained to reveal details of microorganisms since direct viewing under UV light is harmful.

Phase Contrast Microscope

When light is passed through a live biological specimen on a slide, no significant differences in absorption of light by different parts of the specimen can be seen unless the specimen is crowded by pigments such as chlorophyll or melanin. This leads to images with little or no contrast. Staining preparations can provide the much needed contrast but staining

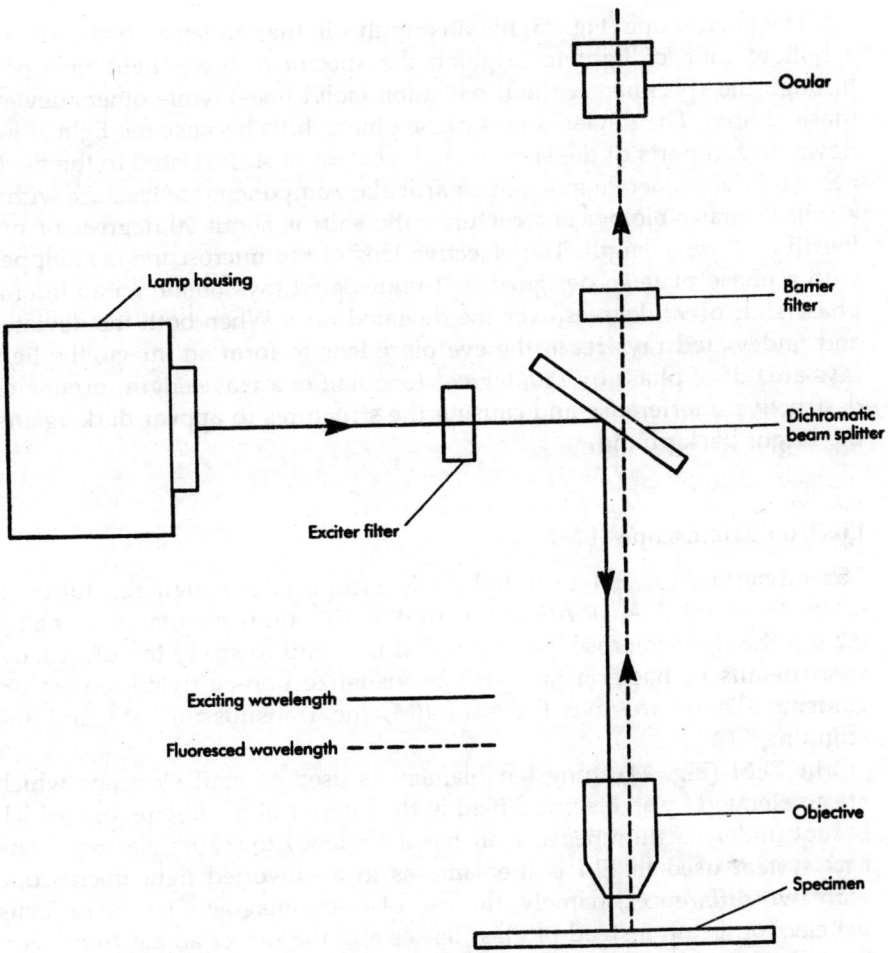

Lamp housing

Ocular

Barrier
filter

Dichromatic
beam splitter

Exciter filter

Exciting wavelength ———————

Fluoresced wavelength – – – – – – –

Objective

Specimen

Fig. 22 Diagrammatic representation of optical system of modern epifluorescence microscope. Light from the high-intensity lamp passes through exciter filters, which
limit transmitted wavelengths to a narrow band appropriate to the fluorochrome
being used. The dichromatic splitter directs exciting wavelengths to the objective,
which focuses them onto the specimen. The longer wavelength emitted as a result
of fluorescence passes the beam splitter, the barrier filter removes wavelengths
of light not caused by fluorescence, and the image is observed directly or
photographed. (Courtesy of Carl Zeiss, Inc., Thornwood, N.Y., from Howard *et
al.* 1994)

can lead to artifacts. Therefore, phase contrast microscopes are designed
to provide contrast to live specimens to detect cell inclusions under natural
conditions. These microscopes are equipped with special interference optics to exploit phase differences in natural biological specimens to reveal
their distinct identities.

The microscope (Fig. 23) has an annulus in the condenser that provides a hollow cone of light to brighten the specimen. Some light rays pass through the specimen without deviation (solid lines) while others deviate (dashed line). The deviated rays cause phase shifts because the light slows down within parts of the specimen, the extent of shift related to the thickness and the refractive index of a particular component or structure within a cell. In many biological structures, the shift is about 90 degrees or one fourth of a wavelength. The objective lens of the microscope is equipped with a phase plate so designed that undeviated rays obtain an additional phase shift of 90 degrees over the deviated rays. When both the deviated and undeviated rays reach the eye piece lens to form an image, the light rays are out of phase by 180 degrees (one half of a wavelength) producing destructive interference and causing the structures to appear dark against the bright background.

Electron Microscope (EM)

The advantage of EM over light microscope is the high resolution it offers to about 3 Å (1 Angstrom unit = 10^{-7} mm) compared to about 0.2 μm the light microscope offers. EM is useful to study the ultrastructural details of bacteria and also to visualize non-cultivable micro-organisms. There are two types of EM, the transmission EM and the scanning EM.

In TEM (Fig. 24), tungsten filament is used to emit electrons which are accelerated by an electrical field in the interior of the microscope which is kept under vacuum because air has a tendency to scatter electrons. The lens system used in EM is the same as in an inverted light microscope with two differences, namely, the use of electromagnetic lenses to focus the electron beam instead of glass lenses and the use of an electronic gun instead of an incandescent lamp. Focusing of the electron beam is done by varying the magnetic field of electromagnetic lenses. The electron beam is concentrated by the use of one or two condenser lenses. An objective lens initially magnifies the image which is further enlarged by the use of one or more projector lenses in the same way the eye-piece lens does in a light microscope. The enlarged image can be flashed on a fluorescent screen which is then photographed. The photograph (electron micrograph) is later interpreted to bring home the characteristic features.

The specimen used for TEM studies has to be processed prior to microscopy. This is done by staining thin specimens (80 nm or less) with heavy metals such as uranium, osmium or lead. When these dense elements are present, electrons are scattered from the beam resulting in dark images on the fluorescent screen or in electron micrographs. Individual cells may be affixed to thin electron-transparent films and stained with heavy metal stains. However, tissues must be fixed in a suitable fixative,

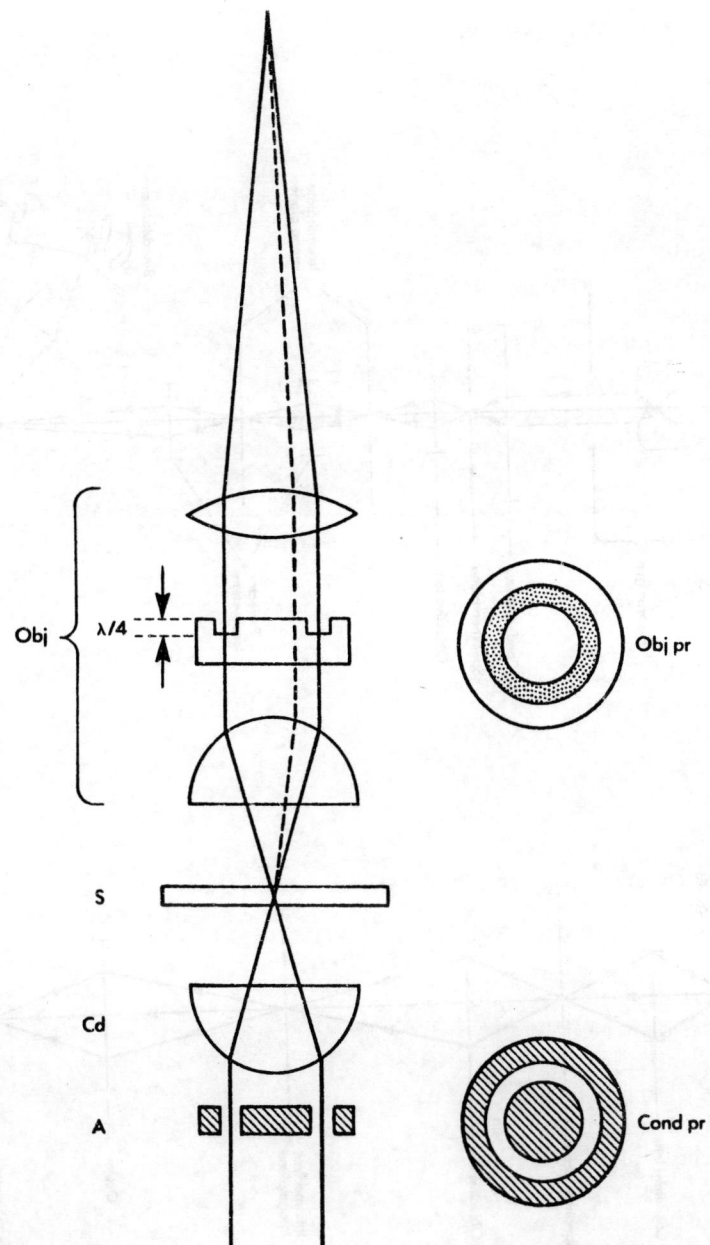

Fig. 23 Basic optical features of the phase contrast microscope displayed in diagram form. Comparison of pathway taken by representative deviated ray *(dashed line)* and undeviated ray *(solid line)* illustrates how the phase plate retards deviated rays by an additional one-fourth wave length ($\lambda/4$), leading to interference and production of contrast in the image. A = annulus; *Cd* = condenser lens; *S* = specimen; *Obj* = phase contrast objective; *Obj pr* = phase ring; *Cond pr* = condenser phase ring. (From Howard *et al.*, 1994).

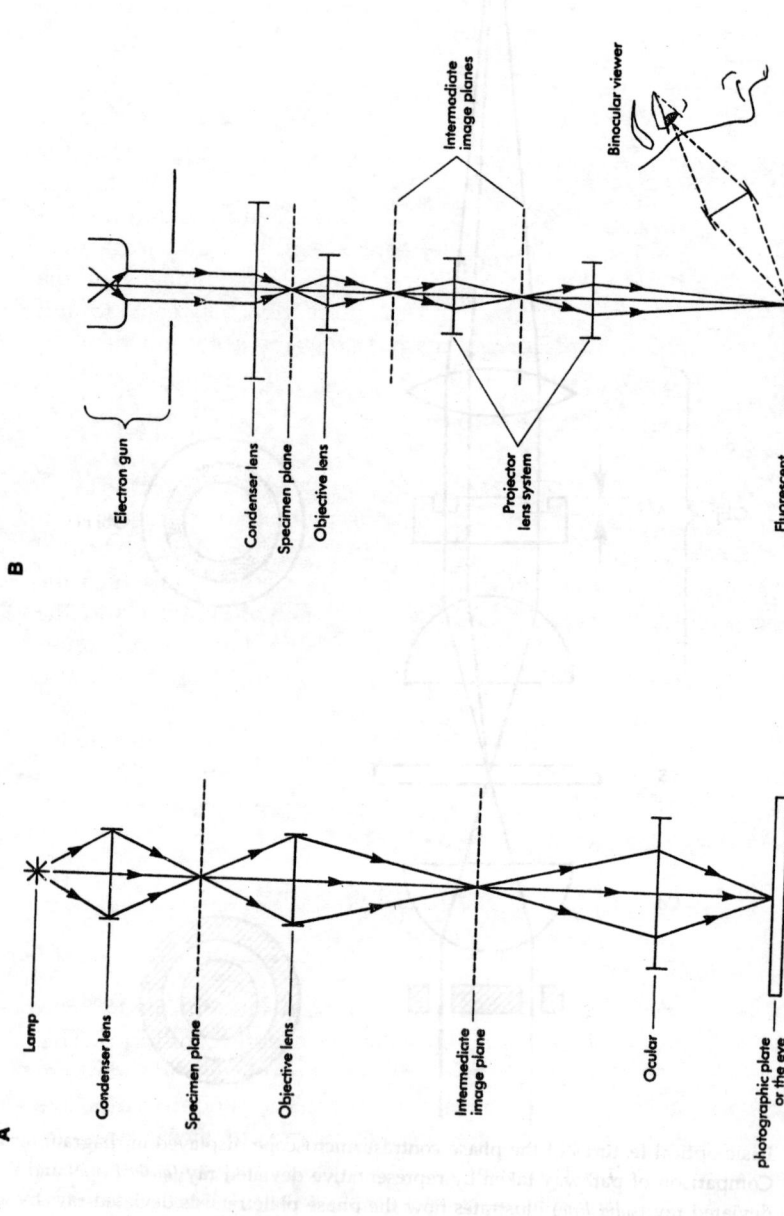

Fig. 24   Comparison of optical systems of, **A**, inverted light microscope and, **B**, transmission electron microscope. Basic layouts are similar, but focusing electron beam requires electromagnetic rather than glass lenses, and an electron gun instead of an incandescent lamp. (From Howard *et al.* 1994).

embedded in plastic and sectioned using diamond or glass knives on an ultramicrotome. These sections are placed on a slide, processed and stained with osmium, uranium or lead to reveal structural details. All these procedures need skill, patience, plenty of time and above all a trained eye to interpret the electron micrographs.

Scanning Electron Microscope

The scanning electron microscope (SEM) operates differently from the transmitted electron microscope (TEM). Specimens (0.5–1 cm blocks) are dehydrated, stuck on a support and vacuum-coated with a metal, for example gold or platinum. A beam of electrons scans back and forth over the surface of the specimen creating three dimensional images of the specimen with a good depth of field and magnification between ×15 and ×100,000. Surface structures of microorganisms can be seen and hence it is particularly useful to view fungal spores. SEM is also useful in viewing internal structure of cells.

Confocal Microscopy

The method of image formation in a confocal microscope is different from a conventional microscope. In the conventional type, the whole specimen is illuminated uniformly and simultaneously along the plane in which the objective lens is focussed, resulting in the reduction of contrast and the resolution of the image. In a confocal microscope the illumination comes from a laser source focussed as a spot in one point of the specimen at a time in a sequential manner and finally a complete in focus image of the entire specimen is obtained. In this way confocal microscope can visualize cellular organelles, cytoskeletal elements, RNA/DNA and provide 3-D pictures of specimens (Fig. 25).

Methods Used in Soil Microbiological Studies

Sterilization Methods

Sterilization is the process of killing microorganisms in media and glassware used in the cultivation of cultures under aseptic conditions. This is accomplished by dry heat, heat under pressure, chemicals and radiation. Hot air ovens are designed to sterilize glassware. The modern autoclave is in principle a sophisticated pressure cooker where the steam pressure is regulated after evacuating air completely from the chamber. Both temperature and pressure have to be checked because pressure alone is of no significance in killing microorganisms. Usually at 15 lbs/square inch pressure when temperature reaches 121°C, 15 minutes retention of media and glassware under these conditions is enough to achieve freedom from

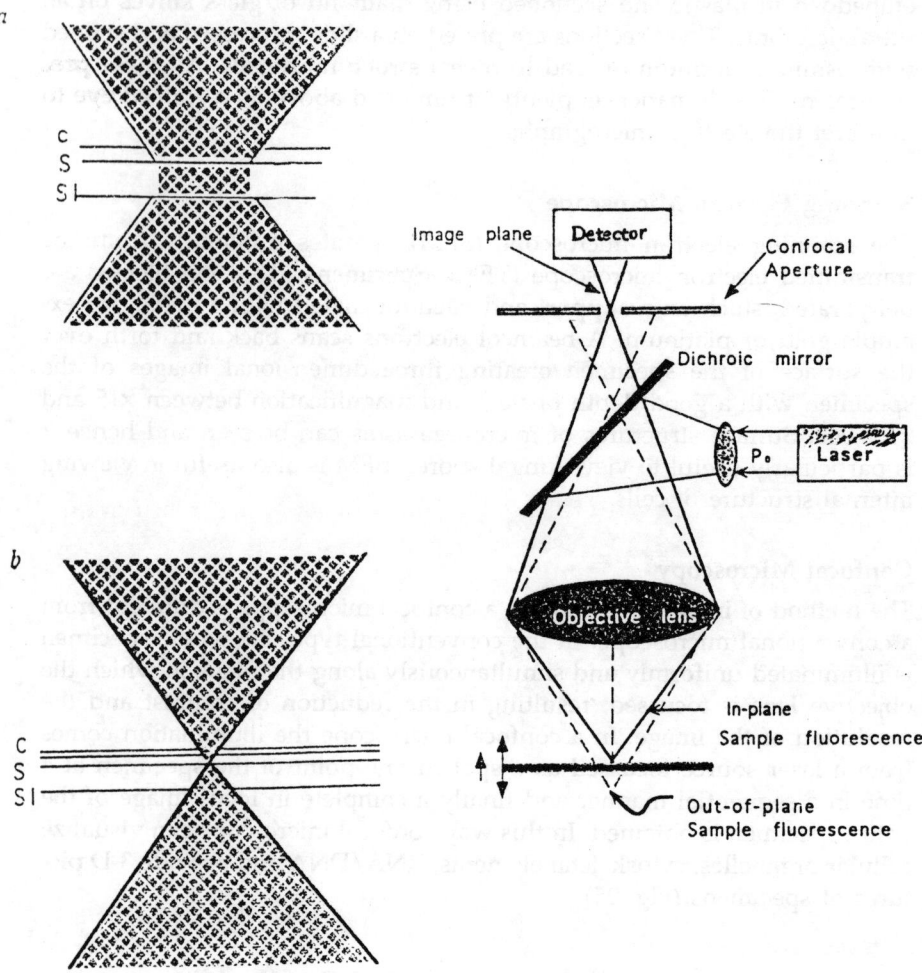

Fig. 25 *Left side:* Illumination of specimen in conventional and laser scanning confocal microscopy. In conventional microscopy (a), the entire depth of specimen is illuminated continuously which results in the detection of out-of-focus and in-focus signals together, causing loss of resolution. In LSCM, the specimen is illuminated sequentially covering specific points at a time (b). These images of points which are devoid of out-of-focus signals are then added to form a complete in-focus image of the specimen (c-coverslip; s-specimen; sl-slide). *Right side:* The confocal principle. A laser beam is directed on to the specimen by a scanning system through a high numerical aperture objective. Induced fluorescent or reflected light is scattered in all directions. Light is collected by the objective lens and directed through the scanning system towards the beam splitter. Imaging aperture allows only in-focus light (straight line) and eliminates the out-of-focus light (broken line). From Singh and Gopinathan, 1998.

microorganisms. Retaining soil samples for 20 minutes successively for two times at 20 lbs pressure/square inch is recommended for sterilization of soil samples and extracts. This procedure allows the removal of spores which cannot resist such treatments. Infact, Tyndallization, named after Tyndall is based on the principle that if media are kept at 60°C for certain period and later the process is repeated after an interval to allow spores to germinate can sterilize vegetative as well as reproductive cells of microorganisms.

Cold sterilization of heat labile liquids is best achieved by the use of ceramic and membrane filters often called millipore filters. These filters are made of cellulose acetate, cellulose nitrate, polycarbonate or polyvinylidene fluoride. Poresizes of 0.4 to 0.2 µm filter out bacteria while filters with pore sizes smaller than these dimensions have been designed to remove viruses.

Laminar flow cabinets are increasingly used to control sterility in a modern laboratory where air is passed through a set of sterilizing filters to provide sterile air currents in a cabinet. The surroundings in the cabinet are free from microorganisms and conducive for aseptic transfer of biological specimens. Laminar flow cabinets coupled with U-V light and disinfectants can achieve complete sterility to handle microbiological procedures under aseptic conditions (Fig. 26).

Obtaining Soil Samples

Soil samples are collected normally at a depth of 15 cm and transferred to clean containers. Three to five samples are taken for each replicate and mixed evenly. From the mixed sample, at least 10–25 g of soil are taken as a representative sample of the particular replicate. For obtaining soil samples from different depths, a special auger can be used.

Soil Dilution and Plate Counts

The soil dilution and plate count method has been most widely used for enumeration and isolation of soil microorganisms. Appropriate soil dilutions (1 ml aliquots) are plated on suitable solid media for this purpose (see Fig. 33 in Chapter 4). To suppress bacteria from appearing on plates and to encourage fungal growth, either rose bengal or streptomycin is incorporated into the media. Thermophilic microorganisms (capable of tolerating higher temperatures) are usually isolated by incubating dilution plates in an incubator adjusted to 55–60°C. Details of the method are given at the end of the chapter on rhizosphere and lists of different media intended for various groups of microorganisms including those capable of degrading substrates such as cellulose and lignin are given in the appendix.

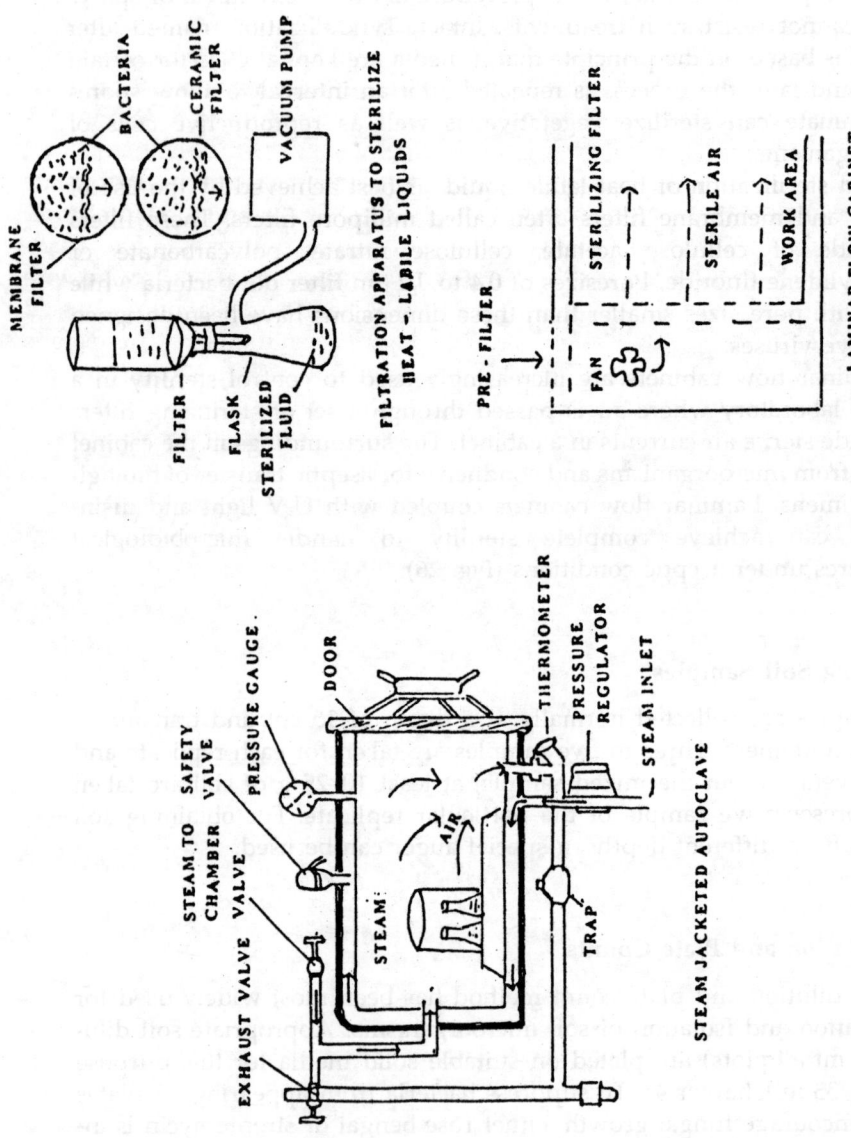

MEMBRANE FILTER

BACTERIA

CERAMIC FILTER

FILTER

FLASK

STERILIZED FLUID

VACUUM PUMP

FILTRATION APPARATUS TO STERILIZE HEAT - LABILE LIQUIDS

PRE - FILTER

STERILIZING FILTER

STERILE AIR

WORK AREA

FAN

LAMINAR FLOW CABINET

STEAM TO CHAMBER VALVE

SAFETY VALVE

PRESSURE GAUGE.

DOOR

THERMOMETER

PRESSURE REGULATOR

STEAM INLET

AIR

STEAM

TRAP

EXHAUST VALVE VALVE

STEAM JACKETED AUTOCLAVE

Fig. 26 Different ways of sterilization of media and work place.

The Streak Plate

Streak plate ensures continuous in situ dilution of a given sample or a colony of bacterium (uniform or mixed) on the surface of solidified agar medium in a Petri plate. Wire loops made of platinum, nichrome or stainless steel mounted on solid metal or wooden handle that are commercially available are used. One such wire loop is passed over the flame of a bunsen burner and air-cooled before use. The sterilized loop is dipped into the specimen to obtain a small quantity of the inoculum which is then placed in the centre of the first two quadrants of a Petri plate. The Petri plate containing the agar medium has to be marked earlier into 4 quadrants on the back of the plate. The initial inoculum is spread in the top two quadrants in the form of a streak back and forth (first streak) using the wire loop. The plate is then rotated 90 degrees by hand and the loop is passed back and forth in the 2nd and 3rd quadrant (second streak). The plate is again rotated 90 degrees and the loop is passed back and fourth for the 3rd time in the 3rd and 4th quadrant (third streak). The objective of making successive streaks is to subject the initial inoculum into serial dilution and all these step-wise procedures are carried out in a sterile hood or inoculation chamber. Streak plate is also useful in picking up single colonies of apparently pure and uniform cultures from a mixture of colonies. Repeated streaking of single colonies will also be helpful in obtaining pure cultures (Fig. 27).

The Soil Plate

In soil dilution and plating method, one disadvantage is that not all the microorganisms adhering to the surface of soil particles are taken into account since it is likely that some microorganisms may still stick to the surface of soil particles in the form of hyphae in spite of shaking soil samples in water to bring them into suspension. To overcome this deficiency, a small quantity of soil (0.005–0.01 g) is directly placed in Petri plates and crushed with a needle followed by pouring of melted and cooled agar medium. In certain cases, a crumb of soil (up to 1 cm) has been used to study soil microflora.

Enrichment Cultures

Winogradsky and Beijerinck developed the technique of enrichment cultures making use of the principle of natural selection. When a heterogeneous inoculum is added to a liquid medium only those organisms which are adapted to the conditions or nutrients available in the medium will start proliferating. A culture medium can easily be devised to cater to the nutritional requirements of a particular group of microorganisms and that group soon dominates the medium to the exclusion of others. Repeated transfers to fresh media finally give pure cultures. By

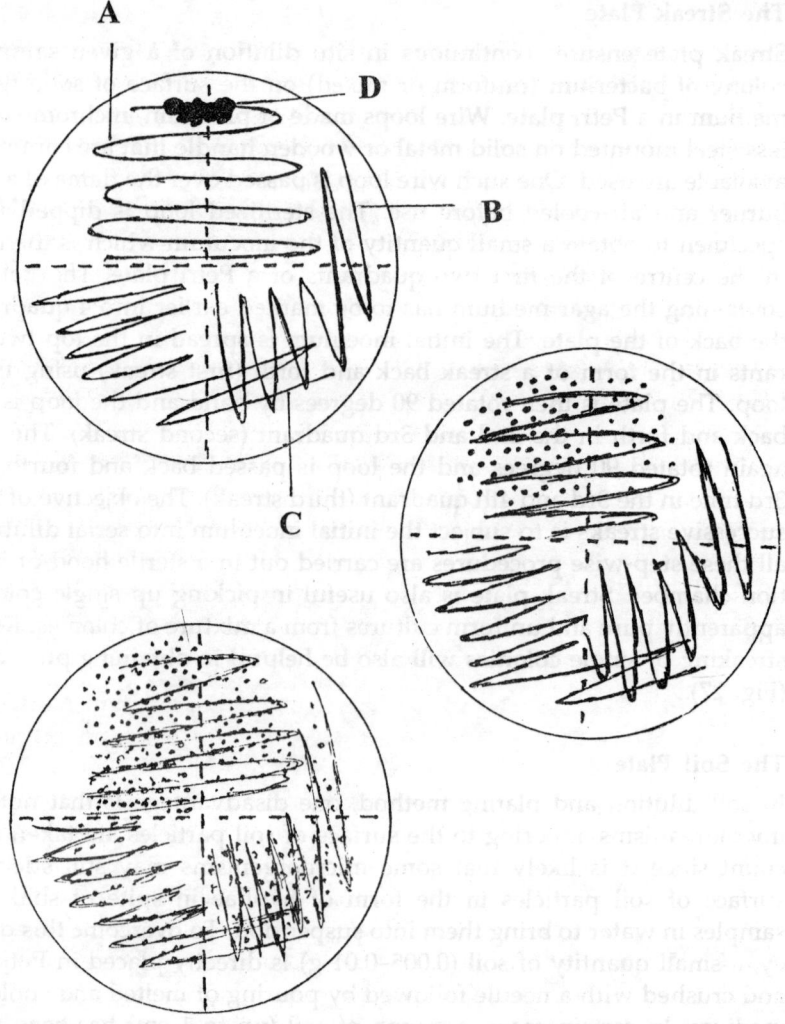

Fig. 27 Streak plates showing the first (A), the second (B) and the third (C) streaks of an initial inoculum (D). Observations are taken of progressive growth of colonies clockwise from 1st quadrant to the 4th quadrant.

means of this technique, it is possible to accomplish the isolation of sparsely occurring microorganisms in soil such as those which oxidize iron and sulphur and utilize nitrate.

The Buried Slide

The method originally devised by Rossi and Chołodny has been variously modified to suit the needs of individual workers. The technique is useful to study qualitative changes in soil microflora under the influence of soil amendments. The principal features of the method are as follows: A clean glass slide is introduced into a previously made slit in soil and left in position for 1 to 3 weeks. The slide is then removed in such a way that one of its sides is not disturbed. This side known as top side is gently washed in a stream of water, air dried and fixed over a low flame. After placing the slide over a steam bath, the top side of the slide is stained with erythrosin or rose bengal, dried and examined under a microscope.

Immersion Plate and Tube

Special methods have been designed for the in situ colonization of fungi in soil by burying agar-coated plates or agar filled and perforated tubes in soils. After desired periods of incubation, the agar surfaces are either examined under a binocular microscope or cut into suitable pieces and the pieces used to isolate microorganisms by the pour plate method. There have been many modifications of this method but the underlying principle is to study the occurrence of microorganisms in a natural way in their habitats.

Direct Microscopic Examination of Soil

The relative preponderance of different types of microorganisms in soil and their interrelationships are best studied by this method originally described by Conn in 1918. One gram of soil is mixed with 9 ml of 0.015% agar solution, shaken and 1/10 of a ml of the suspension transferred to the centre of a glass slide by means of a pipette. The suspension is evenly distributed over a 4 sq cm area of the slide with the help of a needle. When the prepared slide is dry, it is fixed in 0.1 N HCl for a minute, briefly immersed in water (to remove excess acid) and the slide placed on a water bath. The material on the slide is stained with rose bengal or erythrosin and examined under a compound microscope for qualitative studies. For quantitative determinations, the number of microorganisms of different types is counted for each microscopic field. Based on the data obtained, the number of microorganisms per g of soil is then calculated. The original method has been modified by several workers by using a haemocytometer to count the numbers accurately.

Fluorescent Staining

These traditional stains have been replaced by acridine orange (AO) which stains DNA in cells and fluorescein isothiocyanate (FITC) which adsorb to the sulfydryl groups in proteins. These are fluoresent stains and exhibit different colours under U-V light. The AO stain show up as green for living cells and red for dead cells, depending on the ratio of RNA to DNA within the cytoplasm despite the fact that clear differentiation between living and dead cells becomes more difficult in clay soils than in sandy soils. FITC is adequate for bacteria but not for fungi. Fungal cells can adsorb water soluble aniline blue to the B(1–3)–glycan linkages and show up better under U-V light. The calculation of bacterial numbers in soil using fluorescent dyes such as FITC can be done by the following equation, not withstanding limitations.

$$N_g = N_f \frac{A}{a_m} \frac{V_{sm}}{V_{sa}} D \frac{W_w}{W_d}$$

N_g Bacterial numbers in soil per g. of dry soil

N_f Bacteria per field

A   Area (mm2) of smear (or filter)

A_m Area (mm2) of microscope field

V_{sm} Volume (ml) of smear of filter

V_{sa} Volume (ml) of sample

D   Dilution

Ww Wet weight of soil

Wd Dry weight of soil

Chemical Methods to Determine Microbial Biomass

By treating soil with chloroform ($CHCl_3$) vapour, cells are made to leak out their cytoplasm. The carbon content of an appropriate soil extract can be determined later either by conventional methods or by C-14 methods by adding a tracer to soil. Measurement of adenosine triphosphate (ATP) in a given sample of soil using luciferin reaction is another chemical method. The substrate luciferin (pure chemical or from fire fly) can react with ATP and luciferase enzyme in the presence of magnesium to give an enzyme-luciferin-adenosine monophosphate (AMP) intermediate. This intermediate breaks down in the presence of oxygen to liberate free AMP, inorganic phosphorus and light. The light generated is measured by a photometer or scintillation counter and plotted against ATP content from a standard curve. From the data obtained biomass carbon content can be calculated. Measurement of CO_2 evolved in a given soil sample either by simpler chemical methods or the more accurate Warburg or radiorespirometry (involving C^{14}) can also serve to measure microbial biomass.

Soil Percolation Techniques

To measure the potential activities of a soil sample, percolation techniques come in handy. When a specific microbially degradable substrate is incorporated in soil, at least a part of the soil microflora is stimulated both in numbers and activity which determine the speed of degradation and the nature of degraded products. Original models for percolation of soil were proposed by Lees and Quastel and they have undergone many modifications in recent years although the principle remains unchanged.

A column of soil is arranged in a glass tube between two layers of glass wool. The percolation fluid, containing the test substance and saturated with air, is passed continuously through the column at a controlled rate to avoid flooding and is recycled to the top of the column to percolate again. The percolation can be continued for several days or weeks. Samples can be removed from the receiver whenever necessary for evaluation. The technique is also useful for studies on soil respiration.

Soil Enzymes Estimation

Soil enzymes are proteinaceous in nature and are found in soil organic and inorganic colloids. A great variety of enzymes are produced by soil microorganisms. However, there are extracellular enzymes which are not directly linked with soil microorganisms. Soil enzymes are of two types—constitutive such as urease and dehydrogenases commonly produced by microorganisms and inducible such as cellulase produced in the presence of cellulose. Some enzymes found in soil and the reactions they catalcatal-catalcatalyze are shown in Table 4.

Dehydrogenase activity has been measured by several workers and can be a convenient laboratory technique for large-scale comparative studies. Twenty g of fresh soil at 90% of its water holding capacity is mixed with 200 mg of $CaCO_3$ and 2 ml of a 1% solution of triphenyl tetrazolium chloride (TTC). The mixture is incubated for 24 hours. In the absence of O_2, TTC serves as the terminal acceptor for the hydrogen evolved, and through the action of the microbial dehydrogenases, TTC gets reduced to triphenyl formazan (TPF). At 485 mμ, its optical density is proportional to concentration and hence can be read with a spectrophotometer.

Methods for Assaying Antibiosis

Antibiotic activity of one microorganism upon another can often be observed in soil-dilution plates. The organisms involved in this phenomenon are known as antagonists. They can be isolated and tested by several methods (Fig. 28): (1) the antagonist and the test organism are placed on media opposite each other at the periphery of Petri plates, (2) the antagonist is inoculated on media in the form of a line or streak and the test

Table 4 Some enzymes found in soil and the reactions they catalyze (from Paul and Clarke, 1989)

Enzyme	Reaction catalyzed
Oxidoreductases	
Catalase	$2H_2O_2 \rightarrow 2H_2O + O_2$
Catechol oxidase (tyrosinase)	O-Diphenol $+ \frac{1}{2} O_2 \rightarrow O$-quione $+ H_2O$
Dehydrogenase	$XH_2 + A \rightarrow X + AH_2$
Diphenol oxidase	P-Diphenol $+ \frac{1}{2} O_2 \rightarrow P$-quinone $+ H_2O$
Glucose oxidase	Glucose $+ O_2 \rightarrow$ gluconic acid $+ H_2O_2$
Peroxidase and polyphenol oxidase	$A + H_2O_2 \rightarrow$ oxidized $A + H_2O$
Transferases	
Transminase	$R_1R_2\text{-CH-N}^+H_1 + R_1R_4CO \rightarrow R_1R_4\text{-CH-N}^+H_3 + R_1R_2CO$
Hydrolases	
Acetylesterase	Acetic ester $+ H_2O \rightarrow$ alcohol $+$ acetic acid
α- and β-Amylase	Hydrolysis of $\beta(1 \rightarrow 4)$ glucosidic bonds
Asparaginase	Asparagine $+ H_2O \rightarrow$ aspartate $+ NH_1$
Cellulase	Hydrolysis of $\beta(1 \rightarrow 4)$ glucan bonds
Deamidase	Carboxylic acid amide $+ H_2O \rightarrow$ carboxylic acid $+ NH_3$
α- and β-Galactosidase	Galactoside $+ H_2O \rightarrow$ ROH $+$ galactose
α- and β-Glucosidase	Glucoside $+ H_2O \rightarrow$ ROH $+$ glucose
Lipase	Triglyceride $+ 3H_2O \rightarrow$ glycerol $+ 3$ fatty acids
Metaphosphatase	Metaphosphate $\rightarrow$ orthophosphate
Nucleotidase	Dephosphorylation of nucleotides
Phosphatase	Phosphate ester $+ H_2O \rightarrow$ ROH $+$ phosphate
Phytase	Inositol hexaphosphate $+ 6H_2O \rightarrow$ inositol $+ 6$-phosphate
Protease	Proteins $\rightarrow$ peptides $+$ amino acids
Pyrophosphatase	Pyrophosphate $+ H_2O \rightarrow 2$-orthophosphate
Urease	Urea $\rightarrow 2NH_1 + CO_2$

organism inoculated 2–4 cm away from the antagonist, (3) Petri plates are seeded with the test organism on agar and 1 to 3 holes are punched on agar by means of a cork-borer. Alternatively, miniature sterile cups or glass cylinders are placed in contact with the agar surface. The wells formed in the agar or the cavities of miniature glass cylinders or cups are filled with aliquots of dialyzed culture filtrate of the antagonist (the antibiotic producing organism) and the plates incubated for 24 hrs for bacteria and for 2–4 days in the case of fungi at 28°C (± 2°C). A zone of inhibition of growth of the test organism around the wells or cylinders is indicative of the extent of antibiosis produced by the antagonist, (4) for routine screening, small discs of filter paper are soaked in the culture filtrate of the antagonist and placed on agar seeded with the test organism in Petri plates. This is a simplified version of the agar-cup method described earlier.

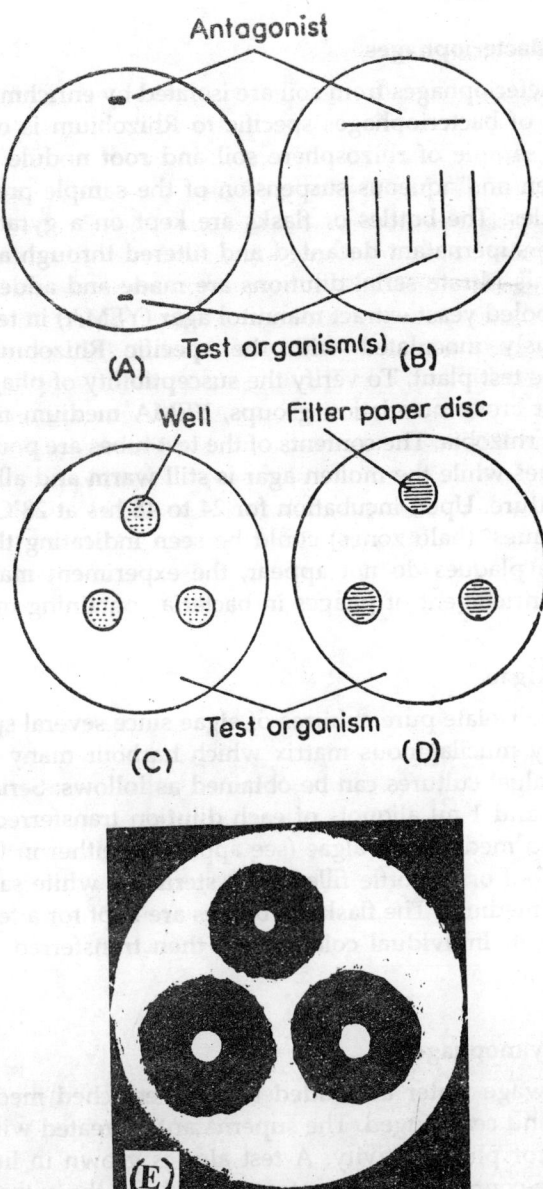

Fig. 28 Methods for asaying antibiotic activity of microorganisms on solid agar medium:
(A) the antagonist and the test microorganism placed on either side of a Petri plate;
(B) the antagonist is streaked across the plate and the test organisms streaked at
right angles to the antagonist; (C) the agar plate is seeded with the bacteria under
test and the wells on agar are filled with the culture filtrate of the antagonist; and
(D) discs of filter paper dipped in the culture filtrate of the antagonist are placed
on agar seeded with the test organism. In (C) and (D), zones of inhibition of growth
of the test organisms will take place as shown in (E).

Isolation of Bacteriophages

In general, bacteriophages from soil are isolated by enrichment techniques. The isolation of bacteriophages specific to Rhizobium is outlined below: A composite sample of rhizosphere soil and root nodule bearing lateral roots are taken and aqueous suspension of the sample prepared in large flasks or bottles. The bottles or flasks are kept on a gyratory shaker for two days, the supernatant decanted and filtered through a sintered glass filter. From this filtrate serial dilutions are made and added to sterilized, melted and cooled yeast extract mannitol agar (YEMA) in test tubes which were previously inoculated with the specific Rhizobium capable of nodulating the test plant. To verify the susceptibility of phages to rhizobia from different cross-inoculation groups, YEMA medium may be seeded with different rhizobia. The contents of the test tubes are poured into sterilized Petri plates while the molten agar is still warm and allowed to set at room temperature. Upon incubation for 24 to 48 hrs at 28°C (± 2°C), characteristic 'plaques' (halo zones) could be seen indicating the presence of phages. If the plaques do not appear, the experiment may be repeated after further enrichment of phages in bacteria containing media.

Isolation of Algae

It is not easy to isolate pure cultures of algae since several species of algae are covered by mucilaginous matrix which harbour many contaminants. However, unialgal cultures can be obtained as follows: Serial dilutions of soil are made and 1 ml aliquots of each dilution transferred into suitable sterilized liquid medium for algae (see appendix) either in flasks plugged with cotton wool or in bottle filled with sterilized white sand moistened with the algal medium. The flasks or bottles are kept for a few weeks near a source of light. Individual colonies are then transferred to agar slants for identification.

Isolation of Cyanophages

Samples of sewage water are added to algae-enriched media, incubated for one week and centrifuged. The supernatant is treated with chloroform and screened for phage activity. A test alga is grown in liquid medium with the phage-containing supernatant to find out the extent of lysis. Alternatively, plaque formation (similar to bacteriophages) can be demonstrated on solid medium which will clearly indicate the presence of cyanophages.

Glass Ring Method for Isolation of Protozoa

Soil protozoa feed on bacteria. Only certain species of bacteria are edible while others are toxic to protozoa. In this method, a culture of a bacterium such as Aerobacter is first grown on a medium capable of supporting good growth. Bacterial cells are taken from this culture and spread in small circles on the surface of solidified water agar (made from 5% sodium chloride and 1% agar) in Petri plates. Suitable soil dilutions are prepared in 0.5% sodium chloride solution and 0.5 ml aliquots of each dilution are transferred into bacterial circles. Subsequently, small glass rings are pressed into agar around the bacterial circles so as to separate different bacterial circles in the same Petri plate. When protozoa grow on the bacterial circles, the bacterial colonies disappear indicating the presence of protozoa. The number of protozoa per gram of soil can be calculated from the number of bacterial circles colonized by the predators at different soil dilutions and by statistically analysing the data.

Determination of Most Probable Number (MPN) of *Nitrosomonas*

Nitrifying bacteria are difficult to isolate but their presence may be detected in soil by suitable tests and their most probable numbers determined. Serial dilutions of 10 g sample of soil are made with sterile water. One ml aliquots of each dilution are transferred to test tubes containing 3 ml sterilized ammonium-calcium carbonate medium [$(NH_4)_2\ SO_4$—0.5 g; K_2HPO_4—1.0 g; $FeSO_4 \cdot 7H_2O$—0.03 g; NaCl—0.3 g; $MgSO_4 \cdot 7H_2O$—0.3 g; $CaCO_3$—7.5 g; Distilled water—1 litre]. One set of tubes without soil dilutions is maintained as controls and the tubes incubated for three weeks at 28°C.

At the end of incubation, three drops of freshly prepared Griess Ilosvay reagent (a—0.6 g sulfanilic acid is dissolved in 70 ml hot distilled water, cooled, 20 ml of concentrated HCl is added and volume made up to 100 ml; b—0.6 g of α-naphthylamine is dissolved in 10 to 12 ml distilled water containing 1 ml concentrated HCl and volume made up to 100 ml; c—16.4 g sodium acetate is dissolved in distilled water and volume made up to 100 ml. The three solutions—a, b and c—are stored in dark bottles and mixed in equal parts before use) are added to the solution in tubes. Observations are made of the development of purplish-red colour within a few minutes. The solutions in the tubes which show up colour indicate the presence of nitrites and hence *Nitrosomonas* (positive tubes). The tubes which do not develop colour may be tested for nitrates by adding a pinch of zinc-copper-manganese dioxide mixture (1 g powdered zinc, 1 g powdered manganese dioxide and 0.1 g copper powder are mixed). If these tubes develop purplish-red colour, they indicate the presence of *Nitrobacter*. Based on the data, the most probable number (MPN) of *Nitrosomonas* may be calculated from a table provided by Alexander and Clark (1965).

MPN for *Nitrobacter*

Nitrite-calcium carbonate medium is used to enumerate *Nitrobacter*. The composition of the medium is as follows: KNO_2—0.006 g; K_2HPO_4—1.0 g; NaCl—0.3 g; $MgSO_4 \cdot 7H_2O$—0.1 g; $CaCO_3$—1.0 g; $CaCl_2$—0.3 g; Distilled water—1.0 litre. Three ml of the medium is transferred to each test tube and sterilized. From the same dilutions prepared for *Nitrosomonas* count, 1 ml aliquot is transferred to five test tubes containing the above medium. One set of uninoculated test tubes is also included and all the tubes are incubated at 28°C for three weeks. At the end of incubation, tubes are tested only for the presence of nitrite using the Griess Ilosvay reagent. Individual tubes are recorded as positive for *Nitrobacter* if they fail to develop the reddish colour. The tubes showing reddish colour are recorded negative for *Nitrobacter*. From the MPN table, the number of *Nitrobacter* in the sample is calculated in the same manner as mentioned for *Nitrosomonas*.

Soil Respiration

When carbon containing substrates are oxidized in soil, CO_2 is evolved. The amount of CO_2 evolved is generally taken as an index of the total activity of the soil microflora. Many simple methods of determining CO_2 are made use of, but Warburg's manometric techniques are more widely adapted for this purpose. Therefore, a knowledge of the working of Warburg's apparatus is essential before using this method.

The amount of soil taken in a manometric flask varies from 1–10 g. Water or buffer solution is added to produce the necessary humidity (generally 60% saturation), 0.2 ml of a 3% potassium hydroxide solution is placed in the central well of the flask together with a small roll of filter paper to increase the surface area for the absorption of CO_2. Flasks are attached to manometers (having Brodie's solution as an indicator) and suspended in the water bath at 30°C. At regular intervals, the changes in the readings on the manometer are noted. The rate of oxygen uptake is expressed as $\mu l/g/hr$.

Molecular Methods in Soil Microbiology (Also see Chapter 8 pp 225, 226)

Conventional methods of soil dilution plating and direct microscopy may not always provide a true picture of microbial diversity in soil because of the existence of some recalcitrant microorganisms which do not readily grow on culture plates (sometimes due to fastidious growth requirements) even after prolonged incubation. In recent years, culture independent molecular microbiological methods have shown great promise in overcoming this limitation. These new methods have also helped in understanding the evolutionary links between microbial groups besides providing help in classifying and quantifying bacteria by in situ analysis. Methods for extraction, purification and quantification of soil DNA have been developed and are being refined continuously. The availability of pure soil DNA has enabled the microbial ecologist to devise strategies in using nucleic acids to obtain a true picture of evolutionary sequence and microbial diversity in a given soil.

Extraction and Purification of Nucleic Acid from Soil

Sediments rich in organic matter and microflora may have to be washed to remove soluble compounds that may interfere with the isolation, purification and separation (into DNA and RNA) of nucleic acids. The release of nucleic acids is accomplished both by chemical and physical methods. Treatments with one or more of the following compounds or using physical means have been adopted by various investigators: lysozyme, heat, proteinase, sodium dodecyl sulfate (SDS) achromopeptidase, hot phenol, guanidine thiocyanate, pronase, acetone, sarkosyl, EDTA, freeze-thaw cycles, freeze-boil cycles, sonication, bead-mill homogenization, microwave heating and mortar mill grinding. DNA yields of the order of 11.8 μg DNA per gram of sediment dry weight have been recorded by a combination of SDS treatment and bead-mill homogenization, reducing the viable counts of left over intact microorganisms to about 2 per cent. The extracted DNA has been successfully used in molecular biology procedures.

DNA Hybridization

The reader is referred to Chapter 16 for an account of structure and function of nucleic acids. The base composition of DNA from bacteria ranges from about 25 to 75 moles per cent guanine plus cytosine (mol % G + C). These values for DNA differ among microorganisms and such differences reflect the existence of several species. On the contrary, if the mol % G + C show similarity between two microorganisms, the conclusion could be that they are either two distinct species or two different species having nearly identical genomes. At high temperature and about 0.4 to 1.0 M NaCl, the base pairs of DNA get separated into single strands and this process is known as melting or denaturation. If this denaturation is carried out with soil DNA (which is a mixture of DNA from different organisms) and the temperature as well as the salt concentration are lowered, the separated single strands begin to reassociate (renature) with complimentary fragments from other DNA that are identical. This process is known as DNA hybridization and the results are expressed as per cent hybridization or relatedness. The values are determined using a spectrophotometer to measure absorbance values. The values indicate the genomic fraction of DNA from one organism that can hybridize or reassociate with DNA from another organism under a given set of experimental conditions. A commonly used method is to label the DNA with radioactive P32, S35 or tritium and distinguish the labelled strands from the unlabelled ones by selective adsorption to hydroxyapatite followed by separation by column chromatography. Alternatively, single stranded DNA can be degraded by specific nuclease and the quantity of labelled DNA present in intact duplex structures measured in a scintillation counter. An older method commonly used involves the immobilization of unlabelled DNA on a nitrocellulose membrane and incubating it with labelled DNA fragments. The labelled fragments can bind to the membrane only if they form duplexes with the immobilized DNA.

Experiments on denaturation of DNA followed by renaturation (hybridization) have demonstrated that generally soil may contain 4×10^4 species of microorganisms per gram of soil.

Amplification by Polymerase Chain Reaction

This technique commonly known as PCR amplifies in vitro specific DNA sequences with the help of a thermostable DNA polymerase enzyme (for example from T. aquaticus known as Taq) and specific primers. Primers are oligonucleotides with 15–20 bases and their sequence is complementary to the target DNA sequence intended for amplification. The first step is to render the double stranded DNA (dsDNA) to a single stranded (ssDNA) by heating which denatures the DNA. When the primers are added, they will attach to the single stranded DNA (annealing step) and the polymerase enzyme begins to synthesize the complementary nucleotide chain beginning at the known primer sequence (extension step). This process can be automated in a thermocycler where products of discrete length are formed exponentially to one million-fold or more, to be detected even with poorly sensitive probes or sometimes without any probing at all by staining an electrophoretic gel of the PCR products prepared by Southern blot method. This PCR technique (Fig. 29) enables the detection of even small number of cells in an environment, cells from museum specimens, forensic materials, human cells containing infectious bacteria and also rhizobia in soils.

Detection and Identification of rRNA Genes

The bacterial ribosome has two subunits. The larger one consists of two RNA species, a 5S molecule which is about 120 bases long and a 23S molecule which is about 3500 bases in length. The smaller sub-unit 16S rRNA gene is about 1500 bases long and is short enough to be sequenced yet providing information for analyses. Certain sequences in

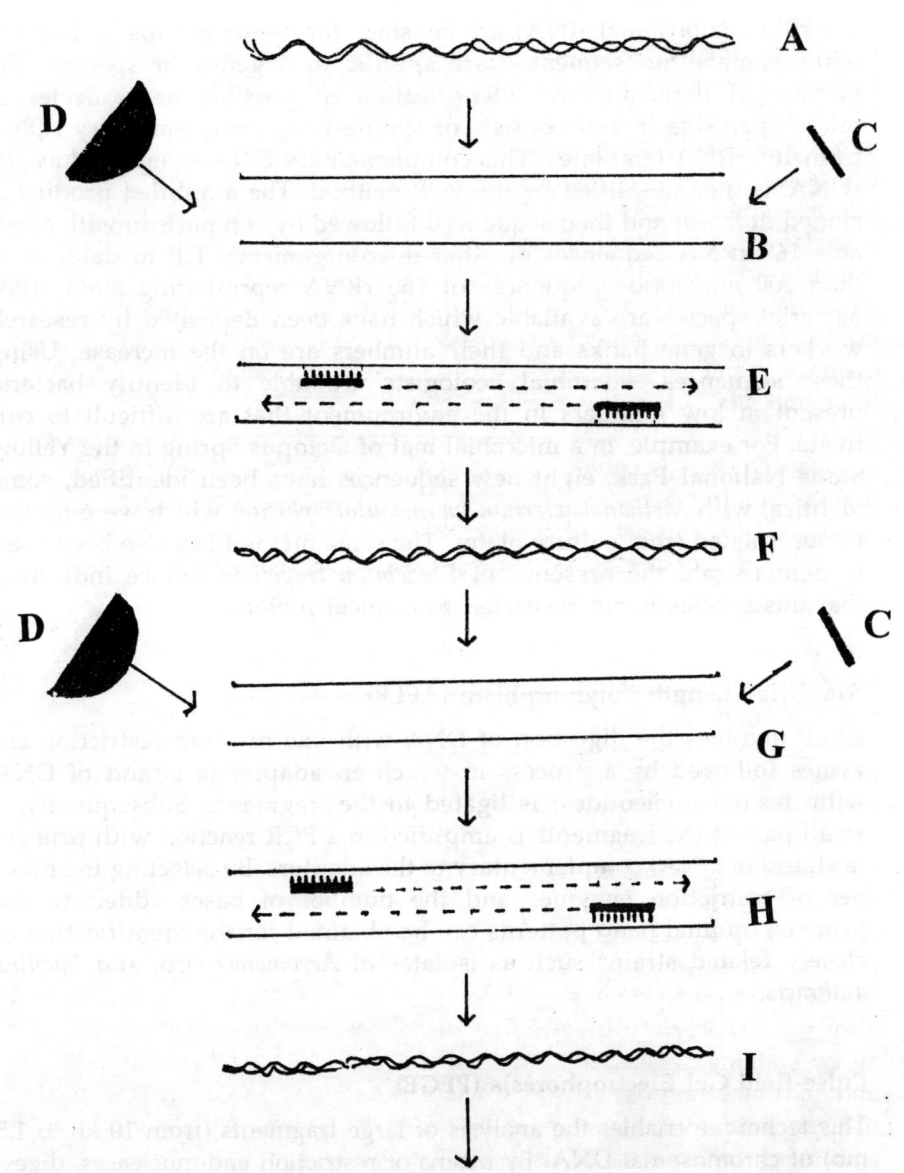

Fig. 29 Polymerase chain reaction (PCR): Double stranded DNA (A) is heated to melt strands into single stranded DNA (B) to which primers (C) and thermostable DNA polymerase (D) have been added to prime the reaction when complementary nucleotide sequences are synthesised (E) on each strand to form again a double stranded DNA (F). The process is repeated as in G, H and I many times to amplify a particular DNA sequence.

16S rRNA (ribosomal RNA) are constant for major groups of bacteria while some other sequences are specific to a genus or species. The method of detecting and identification of possible new species of microorganisms in soil consists of synthesising complementary cDNA from 16S rRNA templates. This complementary DNA designated as 16S rDNA, is then amplified by the PCR method. The amplified product is cloned in *E. coli* and then sequenced followed by comparison with available 16S rRNA sequences of other microorganisms. Till to date, more than 300 nucleotide sequences of 16S rRNA representing about 1000 bacterial species are available which have been deposited by research workers in gene banks and their numbers are on the increase. Using these sequences, microbial ecologists are able to identify bacteria present in low numbers in the environment that are difficult to cultivate. For example, in a microbial mat of Octopus Spring in the Yellow Stone National Park, eight new sequences have been identified, some identical with *Methanobacterium thermoautotrophicum* which were hitherto not isolated from culture plates. The same method has also been used to demonstrate the presence of *Rhizobium tropici in* France indicating that this species is not restricted to tropical regions.

Amplified Length Polymorphism (AFLP)

AFLP involves the digestion of DNA with one or more restriction enzymes followed by a process in which an adapter (a strand of DNA with desired nucleotides) is ligated to the fragments. Subsequently, a small part of the fragments is amplified in a PCR reaction with primers (a strand of DNA) complementary to the adapters. By selecting the number of restriction enzymes and the number of bases added to the primers, optimal band patterns can be obtained for the identification of closely related strains such as isolates of *Aeromonas* spp. and *Bacillus anthracis*.

Pulse-field Gel Electrophoresis (PFGE)

This technique enables the analysis of large fragments (from 10 kb to 1.5 mb) of chromosomal DNA. By means of restriction endonucleases, digestion patterns are obtained which consist of relatively few, well separated fragmented bands. These bands are less ambiguous than patterns generated by conventional electrophoresis. Genomic DNA finger printing patterns by PFGE provide a useful method of distinguishing isolates because the electrophoretic pattern is a reflection of bacterial chromosome organisation. Two or more isolates are presumed to be the same if all the bands in a restriction digest are identical or similar; if they are different, they may belong to different isolates.

Selected References

Akkermans, A.D.L. 1998. Molecular tools for tracking microorganisms in the environment. In Microbial Interactions in Agriculture and Forestry. Eds. N.S. Subbarao and Y.R. Dommergues, pp. 1–18, Oxford and IBH Publishing Co., New Delhi.

Alexander, M. 1961. Introduction to Soil Microbiology. John Wiley & Sons, Inc., New York and London.

Alexander, M. 1982. Most probable number method for microbial populations. Agron. Monogr. 9, 815–820.

Alexander, M. and Clark, F.E. 1965. Nitrifying bacteria. In Methods of Soil Analysis, Part 2. Chemical and Microbiological Properties. pp. 1477–1483. Eds. C.A. Black, American Society of Agronomy, Inc., Madison, Wisconsin, U.S.A.

Alexopoulos, C.J. and Mims, C.W. 1979. Introductory Mycology, Wiley, New York.

Allen, O.N. 1957. Experiments in Soil Bacteriology, 3rd Edition. Burgess Publishing Co., Minneapolis, Minn., U.S.A.

Amarger, N., Bours, M., Revoy, F., Allard, M.R. and Laguerre. 1994. Rhizobium tropici nodulates field-grown Phaseolus in France. Pl. Soil, 161, 147–156.

Amman, R.I., Ludwig, W. and Schleifer, K.H. 1995. Phylogenetic identification and in situ detection of individual microbial cells without cultivation. Microbiol. Rev. 59, 143–169.

Biovin-Johns, V., Bianeshi, A., Ruimy, R., Garcin, J., Daumas, S. and Christen, R. 1995. Comparison of phenotypical and molecular methods for the identification of bacterial strains isolated from a deep sub-surface environment. Appl. Environ. Microbiol. 61, 3400–3406.

Borneman, J., Skroch, P.W., O'Sullivan, K.M., Palus, J.A., Rumjanek, N.G., Jansen, J.L., Nienhuis, J. and Triplett, E.W. 1996. Molecular microbial diversity of an agricultural soil in Wisconsin, Appl. Environ. Microbiol., 62, 1935–1943.

Buchanan, R.E. and Gibbons, N.E., Ed. 1974. Bergey's Manual of Determinative Bacteriology. 8th edition Williams and Wilkins, Baltimore, USA.

Burns, R.G. Ed. 1978. Soil Enzymes. Academic Press, London.

Callaham, D., Del Tredici, P., and Torrey, J.G. 1978. Isolation and cultivation in vitro of the actinomycete causing root nodulation in Comptonia. Science, 199, 899–902.

Carnahan, J.E., Mortenson, L.E., Mower, H.F., and Castle, J.E. 1960. Nitrogen fixation in cell-free extracts of Clostridium Pasteurianum. Biochim. Biophys. Acta, 42, 530–535.

Chase, F.E. and Gray, P.H.H. 1957. Application of Warburg's respirometer in studying respiratory activity in soil. Can. J. Microbiol., 3, 335–349.

Chesters, C.G.C. 1940. A method for isolating soil fungi. Trans. Brit. Mycol. Soc., 24, 352–353.

Chesters, C.G.C. and Thornton, R.H. 1956. A comparison of techniques for isolating soil fungi. Trans. Brit. Mycol. Soc., 39, 301–313.

Cholodny, N. 1930. Über eine neue Methode zur Untersuchung de Bodenmikroflora. Arch. Milrobiol., 1, 620–652.

Clark, G. Ed. 1980. Staining Procedures. H. Williams and Wilkins, Baltimore & USA.

Collins, F.M. and Sims, C.M. 1956. A compact soil perfusion apparatus. Nature. Lond. 178, 1073.

Frobisher, M. 1968. Fundamentals of Microbiology, W.B. Saunders and Co. Philadelphia.

Hardy, R.M.F. and Holstein, R.D. 1977. Methods for measurement of dinitrogen fixation. In A Treatise on Dinitrogen Fixation Sect, Agronomy and Ecology, pp. 451–486. John Wiley and Sons, Inc. New York.

Howard, B.J., Keiser, J.F., Smith, T.F., Weissfeld, A.S. and Tilton, R.C. 1994. Clinical and Pathogenic Microbiology, 2nd Ed., Mosby, St. Louis, USA.

Lane, D.J. 1991. 16S/23S rRNA sequencing. In Nucleic Acid Techniques in Bacterial Systematics. Ed. M. Good Fellow, pp. 115–175, Wiley, New York.

Liesack, W. and Stackebrandt, E. 1992. Occurrence of novel groups of the domain bacteria as revealed by analysis of genetic material isolated from an Australian terrestrial environment. J. Bacteriol. 174, 5072–5078.

Maslaw, J.N., Slutsky, A.M. and Arbiet, R.D. 1993. Application of pulsed field gel electrophoresis to molecular epidemiology. In Diagnostic Microbiology: Principles and Applications, pp. 523–572. Eds. D.H. Persing, T.F. Smith, F.C. Turner and T.J. White. American Soc. for Microbiology, Washington, D.C.

More, M.J., Herrick, J.B., Silva, M.C., Ghiorse, W.C. and Madsen, E.L. 1994. Quantitative cell lysis of indigenous microorganisms and rapid extraction of microbial DNA from sediment. Appl. Environ Microbiol, 60, 1572–1580.

Paul, E.A. and Clark, F.E. 1989. Soil Microbiology and Biochemistry. Academic Press, Inc., San Diego, USA.

Safferman, R.S. and Morris, M. 1963. Algal virus—isolation. Science, N.Y., 140, 679–680.

Safferman, R.S. and Morris, M. 1964. Growth characteristics of the blue-green algal virus LPP. 1. Science, N.Y., 88, 771–775.

Singh, B.N. 1946. A method for estimating the numbers of soil protozoa especially amoebae based on their differential feeding of bacteria. Ann. Appl. Biol., 33, 112–119.

Singh, A. and Gopinathan, K.P. 1998. Confocal microscopy: A powerful technique for biological research. Curr. Sci., 74, 841–851.

Smith, R.F. 1990. Microscopy and Photomicrography, CRC Press, Boca Raton, Florida, USA.

Southern, E.M. 1975. Detection of specific sequences among DNA fragments separated by gel electrophoresis. J. Mol. Biol. 98, 503–507.

Stackebrandt, E. and Goebel, B.M. 1994. Taxanomic note: A place for DNA-DNA reassociation and 16S rRNA analysis in the present species definition in bacteriology. Int. J. Syst. Bacteriol. 44, 846–849.

Waksman, S.A. 1959. The Actinomycetes. Vol. I. Nature, Occurrence and Activities. Williams & Wilkins, Baltimore.

Warcup, J.H. 1950. The soil-plate method for isolation of fungi from soil. Nature, Lond., 116, 117–118.

Ward, D.M., Weller, R. and Bateson, M. 1990. 16S rRNA sequences reveal numerous uncultured microorganisms in a natural community. Nature (Lond.), 345, 63–65.
Woese, C.R. 1987. Bacterial evolution. Microbiol. Rev., 51, 221–271.

4. The Rhizosphere and the Phyllosphere

The Rhizosphere

The region in the vicinity of roots can be distinguished into many microhabitats as shown in Fig. 30. The term 'rhizosphere' was introduced in 1904 by the German scientist Hiltner to denote that region of the soil which is subject to the influence of plant roots. Rhizosphere is characterized by greater microbiological activity than the soil away from plant roots. The intensity of such activity depends on the distance to which exudations from the root system can migrate. The term 'rhizosphere-effect' indicates the overall influence of plant roots on soil microorganisms. It is now clearly established that greater number of bacteria, fungi and actinomycetes are present in the rhizosphere soil than in non-rhizosphere soil and there are innumerable reports in literature to substantiate this fact (Table 5). Several factors such as soil type, its moisture, pH and temperature and the age and condition of plants are known to influence the rhizosphere effect. Apart from the numerical preponderance of microorganisms in the rhizosphere, the rhizosphere effect is also manifest in the occurrence and distribution of bacteria characterized by specific requirements Table 6 of amino acids, B-vitamins and specialized growth factors (nutritional groups). It has also been demonstrated that the rates of metabolic activity of the rhizosphere microorganisms are different from those of the non-rhizosphere soil (Table 7). A wide range of enzymes of plant and microbial origin present in the rhizosphere catalyze the breakdown of organic materials. These enzymes include oxidoreductases, hydrolases, lyases and transferases besides cellulose, dehydrogenases and urease.

Rhizosphere soil can be separated and a soil suspension obtained by shaking roots in aliquots of sterile water from which subsequent dilutions are made. One millilitre of the appropriate dilutions is plated on suitable agar media for enumerating bacteria, actinomycetes and fungi. Likewise, aliquots are transferred to liquid media for counting the numbers of algae, protozoa and bacteria responsible for ammonification, nitrification, denitrification, carbohydrate utilization, cellulose decomposition and nitrogen fixation. Total counts are then expressed on a dry weight basis per gram of soil. Root-free soil or soil samples from uncultivated plots are

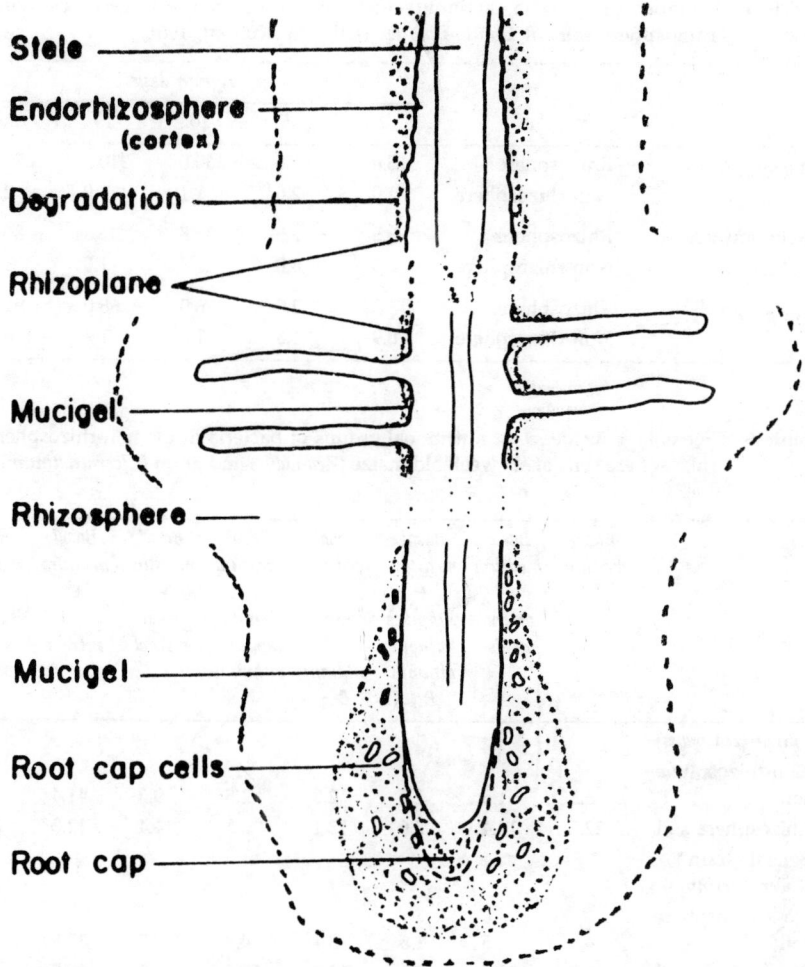

Stele

Endorhizosphere
(cortex)

Degradation

Rhizoplane

Mucigel

Rhizosphere

Mucigel

Root cap cells

Root cap

Fig. 30 A generalized diagramatic view of the root region of a growing plant (from Knowles, 1978).

used as controls to judge the changes in the microbial population due to plant growth. The rhizosphere to soil ratio (R : S) can be calculated by dividing the number of microorganisms in the rhizospehre soil by the number in the soil free from plant growth. Results can also be expressed based on the weight of roots. In many studies, the root sample along with the adhering soil have been subjected to the action of a waring blender with a known volume of sterile water and aliquots plated to determine the microflora from both inside and outside the root cortex.

In many reports another specialized microhabitat has been recognized and defined as the 'rhizoplane' or the 'root surface'. In sampling the root

Table 5   Number of bacteria, actinomycetes and fungi in the non-rhizosphere and rhizosphere soils of *Dolichos lab lab* (Lakshmi Kumari, 1961)

		Age in days				
		1	5	10	15	20
Bacteria ($\times 10^7$)	Rhizosphere	15.0	95.5	260.0	310.8	677.8
	Non-rhizosphere	2.0	2.0	1.1	2.0	2.5
Actinomycetes ($\times 10^6$)	Rhizosphere	5.5	3.5	34.5	95.8	83.3
	Non-rhizosphere	4.5	6.0	1.3	1.0	1.0
Fungi ($\times 10^4$)	Rhizosphere	3.3	2.0	26.0	68.0	91.8
	Non-rhizosphere	0.9	1.6	1.5	1.7	6.8

Table 6   Percentage incidence of nutritional groups of bacteria in the non-rhizosphere and rhizosphere soils of six week old maize (*Zea mays*) and gram (*Cicer arietinum*) (Dey, 1967)

	Basal medium	Basal medium + amino acids	Basal medium + vitamins without vitamin B_{12}	Basal medium + vitamins with vitamin B_{12}	Basal medium + amino acids + vitamins	Basal medium + yeast extract	Basal medium + soil extract	Basal medium + yeast extract + soil extract
Maize (*zea mays*)								
Non-rhizosphere soil	3.5	1.5	1.5	2.3	2.5	9.3	41.4	38.0
Rhizosphere soil	22.5	14.4	7.6	13.1	3.5	14.4	12.3	12.2
Bengal Gram (*Cicer arietinum*)								
Non-rhizosphere soil	4.8	1.5	1.5	3.1	4.8	7.5	37.5	39.3
Rhizosphere soil	21.5	22.5	8.2	13.1	12.5	8.2	10.5	3.5

Table 7   Oxygen uptake (μl) by three composite samples of rhizosphere and non-rhizosphere soils* (from Katznelson and Rouatt, 1957)

	Rhizosphere soil		
Composite sample	Barley	Rye	Control soil
1	359	295	101
2	363	327	101
3	392	294	96
Average	371	305	99

*After nine hours at 30°C; each figure is an average of duplicate determinations.

system for rhizoplane studies, soil adhering to roots is removed and roots subjected to serial washing by sterilized water (10–12 times) until the clean root surface is exposed. When such washed roots are plated, characteristic fungi and bacteria appear on agar plates, thereby indicating that there are certain microorganisms intimately associated with the root surface. Some fungi inhabit the root surface in a mycelial state. They belong to the genera *Mortierella, Cephalosporium, Trichoderma, Penicillium, Gliocladium, Gliomastix, Fusarium, Cylindrocarpon, Botrytis, Coniothyrium, Mucor, Phoma, Pythium* and *Aspergillus*. Fine structure studies on the epithelial layer of plant roots after inoculation with specific bacteria have shown that bacteria get embedded on the surface of the root with the help of the mucilagenous external layer or the 'mucigel' normally present on actively growing root system.

Rhizosphere Effect

Greater rhizosphere effect is seen with bacteria (R : S values ranging from 10 to 20 or sometimes more) than with actinomycetes or fungi. Only negligible changes are noted with regard to protozoa and algae. Literature abounds with reports on quantitative changes in microorganisms in the rhizosphere of crop plants. Qualitative studies, however, reveal some distinct selective influence of the root system. An example of this is the preferential stimulation of Gram-negative non-sporulating rod-shaped bacteria in the root region of many plants. Several genera of bacteria—*Pseudomonas, Arthrobacter, Agrobacterium, Azotobacter, Mycobacterium, Flavobacterium, Cellulomonas, Micrococcus* and others have been reported to be either abundant or sparse in the rhizosphere. From the agronomic point of view, the abundance of nitrogen fixing and phosphate solubilizing bacteria in the rhizosphere of crop plants assumes a natural significance. The preponderance of amino acid requiring bacteria in the rhizosphere of crop plants has been consistently observed by many workers. The preferential stimulation of vitamin requiring bacteria in the rhizosphere has also been emphasized by some workers but the question of the source of B-vitamins in the root region of plants still remains unanswered. Do they accumulate from the activity of microorganisms in soil or are they exuded by the growing root system? Since the quantity of B-vitamins exuded by roots is not as much as that of amino acids, it may be logical to assume that the contribution of microorganisms to the synthesis of B-vitamins in the rhizosphere is higher than that of actively growing root system.

Electron and direct microscopy show that only 4–10 per cent of root surface is colonized by microorganisms in a random fashion depending on the presence of soil organic matter. The proliferation of bacteria takes place at the junction of the epidermal cells indicating that this is an area of maximum root exudation. In this way, root excretions and root derived organic matter provide substrates for microbial proliferation. The microbial ecology in the rhizosphere is also dependent on genotypes of the plant.

Nitrogen Fixation in the Rhizosphere

With the advent of acetylene reduction technique, it has been possible to measure nitrogenase activity in the rhizosphere of many non-legumes. In terms of moles C_2H_4 per g. of dry root per hour, *Brachiaria mutica* showed nitrogenase activity of 150–750 in Brazil and maize seedlings showed activity in France in the range of 100–3000. Similar values have been recorded for various grasses and weeds with differing degrees of variability. Published reports indicate that rhizosphere of these plants harbour nitrogen fixing bacteria of the families Azotobacteriaceae, Spirillaceae, Enteroba .- teriaceae, Bacillaceae, Pseudomonadaceae and Achromobacteriaceae. Colonization by *Azotobacter*, when it occurred, has been shown to be limited in the rhizosphere and practically negligible on the root surface (rhizoplane), probably due to acidity caused by root exudates. Colonization of root surface by *Azospirillum* has been noticed with rather extensive intrusion within root tissues. Sporangia of *Frankia* have been observed in the rhizosphere of *Casuarina* seedlings (Fig. 31), indicating the variety of N_2 fixing microorganisms that are capable of inhabiting the rhizosphere. *Azotobacter chroococcum* and *A. paspali* are known to elaborate gibberellins and cytokin-like substances and these species of nitrogen fixing bacteria happen to be typical colonizers of the rhizosphere of grasses.

In low land flooded rice cultivation, two major subenvironments have come to be recognized—submerged plant parts and the rhizosphere (Fig. 32). Epiphytic bacteria and algae colonize the surface of aquatic weeds. The rhizosphere is a nonphotic environment where redox conditions are determined by the balance of oxidizing and reducing capacities of rice roots and characterized by exuded carbon compounds from roots providing energy sources for microbial growth. Nitrogen fixing microorganisms in the waterlogged rice fields contribute about 40–50 kg N/ha. This fixation is a cumulative effect of *Rhodopseudomonas*, blue-gree algae, both free living as well as symbiotic (photoautotrophs) and *Azotobacter, Beijerinckia, Methylomonas, Clostridium, Desulfovibrio, Klebsiella, Enterobacter, Flavobacterium, Pseudomonas, Azospirillum* and *Rhizobium* (Heterotrophs). The bulk of symbiotic nitrogen fixation in rice rhizosphere of flooded soil comes from root and stem nodules of leguminous green manure species when they are ploughed into the soil.

It is also now known that nitrogen fixing bacteria occur in the rhizosphere, stalks and phyllosphere of sugarcane plants and such bacteria have

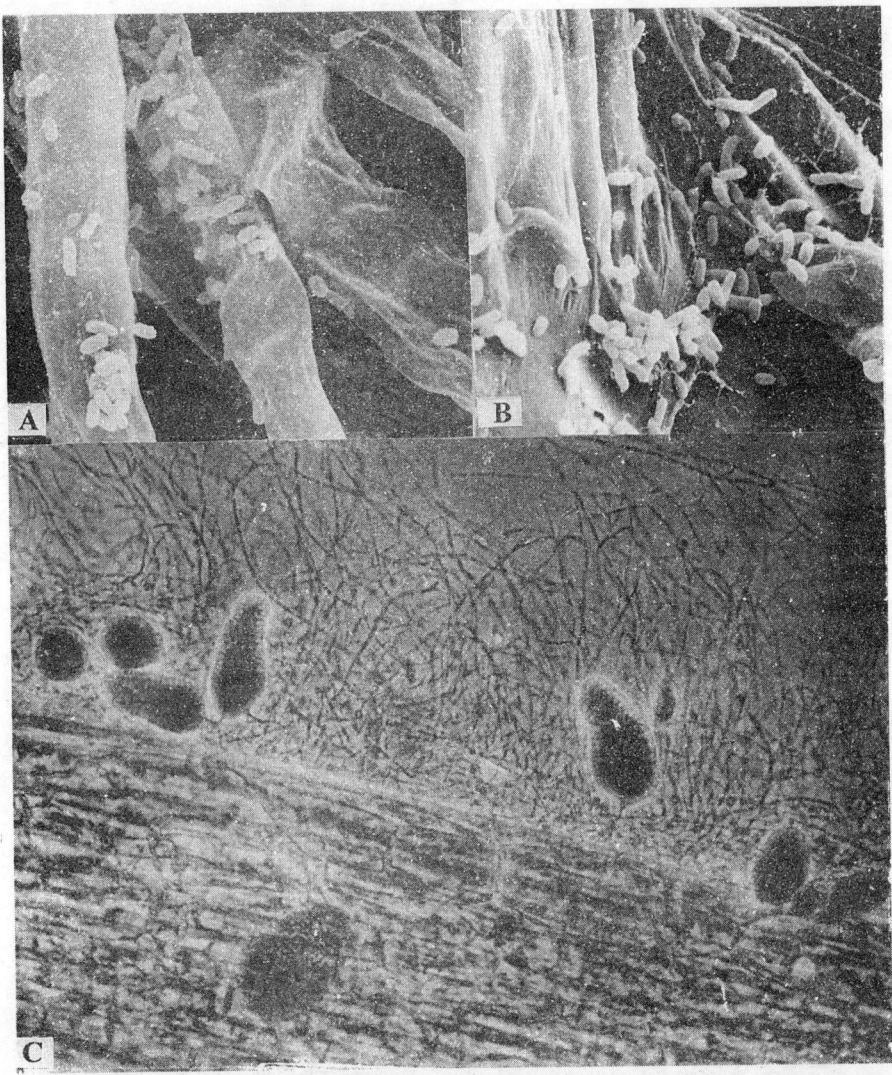

Fig. 31 The occurrence of N₂-fixing microorganisms in the root region of plants: A, B—Scanning electron micrographs of *Azospirillum* bacterial cells on that root surface; C–Sporangia and hyphae of *Frankia* in the rhizosphere of *Casuarina equisetifolia* (from Diem *et al.*, *Can. J. Microbiol.*, 1982, 28, 526–530).

also been reported inside the root cells. The numbers of these bacteria vary among genotypes of sugarcane and the types of soils on which they are cultivated. The following nitrogen fixing bacteria have been isolated from the root region of sugarcane: *Azotobacter vinelandii, Klebsiella pneumoniae, Bacillus polymyxa, Azospirillum brasilense, Derxia gummosa, Enterobacter cloacae* and *Erwinia herbicola* besides many other unidentified isolates. In

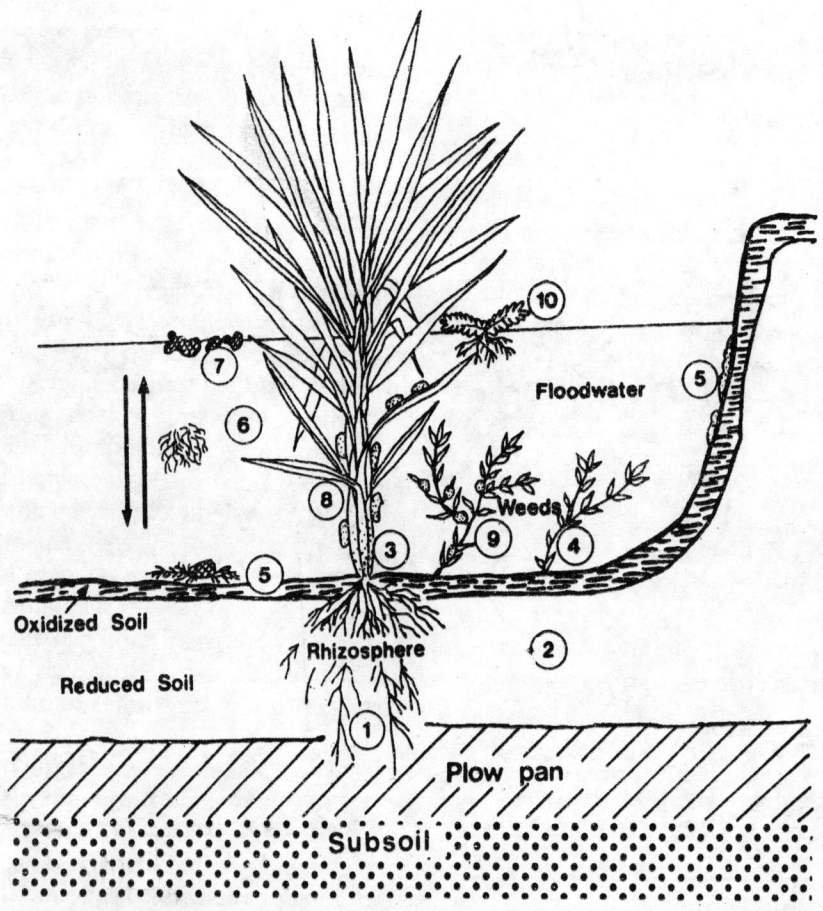

Fig. 32 Diagram of environments and N2-fixing components in a rice field ecosystem. N2 fixing bacteria: 1) associated with the roots, 2) in the soil, 3) epiphytic on rice, 4) epiphytic on weeds. Blue-green algae: 5) at soil-water interface, 6) free floating, 7) at air-water interface, 8) epiphytic on rice, 9) epiphytic on weeds. 10 *Azolla:* (from Roger and Watanabe, 1986).

Brazil, virgin soils supporting the growth of sugarcane receive no applications of chemical nitrogenous fertilizers. Apparently, nitrogen fixing bacteria together with other helper microorganisms that may excrete growth factors or help in lowering oxygen tension, are perpetuated through the continuous practice or vegetative propagation involving planting of sugarcane setts in the organically rich Amazonian soil. Populations of nitrogen fixing bacteria have been found in sugarcane rhizosphere even upto a depth of 120 cm. Conclusive evidences concerning the ability of sugarcane bacteria to fix molecular nitrogen have come forth from field measurements involving acetylene reduction as well as ^{15}N tracer.

Alteration of Rhizosphere Microflora

There are reports of changes in the rhizosphere microflora by (1) soil amendments; (2) foliar application of nutrients; and (3) artificial inoculation of seed or soil with preparations containing live microorganisms, especially bacteria (bacterization).

Many experiments have been done to find out the effects of N, P and K additions on rhizosphere microflora. The results do not lend themselves to any generalization since increase as well as decrease in R : S ratios have been reported as a sequel to fertilizer applications.

Translocation of photosynthates from leaves to roots is a well-known phenomenon as part of the normal metabolic activity of plants. Therefore, it is logical to assume that if materials artificially applied to leaves get into the leaf tissue, translocation may not pose difficult problems. In fact, many workers have reported the recovery of compounds (2,3,6-trichlorobenzoic acid; α-methoxyphenylacetic acid; 2-methoxy-3,6-dichlorobenzoic acid; 2, 4, 5-T; streptomycin) sprayed on leaves from root exudates of plants. Foliar sprays of urea are known to alter the number and nature of microorganisms in the rhizosphere. Extensive studies have been done on induced changes in the rhizosphere microflora by foliar sprays of antibiotics, growth regulators, pesticides and inorganic nutrients in the hope that such an approach may serve as a new tool in biological control of root diseases. However, no definite conclusions or guidelines have emerged from such studies to merit their application under field conditions.

Microbial seed inoculants such as *Azotobacter, Beijerinckia, Rhizobium* or P-solubilizing microorganisms may help in the establishment of beneficial microorganisms in the rhizosphere or in the immediate vicinity of growing roots. Field experiments have shown that counts of *Azotobacter* in wheat rhizosphere increased upon artificial seed inoculation indicating the efficiency of bacterization as a means of altering and improving the rhizosphere microflora.

Associative and Antagonistic Activities in the Rhizosphere

The dependence of one microorganism upon another for extracellular products, chiefly amino acids and growth promoting factors, can be regarded as an associative effect. Many reports indicate that cellular extracts of certain bacteria, fungi and algae increase the growth of other microorganisms in pure culture. Such findings are purely of academic interest unless actual benefits can be demonstrated in soil under field conditions. Russian workers have demonstrated an increase in amino acid content in plants grown in soil inoculated with specific microorganisms. Such observations have also been made with regard to B-vitamins, auxins, gibberellins and antibiotics. Gibberellins and gibberellin-like substances are known to be produced by bacterial genera commonly occurring in the rhizosphere such as *Azotobacter*, *Arthrobacter*, *Pseudomonas* and *Agrobacterium*. The commonly observed increased germination due to *Azotobacter* inoculation may be attributed to growth promoting substances excreted by the bacterium. There is an increase in the exudation of organic acids, amino acids and monosaccharides by plant roots in the presence of microorganisms. Microorganisms also influence root hair development, mucilage secretion and lateral root development of several plants. The fungi inhabiting the surface of roots influence the amount of substances absorbed into the root system. These attributes point out the existence of a two-way movement of metabolites between plants and microorganisms.

Secretion of antibiotics by microorganisms and the resultant biological inhibition of growth of other susceptible microorganisms are demonstrable in soil as well as in pure cultures. Such antagonistic effects are natural to expect even in uncultivated soil and from the agronomic point of view excessive inhibition of *Azotobacter* or *Rhizobium* in the root region may lead to decreased nitrogen fixation or nodulation. On the other hand, co-inoculation of nitrogen fixing *Azotobacter* and *Azospirillum* isolates with *Rhizobium* appears to have beneficial influence in increasing nodule number, nitrogen fixation and yield of soybean, pea and clover.

Root Exudates

Most of the studies on root exudates have been done in plants grown under aseptic conditions. One of the most important factors responsible for rhizosphere effect is the great variety of organic substances available at the root region by way of exudates from roots which directly or indirectly influence the quality and quantity of microorganisms in the root region. The substances exuded by plant roots include amino acids, sugars, organic acids, vitamins, nucleotides and many other unidentified substances. The nature of substances exuded by roots of plants has been summarized in Table 8. The nature and amount of substances exuded are dependent on the species of the plant, age and environmental conditions under which they grow. By the use of $^{14}CO_2$, it has been shown that products of photosynthesis are translocated to the root system and find their way into the rhizosphere in less than 12 hours, clearly indicating the influence of the metabolism of plants in determining the extent of rhizosphere effect. Further, radioactivity in the rhizosphere soil diminishes when samples are taken spatially away from the root system, thereby indicating that rhizosphere effect is centrifugal and would tend to diminish in soil where the ramification of the root system is less.

Table 8 Organic Compounds Detected in Plant Root Exudates (from Bolton *et al.*, 1993)

Class of organic compound	Exudate components
Sugars	Glucose, fructose, sucrose, maltose, galactose, rhamnose, ribose, xylose, arabinose, raffinose, oligosaccharide
Amino compounds	Asparagine, α-alanine, glutamine, aspartic acid, leucine/isoleucine, serine, aminobutyric acid, glycine, cysteine/cystine, methionine, phenylalanine, tyrosine, threonine, lysine, proline, tryptophan, β-alanine, arginine, homoserine, cystathionine
Organic acids	Tartaric, oxalic, citric, malic, acetic, propionic, butyric, succinic, fumaric, glycolic, valeric, malonic
Fatty acids and sterols	Palmitic, stearic, oleic, linoleic, and linolenic acids; cholesterol, campesterol, stigmasterol, sitosterol
Growth factors	Biotin, thiamine, niacin, pantothenate, choline, inositol, pyridoxine, *p*-amino benzoic acid, *n*-methyl nicotinic acid
Nucleotides, favonones, and enzymes	Flavonone; adenine, guanine, uridine/cytidine; phosphatase, invertase, amylase, proteinase, polygalacturonase
Miscellaneous	Auxins, scopoletin, fluorescent substances, hydrocyanic acid, glycosides, saponin (glucosides), organic phosphorus compounds, nematode cyst or egg-hatching factors, nematode attractants, fungal mycelial growth stimulants, mycelium growth inhibitors, zoospore attractants, spore and sclerotium germination stimulants and inhibitors, bacterial stimulants and inhibitors, parasitic weed germination stimulators

The root cap and areas of active growth are primary regions of root exudation and one of major sites of carbon release from seminal wheat roots into the soil happens to be the zone of root elongation. It has been suggested that exudation is either from root tips or regions at which lateral roots emerge from the main root.

Fungistasis

Root exudates influence the proliferation and survival of root infecting pathogens in soil either through soil fungistasis or inhibition of pathogens in the rhizosphere.

The term soil fungistasis is used to explain the inability of non-dormant spores, sclerotia or propagules to germinate even under most favourable conditions of pH, temperature and moisture in soil. The fungistasis could be released or undone by the rhizosphere effect of plants which creates a congenial environment for spore germination.

Sclerotia are perennating structures of certain plant pathogens and are different from spores in structure and behaviour. Sclerotia may be a mass of loosely interwoven hyphae (e.g., *Rhizoctonia solani*) or they may have an organized structure (*Macrophomina phaseoli* and *Verticillium dahliae*). Spores or sclerotia of many pathogenic fungi such as *Rhizoctonia*, *Fusarium*, *Sclerotium*, *Aphanomyces*, *Pythium*, *Colletotrichum*, *Verticillium*, *Phytophthora* and *Plasmodiophora* have been shown to germinate by the stimulus provided by the root exudates of susceptible cultivars of the host plants. The stimulus for germination has been attributed to compounds exuded by plant roots which help to overcome, in some manner, the static nature of dormant reproductive structures in soil. The fungistasis of the sclerotia of *Scelerotium cepivorum* is released by certain volatile stimulators associated with the roots of *Allium*. The volatile compounds have been identified as alkyl sulphides which are produced when alkyl cysteine sulphoxides coming from roots of *Allium* are broken down by soil bacteria. Once germination of spores takes place, the germ tube is less susceptible to soil fungistasis although its further development may be modified. For instance, macroconidia of *Fusarium* may germinate in soil and form germ-tubes but may then form chlamydospores instead of hyphae. Chlamydospore formation is less frequent or postponed in sterile soils indicating the associative effects of other soil microorganisms normally present in unsterilized soil in such a morphogenesis. Indeed, it has been observed that chlamydospore production by *F. solani* is accelerated in the presence of *Bacillus licheniformis*.

Root exudates may provide a food base for the growth of antagonists which could suppress the growth of pathogenic microorganisms in soil. Many instances have been reported where the rhizosphere of resistant varieties harboured more numbers of *Streptomyces* and *Trichoderma* than that of the susceptible varieties. Such observations have been recorded with reference of *Fusarium* wilts of plants. On examination of rhizosphere soil of pigeon-pea (*Cajanus cajan*) planted in seven different soils, it was observed that 13–33% of the rhizosphere isolates of *Streptomyces* from resistant varieties inhibited the growth of *Fusarium udum* the causal organism of pigeon-pea wilt whereas only 6% of the isolates from the rhizosphere of susceptible variety were found to be inhibitory to the wilt causing pathogen. High incidence of *Trichoderma viride* in the rhizosphere of varieties of tomato resistant to *Verticillium* wilt has been noted with its ability to minimize the severity of wilt on susceptible plants in the presence of the pathogen.

Substances in plant root exudates may influence the formation of infection structures such as appressoria as in the case of *Pellicularia filamentosa* and *Rhizoctonia fragariae*, although contrary evidences have also been obtained with regard to many other fungi.

Root exudates containing toxic substances such as glycosides and hydrocyanic acid may inhibit the growth of pathogens. A case in point is the exudation of hydrocyanic acid by the flax variety Bison resistant to *Fusarium*. Although several attempts have been made to correlate hydrocyanic acid content of root exudates with disease resistance, the results remain inconclusive. Thus, no unequivocal evidence has yet been brought forth to indicate the existence of specific compounds inhibitory to the growth and activity of pathogens in different cases of disease resistance.

Virus infections of plants result in stunting and foliar abnormalities which naturally reduce the area of the plant surface for photosynthetic activities. At different stages of disease development in *Dolichos lablab* infected with *Dolichos* enation mosaic virus, gross changes in the total numbers of microorganisms in the rhizosphere take place accompanied by alterations in the incidence of amino acid and vitamin requiring bacteria. The changes induced in the rhizosphere due to virus infection could be restored to normalcy by spraying the leaves with thiouracil or gibberellins which are known to overcome foliar abnormalities and stunting symptoms associated with virus-infected plants.

One of the attributes of root exudates is the possible role they play in neutralizing the soil pH and altering the microclimate of the rhizosphere through liberation of water and carbon dioxide. Such changes may influence infections of roots by pathogenic fungi.

Techniques

The Soil Dilution and Plate Count Method

This method has been generally used to estimate the number of microorganisms in the rhizosphere soil (Fig. 33). From the data obtained, an estimate of the total population of fungi, actinomycetes and bacteria per g of soil can be made. The method of sampling the rhizosphere soil differs among workers and this could be one of the contributory factors for the observed variations in results obtained by different workers. Further, in the soil dilution and plate count method, spore forming fungi such as penicillia and aspergilli overwhelm the non-sporulating fungi especially those of basidiomycetes which thus often get overlooked.

The Soil Plate

This method is an extension or adaptation of Warcup's soil plate method. The rhizosphere soil (0.005–0.01 g) is directly plated on melted and cooled agar medium. Before the agar solidifies, the dishes are rotated to distribute the soil in the agar. The advantage of this technique is that a differentiation of the soil microflora from various regions of the root system could be made. Secondly, mycelial as well as spore forming fungi have equal chances to come up on agar plates.

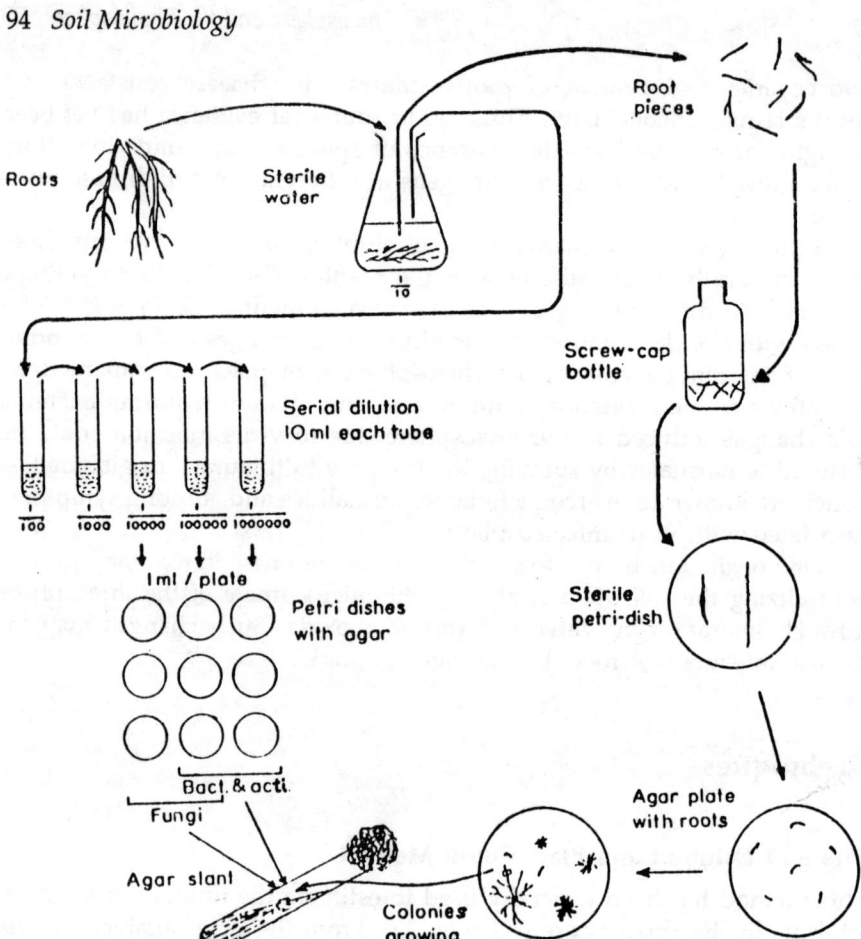

Fig. 33 Serial dilution of soil and plating method for isolation of microorganisms from soil, rhizosphere soil and root surface.

The Soil Box

This technique involves the use of wooden boxes having on one side closely fitting but separate and removable perspex (plexyglass) segments. The roots are observed through the transparent perspex segments and at desired intervals, rhizosphere samples could be obtained to study microorganisms at different depths of the soil.

The Contact Slide Technique

This involves the placing of individual roots with closely adhering soil particles on microscope slides coated with a thin layer of nitro-cellulose dissolved in amyl-acetate. This layer provides an adhesive surface to which the rhizosphere soil sticks. After removal of the roots, the slides are dried, stained with phenolic aniline-blue, washed with water and mounted in lactophenol. The advantage of this technique is the facility it provides for *in situ* observations of the rhizosphere microflora.

Artificial Rhizosphere

This can be established by placing collodion membrane sacs containing root exudates in soil and observing the effect of exudates (passing through the semipermeable membrane) on the nature of microflora colonizing the artificial rhizosphere.

Model Systems

Cultures of selected bacteria are used to inoculate media supplemented with root extracts or root exudates to find out associative and antagonistic effects. The results of such experiments must be interpreted with caution since the technique appears to be an over-simplification of conditions met with in natural soils.

The Use of Labelled Nutrients

Radioactive carbon, phosphorus and sulphur have been used to find out the rate and extent of the transport of photosynthetic products from shoots to the root system and their ultimate liberation into the root medium as exudates. Conversely, the role of root surface fungi in the mobilization of nutrients to the root system and subsequently to the tops of plants can also be studied. In such studies plants are grown under aseptic conditions and roots infected with specific root surface fungi. The tops of plants are enclosed in polythene bags or special chambers containing radioactive carbon dioxide and root exudates or rhizosphere soil extracts assayed for radioactivity in relation to time. By growing plants for longer periods in chambers containing $^{14}CO_2$, it would be possible to detect qualitatively and quantitatively, the nature and amounts of amino acids, organic acids and sugars liberated into the surrounding medium, by means of paper chromatography followed by autoradiography.

Fluorescence Microscopy

Suitable dilutions of soil suspensions are mixed with 1% Difco Bactoagar solutions and appropriate aliquots (0.01 ml) transferred to microscopic glass slides by means of micropipettes. The glass slides may be etched suitably so that required amounts of the agar suspension get evenly and firmly distributed on the surface. The slides are dried uniformly and stained for 2 minutes in a very dilute solution of acridine orange, care being taken to remove excess of the stain by dipping in water. Acridine orange stain reacts with deoxyribonucleic acid of living bacteria and fluoresce red under a fluorescence microscope. Dead bacteria show green-fluorescence upon staining, a characteristic feature also shared by some Gram-negative bacteria which in fact has posed difficulties in differentiating living bacteria from dead ones by the use of this technique. Other factors such as the concentration of the dye, the time of exposure and pH of the medium govern the efficacy of this technique.

Fluorescent Antibody

The technique is based on the specificity of antigen-antibody reaction which is widely used in medical microbiology and pathology. Small amounts of flurorescence can be detected in bacteria if a fluorochrome dye is attached to them. This is accomplished by coupling an antibody (Ab) to a fluorescent dye and allowing the complex to react with its antigen (Ag). The microscopic preparations are then examined under a fluorescence microscope. Commonly known as FA technique, this precise technique has paved the way for an elegant method of observing or detecting a microorganism in its natural environment. Special skill and equipments are needed for its successful operation which can be briefly outlined as follows: The antigen (Ag) providing organism is injected into rabbits and serum collected by standard procedures. The serum is fractionated to give gamma globulin which is dialysed and diluted to 1% protein level. A fluorochrome dye, fluorescein isothiocyanate (FITC) is conjugated to immune globulin under carefully controlled pH and buffer conditions. The conjugated complex is separated from the dye on Sephadex column and dilutions prepared until the desired dilution is obtained which will act as the correct indicator of good antibody activity avoiding non-specific background staining. The FA thus prepared can be preserved for many years at 20°C.

The soil sample or the sample of any ecological microhabitat to be studied (natural material) is placed on a microscopic slide. The latter may as well as allowed to be in contact with the material for long periods to obtain a 'contact slide'. The slide is heated and a gelatin-rhodamine isothiocyanate (RhITC) conjugate is applied to the slides with the object of not only suppressing non-specific antibody absorption but also to serve as a counterstain. FA is finally applied to the dried gel-RhITC layer and incubated, for reaction for one hour. The preparation is washed in buffer, air dried and examined under fluorescence microscope.

One proven example of the application of FA technique, was the detection of an efficient strain of *Rhizobium japonicum* as brightly stained rhizobia-like rods in field soils containing myriads of *R. japonicum* strains (Fig. 34). The survival and competition of rhizobial strains in soils have also been studied by the use of this technique. Despite many inherent limitations, the technique has many potentialities in studies on soil ecology

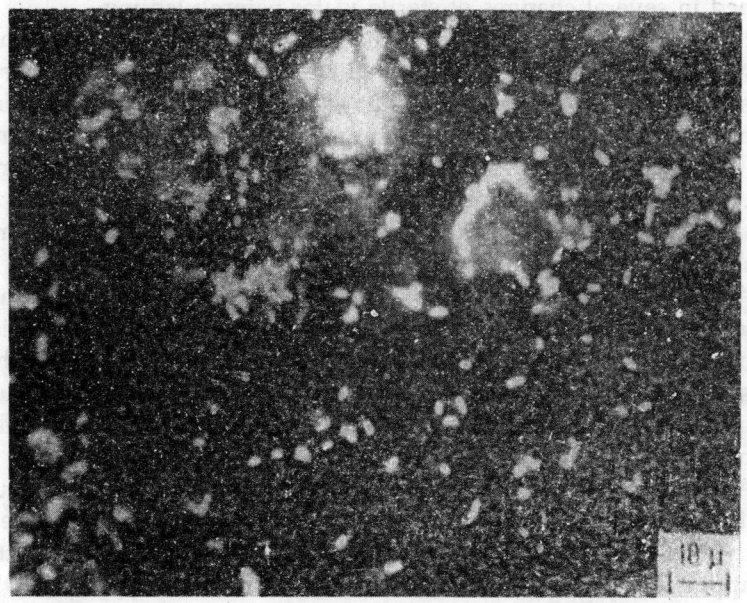

Fig. 34 Specific detection of *Rhizobium japonicum* strain 61A72 by immunofluorescence. Preparations made from slides that had been in contact with field soil inoculated with *R. japonicum* 61A72. In some areas cells are seen in association with soil particles (from B.B. Bohlool and E.L. Schmidt, *Science*, 162: 1012, 1968).

in laboratories where the required facilities for this type of work are available.

Techniques in the Collection of Root Exudates

From time to time methods of raising plants aseptically have been described in the literature and some of them are illustrated in Fig. 35. Many of them can be used successfully to collect root exudates under aseptic conditions. Initially, seeds are surface sterilized with any one of the following chemicals: mercuric chloride, calcium or sodium hypochlorite, freshly prepared chlorine water, hydrogen peroxide, sulphuric acid and ethyl alcohol. An ideal choice for surface sterilization of seeds with soft seed coat is chlorine water, prepared afresh each time until the water receiving chlorine gas turns yellow. Concentrated sulphuric acid has been used with success for seeds having tough seed coat such as cotton and clovers. It is essential to use dry test tubes or beakers while surface sterilizing seeds with sulphuric acid and it is also advisable that subsequent washings with sterile water must be made as quickly as possible so as to prevent damage to seed by excessive heat generated by the addition of water to acid. In all cases, the surface sterilized seeds have to be necessarily washed in several changes of sterile water before planting.

It is emphasised that all parts of the equipment involved in the techniques are sterilized prior to use. Similarly, all the transfers and handling are done in a sterile room under aseptic conditions to prevent microbial contamination.

In the technique referred to in Fig. 35A, a seedling is raised initially in a test tube and then transferred to a one litre capacity Erlenmeyer's flask filled with the nutrient solution and stoppered with a cotton plug in such a way that the sterilized cotton plug separates the root system from the shoot system.

A three-neck distilling flask (capacity 300 ml) serves as a root chamber in the assembly referred to in Fig. 35B. The centre neck of the flask is used for growing plants in a planting tube inserted into the neck while the side tubes are used as inlet and outlet for aeration. The reservoir of nutrient solution is connected to one of the side opening of the root chamber by means of rubber and glass tubing.

An ordinary glass jar with a metal screw type lid serves the purpose for the technique referred to in Fig. 35C. A hole (3.5 cm diameter) is punched in the centre of the metal lid and an aluminium tube fitted on the upper side of the hole. On the inside of the screw type lid of the jar, a disc of aluminium sheet with a central hole (similar to the one on the screw type lid) is placed after sandwiching a piece of cotton gauze in between the lid and the disc. A gauze wrapped cotton plug is inserted into the aluminium tube fitted on top of the lid. The entire apparatus is assembled as shown in the figure after pouring nutrient solution into the jar. The seedling is raised in the aluminium tube and when it grows to a sufficient height, the cotton plug is removed and a small amount of steril-ized sand or vermiculite is placed around the seedling to separate the root system from the shoot system.

One end of a glass vessel open at both ends is plugged with cotton wool over which a small amount of sand is placed. This serves as a seed-ling chamber in the technique referred to in Fig. 35D. The entire vessel is placed in a two-litre conical flask in such a way that the seedling chamber is half-way immersed in the nutrient solution and the growing plant would then be automatically fed with nutrient by the capillarity action of the cotton gauze and the sand. A mechanical aerating device is also shown in the figure.

In the technique referred to in Fig. 35E, a glazed pot or jar is used to grow plants with arrangement for drainage of excess nutrient by means of a flask attached to the lower end of the pot. A large glass cylinder, stoppered at the top with a large cotton plug and sealed with paraffin is placed over the glazed pot. An aluminium tube is made to pass through the cotton plug, then through the glass cylinder and end up as a loop on

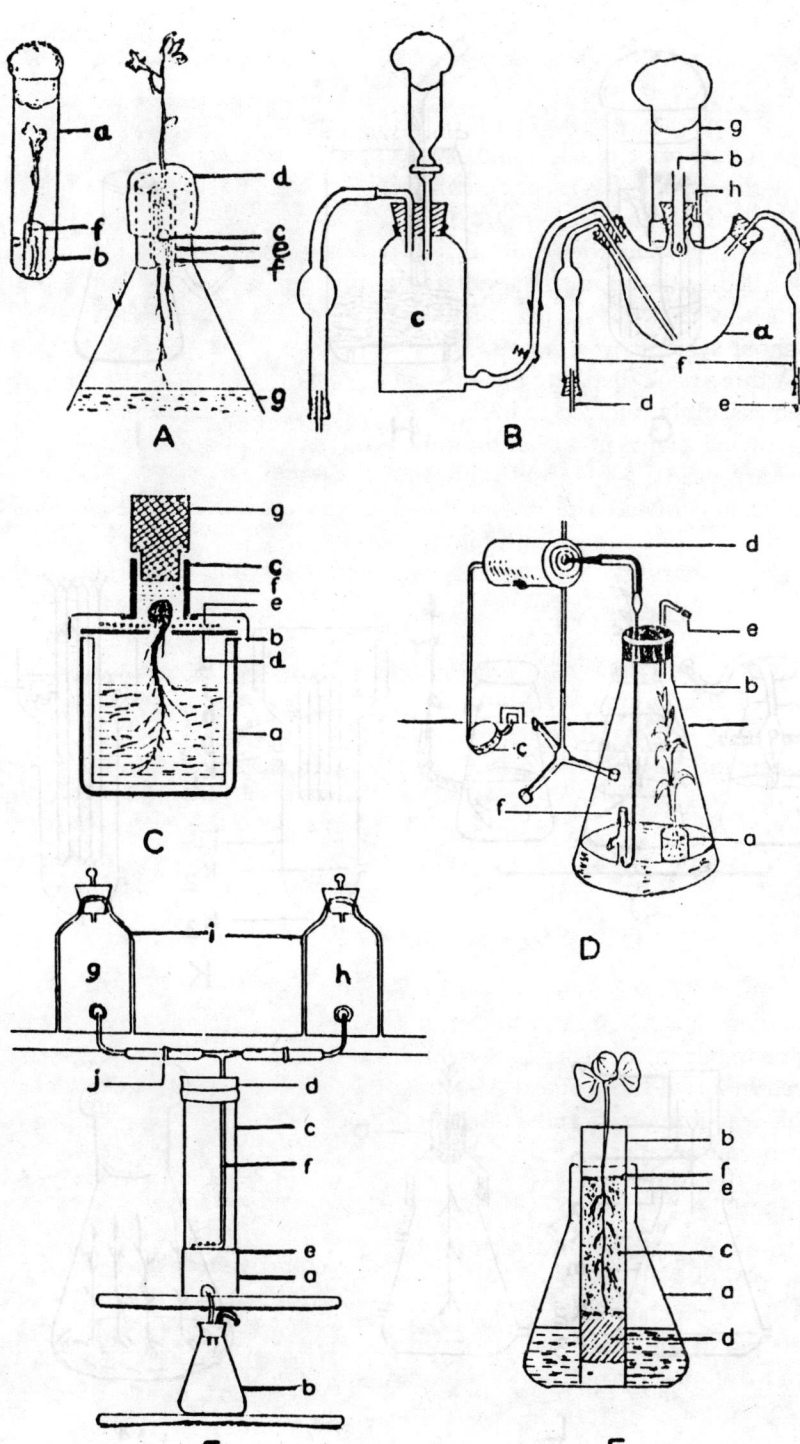

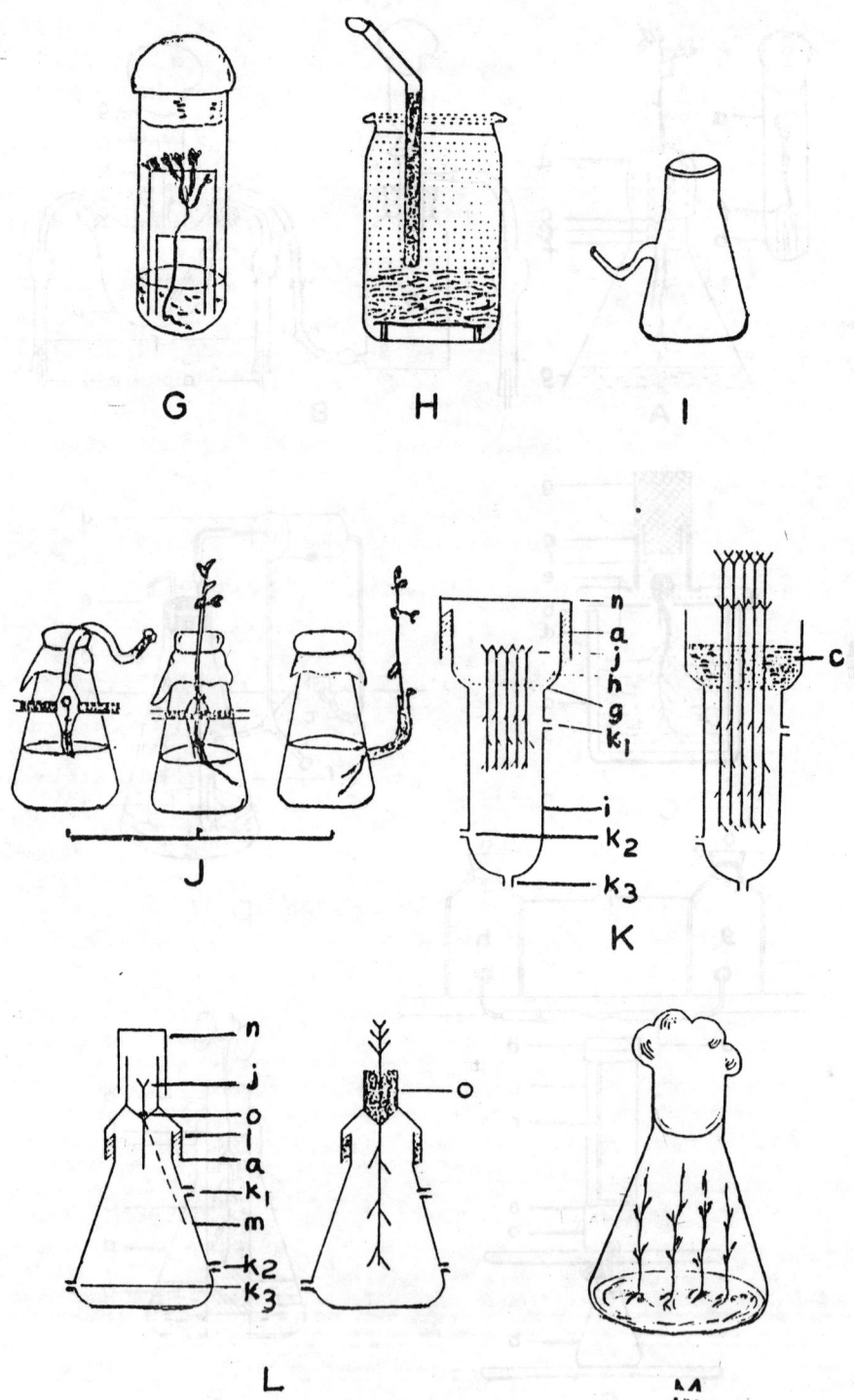

G   H   I

J   K

L   M

the soil jar. The loop portion of the aluminium tube is drilled with holes. The upper end of the tubing is connected to two aspirator bottles which serve as reservoirs of nutrient. Seeds are sown on sand in the glazed pot and plants are raised by releasing nutrient by means of stoppers in the aspirator bottle assembly.

Seedling can be grown in natural soil and root exudates collected if the technique referred to in Fig. 35F is used. Soil is filled in a glass tubing open at one end and closed at the other end with glass wool. The glass wool end is placed in a flask containing water and the open end is fitted into the neck of a flask by means of an annular ring of cotton. Seeds are sown on the open side of the tube and when the seedling emerges, sterilized sand is placed around the stem to prevent aerial contamination.

In the technique referred to at Fig. 35G, a boiling tube is filled with nutrient solution and a microscope slide and a cover slip dipped in it are used to raise small seedlings. Approximately 0.2 ml of cooled agar is pipetted over one-half of the slide and a previously germinated seed is placed in position on the agar surface followed by the placement of the

---

Fig. 35 Methods for aseptic culturing of plants: A—after Knudson and Smith, 1919, *Bot. Gaz.*, *68*, 460–466—a) sterile culture tube, b) 1% agar, c) Erlenmeyer's flask, d) sterile cotton, e) outer tube, f) germination tube, g) nutrient solution; B—after Reuszer, 1949, *Proc. Soil Sci. Amer.*, 14, 175–179—a) root chamber, b) planting tube, c) nutrient solution reservoir, d) to air supply, e) air outlet, f) air filters, g) hood, h) cotton plug; C—Cross sectional diagram of jar unit, after Blanchard and Diller 1950, *Plant Physiol.*, *25*, 767–769—a) glass container, b) lid, c) aluminium cylinder, d) aluminium disc, e) cotton gauze, f) vermiculite, g) gauze-wrapped cotton plug; D—after German and Bowen, 1951, *Plant Physiol.*, *26*, 840–842—a) glass vessel, open top and bottom containing gravel, b) two-litre wide mouth Erlenmeyer pyrex flask, c) air-pump, d) Koby air filter, e) glass wool fitted outlet tube, f) air-bubble lift pump; E—after Kathrein, 1951, *Plant Physiol.*, *26*, 843–847—a) 4" glazed soil jar, b) drainage collection flask, c) 100 mm glass cylinder, d) cotton plug, e) paraffin seal, f) aluminium watering tube, g) nutrient solution, h) distilled water, i) 20-litre aspirator bottle; F—after Bhuvaneswari and Sulochana, 1955, *Curr. Sci.*, 24, 376–377—a) pyrex container, b) pyrex tube open at both ends, c) a layer of soil, d) glass wool, e) an annular ring of cotton, f) a layer of acid washed sand; G—after Fahraeus, 1957, *J. Gen. Microbiol.*, *16*, 374–381—culture tube containing slide with a few-week old seedling, held in place by a cover slip; H—after Wieringa and Bakhuis, 1957, *Pl. Soil. 8.*, 254–262—jar for sterile culture of plants; I—after Waris, 1958, *Physiol. Pl. 11*, 627–630—S-tube flask; J—after Waris, 1950, *Physiol. Pl. 11*, 627–630—cross-tube flask with sigmoid tube and pea seedlings in different views; K and L—after Subba Rao and Bailey, 1961, *Can J. Bot.*, *39*, 1747–1758—methods for growing multiple and single plants respectively in two stages of operation—h) cup to hold the chromel wire mesh, i) vessel to contain nutrient solution, j) plant(s), k_1) inlet for nutrient, k_2) inlet for aeration, k_3) exit for nutrient, l) lid with an upper limb to hold the shoot, m) orifice, n) lid, o) glass beads, and a) cotton gauze; M—after Andal *et al.*, 1956 *Nature*, Lond., *173*, 1063—flask containing filter paper on which seedlings could be raised aseptically.

cover slip on the radicle. The root system grows in between the glass slide and the cover slip.

An ordinary glass jar is again used as a container for growing plants aseptically in the technique referred to in Fig. 35H. The jar is filled with sand and a bent glass tube, also filled with sand, is inserted into the sand. Seeds are sown on sand in the jar and when the seedlings are sufficiently tall, the surface of the sand is covered with a thin layer of paraffin covered glass sand. The bent glass tube serves as a means to provide nutrient solution to plants as and when required.

A conical flask with a sigmoid side tube (Fig. 35I) is intended to grow plants wherein surface sterilization of seeds is also accomplished. A seed is placed in the sigmoid tube and a small quantity of surface sterilizing agent poured into it by means of a pipette. By tilting the flask to one side, the sterilant could be poured out and seed rinsed with sterile water before pushing the seed into the flask for subsequent growth.

Cross-tube flasks are shown in different views in Fig. 35J. The two openings of the horizontal portion of the tube system are plugged with cotton. The upper opening of the longitudinal stem is connected to a sigmoid tube by means of a short rubber tube, while the lower opening is continuous with the flask. The open end of the sigmoid tube is covered by cotton plug. The sigmoid tube is intended for surface sterilization and subsequent transfer of seed into the basal portion of the longitudinal stem. A seed is introduced into the sigmoid tube, the cotton plug removed and the surface-sterilizing agent poured over the seed. The plug is reintroduced and after half an hour the sigmoid tube is tilted in a manner so as to draw out the sterilizing agent through the cotton plug. This is achieved by pressing dry cotton wool against the wet cotton plug. Subsequently, the sterile seed is conducted by tilting the set-up suitably until the seed is lodged in the basal portion of the longitudinal stem. On germination, the root will grow into the flask and the stem of the seedling passes through the longitudinal stem of the cross-tube system. At this stage of plant growth, the cotton plugs in the horizontal tube are pushed inwards so that they hold the stem in position. The sigmoid tube and the rubber tubing are then removed.

Many seedlings can be grown in the technique referred to in Fig. 35K. The main vessel has a wide opening at the top with narrow side openings for aeration and drainage. The neck of the wide opening has a chrome wire mesh held in position by a glass cup and this in turn is closed by a glass lid. Seeds are sown on the chrome wire mesh and nutrient solution poured into the main stem of the vessel. When the seedlings grow sufficiently tall, the lid is removed and glass beads are poured on the wire mesh to prevent aerial contamination.

A similar technique (Fig. 35L) for growing single plants takes advantage of a flask, whose neck is fitted with a special lid having a small hole in the middle as shown in the figure. As usual, the apparatus has outlets for aeration and passage of nutrients. A previously germinated seed is held gently by a pair of forceps and introduced into the hole and kept in position by means of two glass beads. When the seedling establishes, more glass beads are put in position to prevent aerial contamination.

Two superimposed acid washed filter papers and placed in flasks (Fig. 35M) and moistened with nutrient solution. Seeds are sown on the filter paper and when seedlings have established the filter papers are removed, ground with acid washed sand and extracted for recovery of root exudates. This appears to be the simplest technique so far described.

All the techniques have been designed to grow plants free of microorganisms and therefore periodical checking of the nutrient solution by plating it on agar media is a necessary step in maintaining the efficiency of the technique employed. The techniques could be divided into two lots: (1) where root and shoot systems grow in the same environment, and (2) where both the systems are separated so as to grow the root system aseptically and the shoot system free in the air. Cotton wool, special wax and glass beads have been used to achieve the separation between the root and shoot systems.

Plant Growth Promoting Rhizobacteria

The use of the term rhizobacteria implies the ability of certain bacteria to colonize the rhizosphere very aggressively. *Pseudomonas* spp. are receiving world-wide attention under the broad general category known as plant growth promoting rhizobacteria (PGPR). The bacteria exhibit fluorescence under u-v light and hence are also known as fluorescent pseudomonads. Initial observations were based on the ability of *P. fluorescence* and *P. putida* when applied to potato seed pieces improved the growth of potatoes. Subsequent field studies have revealed that these species and similar bacteria increased the yield of potato (5.33 per cent), sugarbeet (4–8 t/ha) and radish (6–144 per cent of root weight).

Soil-borne pathogens may be distinguished as major and minor ones. The major ones include the well-known *Phytophthora* and *Fusarium* causing root rots and vascular wilts. The minor pathogens cause damage to young tissues of roots with no visual symptoms. The minor pathogens also include, according to recent thinking, certain non-parasitizing deleterious rhizosphere microorganisms (DRMO) which include deleterious rhizobacteria (DRB) and deleterious rhizofungi (DRF). Some soils are conducive to soil-borne diseases whereas others are suppressive to diseases. The reasons behind these observations may lie in the soil structure and or microbial composition.

Many genera of soil bacteria have shown great potentiality as biocontrol agents operating not merely by secreting antibiotics but by means of other mechanisms. These genera include *Actinoplanes, Agrobacterium, Amarphosporangium, Arthrobacter, Cellulomonas, Bacillus, Azotobacter, Enterobacter, Erwinia, Flavobacterium, Micromonospora, Rhizobium, Bradyrhizobium, Serratia, Streptomyces* and *Xanthomonas. Agrobacterium radiobacter* strain 84 is an excellent example of a biocontrol agent controlling crown gall disease caused by *Agrobacterium tumefaciens. Bacillus subtilis*, capable of producing endospores and tolerating heat can suppress major and minor soil-borne diseases of carrots, oats and groundnut.

In Netherlands, it has been shown that the frequency of potato cultivation in the same field has a bearing on the yield of potato tubers. When the crop was grown in the same field every third year, yields were reduced by 10–15 per cent from what would be expected if the crop was grown once in six years. The severity of decline in yield appeared to be progressive when yield decrease reached 30 per cent if potato was cropped continuously in the field. Fluorescent pseudomonads are believed to improve the growth of plants by colonizing the root region aggressively and thus preempt the establishment of DRMO on roots, especially those which produce growth inhibiting cyanide. No such growth promotion was possible in plots where no potato was cultivated probably due to the absence of factors which stimulate the production of toxic substances. Fluorescent pseudomonads have been shown to suppress major plant pathogens like the take-all, a root disease of wheat caused by *Gaeumannomyces graminis* var *tritici*, by 11–17 per cent.

About 10 per cent of bacteria in the rhizosphere appear to be aggressive in reducing the population of DRMOs and there appears to be no relationship between *in vitro* inhibition and *in vivo* suppression effects. The field beneficial effects are dependent on soil temperature, pH, moisture and clay content which influence the survival of PGPRs in the rhizosphere. This ecological competence may diminish by repeated subculturing *in vitro*, possibly related to loss of cell surface structure or reduction in antibiotic and siderophore production as explained hereunder.

It has been postulated that hydrocyanic acid (HCN) produced by many DRMOs reduces potato root growth and is responsible for the decreased yield of potato tubers. Pseudomonad PGPRs increase potato yields, according to one theory, by reducing HCN production by DRMOs, through siderophore-mediated competition for Fe(III), which is required for HCN production. The part played by HCN in reducing yield of potato has become debatable because evidences have also been presented to demonstrate that HCN infact proved beneficial in biological control in other root diseases such as take-all of wheat and black rot of tobacco. Therefore, more critical studies in future may clarify the situation.

Three possible mechanisms have been suggested to explain the beneficial effects of PGPRs in enhancing production. They are competition for substrate and niche exclusion, production of siderophores and antibiotics. However, more than one mechanism may operate for mediating a biological control. Fluorescent pseudomonads 'mop up' nutrients in the rhizosphere because of their versatility in growth and nutrient absorption. The points of emergence of lateral roots are favourite spots for DRBs and PGPRs appear to compete for these spots very effectively.

Siderophores

Siderophores are low molecular weight, high affinity iron chelators that transport iron into bacterial cells. Fluorescent pseudomonads produce yellow-green, fluorescent siderophores which specifically recognize and sequester the limited supply of iron in the rhizosphere and thereby reduce the availability of this trace element for the growth of the pathogen. The availability of iron in soil decreases with increase in pH and therefore PGPRs function better in neutral and alkaline soils than in acid soils. The demonstration that siderophore minus mutants are less suppressive to pathogens in the rhizosphere than parental strains appear to convince the role of siderophores in pathogen suppression. As stated earlier, an example of the involvement of antibiotics is the role of Agrocin-84 produced by *Agrobacterium radiobacter* in controlling the crown-gall symptoms of plants caused by *Agrobacterium tumefaciens*. The phenazin-type antibiotic produced by *Pseudomonas fluorescence* in the control of take-all disease of wheat has also been cited as another example.

Several investigations have shown that fluorescent pseudomonads in the rhizosphere produce yellow-green fluorescent pigment. Some strains, particularly B10 inhibit the growth of *Erwinia caratovora* which causes the soft rot disease of potato. In the presence of iron, no beneficial effect of *Pseudomonas* inoculation was observed when the soil was amended with Fe in the form of FeEDTA (Ethylenediaminetetraacetatoferrate) in spite of effective colonization in the rhizosphere. A yellow-green pigment called 'pseudobactin' isolated from this fluorescent pseudomonad also did not exhibit beneficial effect when iron was sequestered (bound) in the form of 'red-brown ferric pseudobactin' whereas pseudobactin by itself was effective. These results imply that the siderophore pseudobactin deprived *E. caratovora* of Fe, by scavenging the element available in the vicinity and thus reduce disease severity by minimizing the virulence of the pathogen.

In California, when FeEDTA was added to a *Fusarium* suppressive soil where no inoculation was done to flax seedlings with its pathogen *Fusarium oxysporum* f. sp. *lini* 90% of the seedlings survived. On the other hand, in the same suppressive soil with the pathogen in the presence of EDTA, only 47% of the seedlings survived. However, the presence of the pathogen alone in the same soil had adverse reaction resulting in the survival of only few seedlings. These results reflect the possibility that microorganisms present in the soil produce siderophores which have the affinity for the limited available iron in the root milieu and thus depriving the pathogen of this vital element.

A similar experiment carried out in pathogen conducive soil also proved that sequestering the limited Fe was the reason behind the reduction in the severity of the disease. The experiment involved the addition of strain B10 of *Pseudomonas* sp. or siderophore pseudobactin to a pathogen conducive soil, infested with *Fusarium oxysporum* F. *lini* and growing flax seedlings. Pseudobactin is a linear hexapeptide requiring at least five gene clusters with a minimum of five genes for its biosynthesis. The addition of PGPR or its siderophore pseudobactin increased the survival of flax seedlings to 87 to 90 per cent when compared to other treatments where ferric pseudobactin or FeEDTA plus *Pseudomonas* sp. strain B10 were added bringing down seedling survival to 48 to 50 per cent. Similar results were also obtained with the take-all disease of wheat caused by *Gaeumannomyces graminis* var. *tritici* by the addition *Pseudomonas* sp. strain B10. Some of these results obtained either with *Pseudomonas* or siderophores with and without EDTA have been summarised in Table 9.

Genetic Manipulation

The approach that has received much attention has been the transfer of genetic loci encoding the pathways for the synthesis of antifungal metabolites such as phenazines, phloroglucinols, oomycin(A), pyoluteorin, pyrrolinitrin and HCN. For example a strain of *P. fluorescens* Hv37a was constructed by placing the biosynthetic genes for the production of oomycin A under the control of the constitutive *tac* promoter from *Escherichia coli*; this recombinant strain suppressed *Pythium* damping-off cotton to a greater extent than the parent strain. Another example is the overproduction of phenazine-1-carboxylase or phloroglucinol by the introduction of plasmids possessing the respective biosynthetic loci into a number of different strains of fluorescent *Pseudomonas* spp; the recombinant strains had better ability to suppress the take-all disease of wheat than the parent strain. One more example is the use of Tn5 mutagenesis approach to generate mutants of *P. fluorescens* CN 12 with enhanced level of *in vitro* antibiosis towards the take-all disease organisms of wheat; in four years field trials, the mutant strain provided significantly greater control of the take-all disease than the parent strain.

Another strategy in genetic manipulation has been the introduction of the ability to catabolize a novel substrate so that the recombinant strain, when added to soil with the novel substrate, gets the ability to colonize the root region preferentially over indigenous microflora. For example,

Table 9 The influence of *Pseudomonas* sp. (strain B10) or its siderophore with or without FeEDTA on the intensity of wilt of flax incited by *Fusarium oxysporum* f. sp. *lini* or take-all of barley incited by *Gaeumannomyces graminis* in disease suppressive or conducive soils of California (summary of data from Leong., J., *Ann. Rev. Phytopathol.*, 24, 187–209)

Soil	Pathogen Present/ Absent	Treatment	Per cent of seedling survival Flax	Per cent of seedling survival Barley
Disease suppressive soil	Present	H₂O	82	83
	Present	50 μm FeEDTA	47	38
	Absent	50 μm FeEDTA	90	85
Disease Conducive Soil	Present	H₂O	48	27
	Present	50 μm FeEDTA	52	25
	Absent	50 μm FeEDTA	92	87
	Present	Strain B10	87	88
	Present	Strain B10 + 50 μm FeEDTA	48	25
	Present	50 μm Pseudobactin (Siderophore)	90	73
	Present	50 μm Ferric Pseudobactin	50	20

strain of *Pseudomonas* R20 was genetically engineered to possess the plasmid NAH 7, which encodes the enzymes for salicylate degradation; when this recombinant strain [R 20 (pnAH 7)] was added to soil with salicylate, its population was two fold in sugarbeet soil when compared to the parent strain, thus affording better chance of operating in the root region to bring about reduction in the severity of DRMOs.

Integrated Approach

Many reports have come on the dual application of fluorescent pseudomonads with species of *Trichoderma* or other fungi possessing disease controlling abilities that are mutually compatible. Combinations of biocontrol agents and chemical control agents (fungicides) provide yet another approach to minimize fungal infections of seedlings caused by *Fusarium*, *Pythium*, *Rhizoctonia*, *Cylindrocarpon* and *Cylindrocladium*.

Commercialization of PGPRs

A product by name QUANTUM 4000 is being marketed by Gustafson Inc. Dallas, Texas as a growth promoter on peanut (groundnut) and cotton. The product contains *Bacillus subtilis* strain GBO3, which is a derivative of strain A13. Other products may follow once procedures are standardized for mass multiplication together with strategies for carriers and quality control. Prior to this, the repeatability of success under field conditions have to be established with regard to fluorescent pseudomonads;

quite often, strains that succeed under green house conditions fail to do so in the field. The use of multiple strains of fluorescent pseudomonads has shown success in the control of diseases such as take-all of wheat and *Fusarium* wilt of radish. In northwest China, encouraging results in the control of take-all disease of wheat have come from field trials covering 4000 ha during 1991–94 by using *P. fluorescens* CN12 and Tn5 derivatives, in diverse sites exhibiting varying environmental conditions; the yield increases of wheat due to rhizobacterial inoculation varied from 16 to 64 per cent.

Novel Approaches (Biotechnology of the Rhizosphere)

An approach that merits consideration is the use of rhizosphere microorganisms in the control of weeds. For example, the graminaceous downy brome is a weed in wheat cultivation that grows as fast as the main crop. Several isolates of bacteria from the rhizosphere of the weed were found to inhibit the growth of the weed and not of the wheat crop. In a field trial, the efficacy of a strain of *Pseudomonas* spp. from the rhizosphere in controlling the growth of downy brome weed has been demonstrated in the USA. This approach needs to be followed up with other weeds of many important crop plants.

Manipulation of the rhizosphere to enhance specific microorganisms to achieve *in situ* bioremediation of pesticides has been attempted in some cases. For instance, in flooded soil planted with rice, the pesticide parathion was found to be mineralized to a greater extent than in un-flooded soil planted with rice; moreover, merely flooding soil without planting rice mineralized the pesticide to a marginal extent thereby indicating that rhizosphere of rice harbour microorganisms capable of enhancing the mineralization of parathion that was attributed to a greater biomass of root as well as shoot system. In another study, the extent of degradation of several polycyclic aromatic hydrocarbons was enhanced in soil planted with deep-rooted prairie grasses over that of unplanted soil. Similarly, the biodegradation of trichloroethylene has been shown to be significantly higher in the rhizosphere soil in comparison with the non-rhizosphere soil, an outcome that was attributed to increased root biomass. Despite the fact that it is too early to contemplate inoculation technologies with specific microorganisms by seed coating to enhance biodegradation of xenobiotics, it can be presumed that planting high density rooting plants in contaminated soil may contribute towards scavenging contaminants in soil.

The Phyllosphere

Plant parts, especially leaves are exposed to dust and air currents resulting in the establishment of a typical flora on their surface aided by the cuticle, waxes and appendages which help in the anchorage of microorganisms. These microorganisms may die, survive or proliferate on leaves depending on the extent of influence of the materials in leaf diffusates or exudates. Leaf diffusates or leachates have been analysed for their chemical constituents. The principal nutritive factors are amino acids, glucose, fructose and sucrose. If the catchment areas on leaves or leaf sheaths are significantly substantial, such specialized habitats may provide niches for nitrogen fixation and secretion of substances capable of promoting the growth of plants.

The leaf surface has been termed 'phylloplane' and the zone on leaves inhabited by microorganisms as 'phyllosphere' (Fig. 36). The Dutch microbiologist Ruinen coined the word 'Phyllosphere' from her observations on Indonesian forest vegetation where thick microbial epiphytic associations exist on leaves. The dominant and useful microorganisms on the leaf surfaces in the forest vegetation in Indonesia happened to be nitrogen-fixing bacteria such as *Beijerinckia* and *Azotobacter*. In general, apart from nitrogen-fixing bacteria like *Azotobacter*, other genera such as *Pseudomonas*, *Pseudobacterium*, *Phytomonas*, *Erwinia*, *Sarcina* and other unidentified ones have been encountered on plant surface, especially on leaf surface. In Puerto Rico, nitrogen-fixing blue-green algae such as *Anabaena*, *Calothrix*, *Nostoc*, *Scytonema* and *Tolypothrix* have been encountered on plant surfaces in moss forests.

Some of the fungi and actinomycetes recorded on the plant surface are: *Cladosporium, Alternaria, Cercospora, Helminthosporium, Erysiphe, Sphaerotheca, Podospora, Uncinula, Sporobolomyces, Bullera, Cryptococcus, Rhodotorula, Torula, Torulopsis, Oidium, Puccinia, Melanospora, Saccharomyces, Candida, Tilletia, Tilletiopsis, Penicillium, Cephalosporium, Fusarium, Periconia, Darluca, Rhynchosporium, Spermospora, Aureobasidium, Colletotrichum, Metarrhizium, Myrothecium, Verticillium, Pithomyces* (potential skin infecting genus), *Mucor, Cunninghamella, Fusarium, Aspergillus, Curvularia, Rhizopus, Trichoderma, Heterosporium, Stachybotrys, Syncephalastrum, Actinomyces* and *Streptomyces.*

Biochemical Reactions in the Phyllosphere

Leaf surface microorganisms may perform an effective function in controlling the spread of air-borne pathogens inciting plant diseases. The presence of spores of a pathogen on the surface of leaves and pods results in the formation of a substance referred to as 'Phytoalexin'. Alternatively, the phytoalexin may be normally present in plants and the concentration of such a substance may rise markedly in response to microbial infection. The term phytoalexin is derived from Greek *phyto* meaning plant and *alexin* meaning warding-off compound. The fungal spores produce a chemical substance or substances which are active in inducing the production of

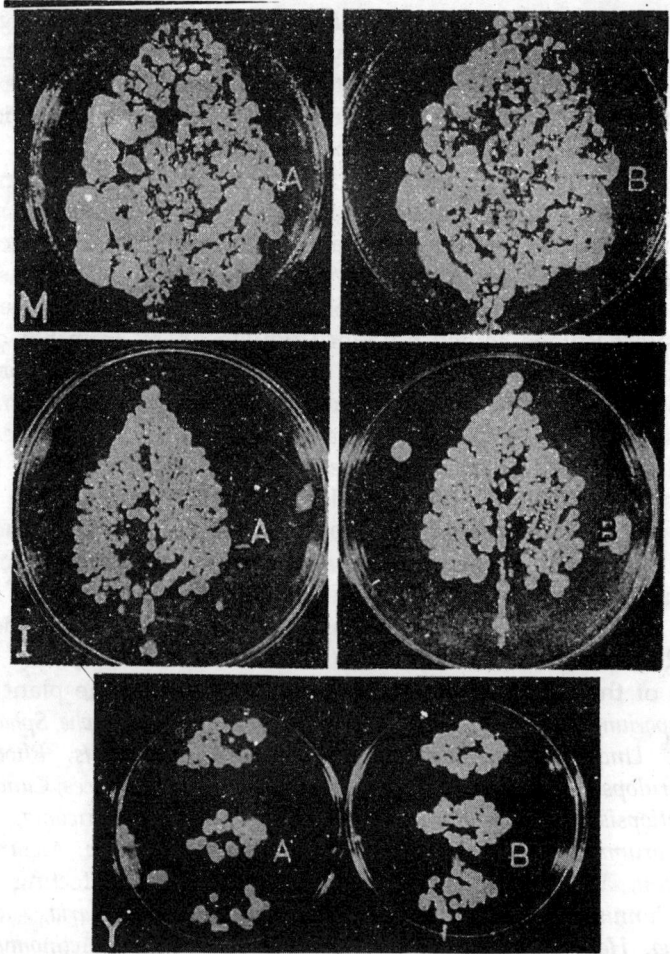

Fig. 36 Showing distribution of bacterial colonies on ventral (A) and dorsal (B) surface of mature (M), immature (I), and young (Y) leaves of *Lantana Camara*. (From K.G. Mukherji and N.S. Subba Rao, 1982.)

phytoalexins by the host as a defence reaction. The fungal metabolite, even in the absence of spores, may also independently induce phytoalexin formation. However, the concept of phytoalexin has been brought under a wide coverage so as to include all chemical compounds contributing to disease resistance in response to injury, physiological stimuli, the presence of infectious agents and their products. In keeping with this view, some of the known phytoalexins are summarized in Table 10. Resistance to disease causing microorganisms has also been attributed to fungistatic compounds secreted by leaves such as malic acid from leaves of *Cicer arietinum,* phenols from apples and waxy materials (α-hexanol) from leaves of *Ginko*.

The name 'elicitor' has been commonly used to denote the compounds which induce the synthesis of phytoalexins. There are biotic elicitors which include complex polysaccharides from fungal and plant cell walls, lipids, microbial enzymes and polypeptides. Others are abiotic which range from heavy metal salts to detergents, autoclaved ribonuclease, cold and U-V light.

Epiphytic microorganisms are known to synthesize indole acetic acid. A more interesting and useful biological function is the fixation of nitrogen by microorganisms inhabiting the leaf surface (Table 11). Bacteria of the genera *Escherichia, Brevibacterium, Bacillus, Diplococcus, Pseudomonas, Flexibacterium, Rhizobium, Beijerinckia, Azotobacter, Xanthomonas* and

Table 10 Some of the known phytoalexins, their incitants and specific hosts

Plant	Incitants	Compounds produced (Phytoalexins)
Potato (*Solanum tuberosum*)	*Phytophthora infestans*	Chlorogenic acid, Caffeic acid, Scopolin, *a*-Solanine, *a*-Chaconine, Solanidine, Rishitin and Phytuberin
Pea (*Pisum sativum*)	*Aschochyta pisi, Penicillium expansum*	Pisatin
Green bean (*Phaseolus vulgaris*)	*Monillia fruticola, Rhizoctonia solani, Colletotrichum lindemuthianum*	Phaseolin
Carrot (*Daucus carota*)	*Ceratocystis* spp.	Chlorogenic acid and 6-Methoxy-mellein (MM)
Soybean (*Glycine max*)	*Phytophthora sojae* and other fungi	Hydroxyphaseolin
Sweet potato	Infection, injury and treatment to chemicals	Chlorogenic acid, Caffeic acid, Scopoletin, Esculentin, Umbelliferone and Ipomeamarone
Cotton (*Gossypium* spp.)	*Verticillium albo-atrum*	Gossypol
Broad bean (*Vicia faba*)	*Botrytis* spp.	Wyerone
Alfalfa (*Medicago sativa*)	*Helminthosporium turcicum, Colletotrichum phomoides, Ascochyta imperfecta, Cylindrocladium scoparium, Colletotrichum trifolii, Uromyces striatus*	As examples—Sativol, Medicagol, Coumesterol and other glycosides
Red clover (*Trifolium subterraneum*)	*Sclerotinia trifoliorum*	Formononetin, Biochanin A, Trifolirhizin, Medicarpin
Apple	*Venturia inaequalis*	Phloridzin and Phloretin oxidation products
Tobacco	*Psudomonas solanacearum, Nectria galligena*	Scopoletin, Scopolin Benzoic acid

Table 11 Nitrogen fixation by bacteria isolated from phyllosphere (Bhurat, 1969)

Crop	Bacteria	Nitrogen fixed (mg per g of sucrose consumed)
Wheat	Achromobacter iophagus	13.4
	Pseudomonas atrofaciens	10.8
	Cellulomonas galba	13.6
	Pseudomonas seminum	7.6
	Cellulomonas cellasea	9.7
Pea	Achromobacter iophagus	8.7
	Pseudomonas calcis	10.8
	Achromobacter xerosis	13.4
	Cellulomonas uda	15.0
	Bacillus licheniformis	6.9

Micrococcus have been isolated from the phyllosphere of maize, cowpea, sugarcane and gram and some of them have proved to be potential nitrogen fixers (Table 10).

Notwithstanding all the observations and implications of phyllosphere microorganisms in plant disease manifestation, the exact role of these microorganisms in the nutrition of plants has remained largely conjectural in spite of the demonstration that many epiphytic bacteria grow on nitrogen-free media and fix atmospheric nitrogen to varying degrees. Experiments have also been done under laboratory conditions to demonstrate nitrogen fixation in the phyllosphere of several plants by the use of ^{15}N and the quantitative data obtained are so divergent that one is led to believe that fixation of nitrogen is a very variable phenomenon on the plant surface. Perhaps, the use of ^{15}N in such studies with known bacteria on the leaf surface of aseptically grown plants may be a rewarding exercise. Experiments have also to be designed to study the fate of biologically fixed nitrogen in the phyllosphere. In recent years, spraying of leaves of crop plants with aqueous solutions of sucrose or with bacterial suspensions has resulted in enhanced growth and yield of certain legumes and cereals in pot trials. Apparently, such sprays may have intensified the biochemical events on the phyllosphere towards the beneficial side. These observations have to be necessarily evaluated under field conditions so as to exploit the phyllosphere phenomenon towards improvement of agricultural output (also See pages 358–360 on Frost Control Biotechnology).

Selected References

Alexander, M. 1971. Biochemical ecology of microorganisms. *Ann. Rev. Microbiol.* 25, 361–392.

Andal, R., Bhuvaneshwari, K. and Subba Rao, N.S. 1956. Root exudates of paddy. *Nature*, Lond., 178, 1063.

Baker, K.F. and Cook, R.J. 1974. *Biological Control of Plant Pathogens.* W.H. Freeman and Co., Sanfrancisco.

Bhurat, M.C. 1969. *Studies on Phyllosphere Organisms.* Doctoral Thesis, IARI, New Delhi.

Bhuvaneshwari, K. and Subba Rao, N.S. 1957. Root exudates in relation to the rhizosphere effect. *Proc. Indian Acad. Sci.*, 45B, 299–301.

Bohlool, B.B. and Schmidt, E.L. 1970. Immunofluorescent detection of *Rhizobium japonicum* in soils. *Soil Sci.*, 110, 229–236.

Bolton, H. Jr., Fredrikson, J.K. and Elliot, L.F. 1993. Microbiology of the rhizosphere. In *Soil Microbial Ecology*, pp. 27–63. Ed. F.B. Metting, Jr., Marcel Dekker, Inc., New York.

Bowen, G.D. and Rovira, A.D. 1976. Microbial colonisation of plant roots. *Ann. Rev. Phytopathol.* 14, 121–144.

Burges, H.D. Ed. 1998. *Formulation of Microbial Biopesticides, Beneficial Microorganisms and Nematodes*, Chapman and Hall, London.

Coley Smith, J.R. and Cooke, R.C. 1971. Survival and germination of fungal sclerotia. *Ann. Rev. Phytopathol.*, 9, 65–92.

Darvill, A.G. and Albersheim, P. 1984. Phytoalexin and their elicitors. A defense against microbial infection in plants. *Ann. Rev. Plant Physiol.*, 35, 243–275.

Dey, B.K. 1967. Doctroal thesis, IARI, New Delhi.

Dickinson, C.H. and Preece, T.F. 1976. (Eds.) *Microbiology of Aerial Plant Surfaces*, Academic Press, London.

Ebel, J. 1986. Phytoalexin synthesis: The biochemical analysis of the induction process. *Ann. Rev. Phytopathol.*, 24, 235–264.

Fletcher, L.M. and Rhodes-Roberts, M.E. 1979. The bacterial leaf nodule association in *Psychotria*, pp. 99–118 in *Plant Pathogens*, Ed. D.W. Lovelock, Society of Applied Bacteriology. Technical Series No. 12, Academic Press, London.

Hiltner, L. 1904. Iiber neuere Erfahrungen ünd probleme aufdem Gebiet der Bodenbackteriologie ünd unter besonderer Berucksichtingung der Gründünging ünd Brache. *Arb. Dtsch. Landw. Ges.*, 98, 59–78.

Jackson, R.M. 1960. Soil fungistasis and rhizosphere. In *The Ecology of Soil Fungi*, pp. 168–176, Eds. D. Parkinson an J.S. Waid, Liverpool Univ. Press, Liverpool.

Katznelson, H. and Rouatt, J.W. 1957. Studies on the incidence of certain physiological groups of bacteria in the rhizophere. *Can. J. Microbiol.* 3, 265–269.

Katznelson, H. and Rouatt, J.W. and Payne, T.M.B. 1954. Liberation of amino acids by plant roots in relation to desiccation. *Nature*, Lond., 174, 1110–1111.

Kerr, A. 1982. Biological control of soil-borne microbial pathogens and nematodes. pp. 429–463. In *Advances in Agricultural Microbiology* Ed. N.S. Subba Rao Butterworths, UK.

Kloepper, J.W., Leong, J., Teintze, M. and Schroth, M.N. (1980a). *Pseudomonas* Siderophores: a mechanism explaining disease suppressive soils, *Current Microbiology* 4, 317–320.

Kloepper, J.W., Leong J., Teintze, M. and Schroth, M.N. (1980b). Enhanced plant growth by siderophores produced by plant growth promoting rhizobacteria. *Nature*, 286, 885–886.

Knowles, R. 1978. Free-living bacteria. In *Limitations and Potentials for Biological Nitrogen Fixation in the Tropics*, pp. 25–40, Eds. J. Dobereiner, R.H. Burris and A. Hollaender, Plenum Press, New York.

Krasilinikov, N.A. 1958. *Soil Microorganisms and Higher Plants.* Academy of Sciences, USSR, Moscow.

Kuc, J. 1966. Resistance of plants to infectious agents. *Ann. Rev. Microbiol.,* 20, 337–370.

Kuc, J. 1972. Phytoalexins, *Ann. Rev. Phytopathol.,* 10, 207–232.

Lakshmi Kumari, M. 1961. *Rhizosphere Microfloras and Host-Parasite Relationships,* Doctoral Thesis, Univ. Madras.

Last, F.T. and Warren, R.C. 1972. Non-parasitic microbes colonising green leaves: Their form and functions, *Endeavour,* 31, 143–150.

Leong, J. 1986. Siderophores: Their Biochemistry and possible role in the biocontrol of plant pathogens. *Ann. Rev. Phytopathol.,* 24, 187–209.

Lochhead, A.G. and Chase, F.E. 1943. Qualitative studies of soil microorganisms. V. Nutritional requirements of the predominant bacterial flora. *Soil Sci.,* 55, 185–195.

Lersten, N.R. and Horner, H.T. Jr. 1976. Bacterial leaf nodule symbiosis in angiosperms with emphasis on Rubiaceae and Myrsinaceae, *The Bot. Rev.,* 42, 145–214.

Mazzola, M. 1998. The potential of natural and genetically engineered fluorescent *Pseudomonas* spp. as biological control agents. In *Microbial Interactions in Agriculture and Forestry,* vol. I, pp. 193–217, Eds. N.S. Subba Rao and Y.R. Dommergues, Oxford and IBH Publishing Co. New Delhi.

McDougall, M.B. and Rovira, A.D. 1965. Carbon-14 labelled photosynthate in wheat root exudates. *Nature,* Lond., 207, 1104–1105.

McDougall, B.M. and Rovira, A.D. 1970. Sites of exudation of ^{14}C labelled compounds from wheat roots. *New Phytol.,* 69, 999–1003.

Mukherji, K.G. and Subba Rao, N.S. 1982. Plant surface microflora and plant nutrition, pp. 111–138 in *Advances in Agricultural Microbiology,* Oxford and IBH Publishing Co., New Delhi.

Pal, V. and Jalali, I. 1998. Rhizosphere bacteria for biocontrol of plant disease. *Ind. J. Microbiol.,* 38, 187–204.

Preece, T.F. and Dickinson, C.H. 1971 (Eds.) *Ecology of Leaf Surface Microoganisms.* International Symposium, Academic Press, New York.

Rovira, A.D. 1965. Interactions between plant roots and soil microorganisms. *Ann. Rev. Microbiol.,* 19, 241–266.

Rovira, A.D. 1965. Plant root exudates and their influence upon soil microorganisms. In *Ecology of Soil-borne Plant Pathogens—Prelude to Biological Control,* pp. 170–186, Eds. K.F. Baker and W.C. Snyder, Univ. California.

Ruinen, J. 1956. Occurrence of *Beijerinckia* species in the "Phyllosphere". *Nature,* London, 177, 220–221.

Ruinen, J. 1961. The phyllosphere, 1. An ecological neglected milieu. *Pl. Soil,* 15, 81–109.

Ruinen, J. 1974. Nitrogen fixation in the phyllosphere. In *The Biology of Nitrogen Fixation,* pp. 121–161. Ed. A. Quispel, North Holland Publishing Co., Amsterdam.

Ruschel, A.P. and Vose, P.B. 1984. Biological nitrogen fixation in sugarcane. In *Current Developments in Biological Nitrogen Fixation,* pp. 219–235. Ed. N.S. Subba Rao, Oxford and IBH Publishing Co. New Delhi.

Ryder, M.H., Stevens, P.M. and Bowen, G.D. Eds. 1994. *Plant Productivity with Rhizosphere Bacteria*, CSIRO Division of Soils, Glen Osmond, South Australia.

Schippers, B., Bakker, A.W., Bakker, P.A.H.M. 1987. Interactions of deleterious and beneficial rhizosphere microorganisms and the effect of cropping practices. *Ann. Rev. Phytopathol.*, 25, 339–458.

Schmidt, E.L. 1973. Fluorescent antibody techniques for the study of microbial ecology. *Bull. Ecol. Res. Comm.* (Stockholm), 17, 67–76.

Subba Rao, N.S. 1965. Aseptic culturing of plants. *J. Indian Bot. Soc.*, 44, 110–121.

Subba Rao, N.S. and Bailey, D.L. 1961. Rhizosphere studies in relation to varietal resistance or susceptibility of tomato to *Verticillium* wilt. *Can. J. Bot.*, 39, 1747–1758.

Subba Rao, N.S., Bidwell, R.G.S. and Bailey, D.L. 1961. The effect of rhizoplane fungi on the uptake and metabolism of nutrients by tomato plants. *Can. J. Bot.*, 39, 1759–1764.

Subba Rao, N.S., Bidwell, R.G.S. and Bailey, D.L. 1962. Studies of rhizosphere activity by the use of isotopically labelled carbon. *Can. J. Bot.*, 40, 203–212.

Vancura, V. 1964. Root exudates of plants. I. Analysis of root exudates of barley and wheat in their initial phases of growth. *Pl. Soil*, 21, 221–248.

Walker, N. Ed. 1975. *Soil Microbiology—A Critical Review*, Butterworths, London.

Watson, A.G. and Ford, E.J. 1972. Soil fungistasis—A reappraisal. *Ann. Rev. Phytopathol.*, 10, 327–348.

Wenhua, T., Cook, R.J. and Rovira, A.D. Eds. 1996. *Advances in the Biological Control of Plant Diseases*, China Agricultural University, Beijing.

5. Nitrogen Fixation in Free-living and Associative Symbiotic Bacteria

Nitrogen-fixing Bacteria

The free-living bacteria having the ability to fix molecular nitrogen can be distinguished into obligate aerobic, facultative aerobic and anaerobic organisms. Obligate aerobic bacteria belong to the genera *Azotobacter*, *Beijerinckia*, *Derxia*, *Achromobacter*, *Mycobacterium*, *Arthrobacter* and *Bacillus*. Among the facultative anaerobic bacteria are the genera *Aerobacter*, *Klebsiella* and *Pseudomonas*. Anaerobic nitrogen-fixing bacteria are represented by the genera *Clostridium*, *Chlorobium*, *Chromatium*, *Rhodomicrobium*, *Rhodopseudomonas*, *Rhodospirillum*, *Desulfovibrio* and *Methanobacterium*. In some of these genera, nitrogen fixation takes place in a photoautotrophic manner by virtue of the presence in them of photosynthetic pigments as exemplified by the well-known genus *Rhodopseudomonas*. On the other hand, the genus *Desulfovibrio* fixes nitrogen in the process of reducing sulphates.

Bacteria of the family *Azotobacteraceae* constitute the majority of heterotrophic free-living nitrogen-fixing bacteria. They are grouped into three genera—*Azotobacter*, *Beijerinckia* and *Derxia*. Several species of *Azotobacter* are recognized as—*A. chroococcum* (Fig. 37), mainly occurring in neutral or alkaline soils; *A. agilis*, an aquatic species; *A. vinelandii* and *A. beijerinckii* originally isolated from North American soils: *A. insignis*, isolated from Indonesian water samples; *A. macrocytogenes* isolated from Danish soils; and *A. paspali* from the rhizosphere of *Paspalum* spp. originally isolated from Brazilian soils. While the genus *Beijerinckia* has three species—*B. indica* (the earlier *Azotobacter indicum*), *B. mobile* and *B. fluminensis*, the genus *Derxia* has only one, *D. gummosa*.

Cell-size, flagellation, pigmentation and production of extra-cellular slime are considered as diagnostic features of these bacteria in distinguish-

---

Fig. 37 Nitrogen fixing *Beijerinckia* and *Azotobacter*: A and B—*Beijerinckia* colonies showing profuse gum formation; C—Four different strains of *A. chroococcum* showing melanin pigment; D—*Azotobacter* cells magnified; E—*A. chroococcum* inoculation increases the growth of barley plants in pots; F—*Beijerinckia* cells magnified under a phase contrast microscope. (A, B, D and F—from Hegazi in an FAO Bulletin by Y.A. Hamdi, 1982; C and E—Courtesy Dr. S.T. Shende).

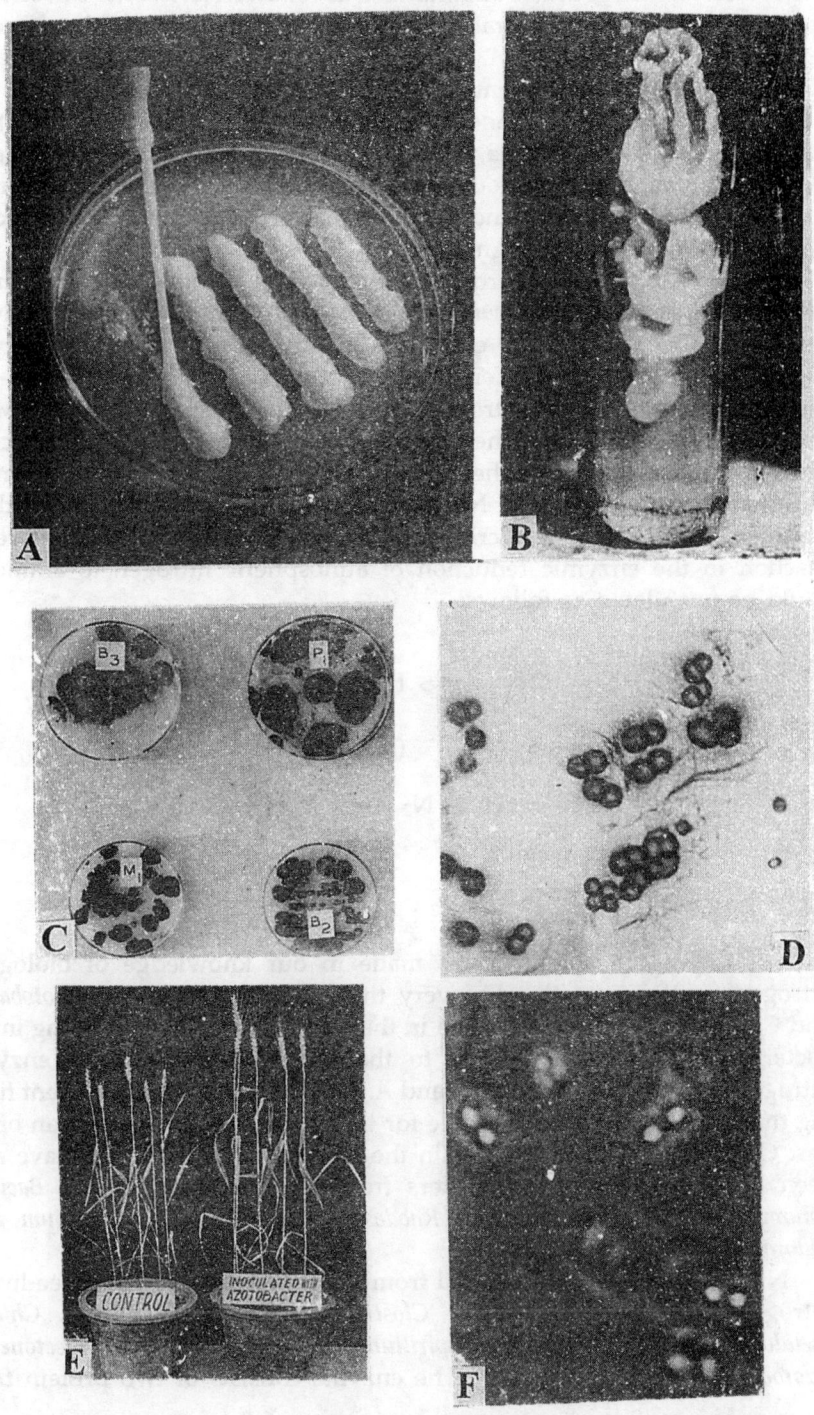

ing species, which could be summarized as follows: *Azotobacter chroococcum* (peritrichous flagella, moderate slime and black-brown insoluble pigment): *A. vinelandii, A. paspali* and *A. agilis* (peritrichous flagella, little to moderate slime and green, fluorescent and soluble pigment): *A. beijerinckii* (no flagella, moderate slime and yellow-light brown insoluble pigment); *A. macrocytogenes* (polar flagella, abundant slime and pink soluble pigment); *Beijerinckia indica* (peritrichous flagella, abundant slime and light rust-brown insoluble pigment) and *Derxia gummosa* (polar flagellum, abundant slime and yellow-brown pigment).

Among the different nitrogen fixing bacteria, *Clostridium pasteurianum* and *Azotobacter (A. chroococcum* and *A. vinelandii)* are the most intensively investigated genera. Earlier evidences mainly pointed out the ability of these bacteria to fix atmospheric nitrogen when cultured on a nitrogen-free medium. The amount of nitrogen fixed is usually estimated by the well-known Kjeldahl method. The use of ^{15}N tracer and acetylene reduction method have however enriched our knowledge regarding the biochemical pathway between N_2 and NH_3 but the exact nature of intermediate products have eluded even critical investigators. Nevertheless, the overall reaction in the enzymic reduction of atmospheric nitrogen to ammonia could be postulated as follows:

$$N_2 \xrightarrow[2e]{2H^+} HN = NH \xrightarrow[2e]{2H^+} H_2N\!-\!NH_2 \xrightarrow[2e]{2H^+} 2NH_3$$

(Dinitrogen) (Di-imide)   (Hydrazine)   (Ammonia)

$$[6e + 6H^+ + N_2 \longrightarrow 2NH_3]$$

Nitrogenase

One of the significant advances made in our knowledge of biological nitrogen fixation was the discovery that cell-free extracts of *Azotobacter* and *Clostridium* could fix nitrogen in the same way as the free-living intact bacterial cells. The finding led to the initial isolation of the enzyme nitrogenase from *C. pasteurianum* and *A. vinelandii* and the subsequent finding that the enzyme is responsible for the adsorption and reduction of N_2 gas. Cell-free extracts which retain the capacity to fix nitrogen have also been obtained by various workers from *Mycobacterium flavum, Bacillus polymyxa, Klebsiella pneumoniae, Rhodospirillum rubrum, Chromatium* and *Chloropseudomonas ethylicum*.

Nitrogenase has been isolated from the following genera of free-living nitrogen-fixing microorganisms: *Clostridium, Bacillus, Klebsiella, Chloropseudomonas, Chromatium, Rhodospirillum, Anabaena, Gloeocapsa, Plectonema, Azotobacter* and *Mycobacterium*. The enzyme consists of two protein fractions—the Mo-Fe containing protein (mol. weight 220,000–270,000) and the Fe containing protein (mol. weight 55,000–66,800). Active nitrogenase can be reconstituted by the addition of purified Mo-Fe and Fe proteins of different microorganisms. For examples, proteins of *Klebsiella pneumoniae* and *Bacillus polymyxa* and those of blue-green algae and photosynthetic bacteria have been combined to reconstitute active nitrogenases, capable of reducing acetylene to ethylene.

The Mo-Fe protein has been designated as dinitrogenase because nitrogen binds to the protein moiety whereas the Fe protein has been referred to as the dinitrogen reductase since the second moiety serves the specific function of reducing the Mo-Fe protein. During catalysis by nitrogenase, protons and nitrogen compete for electrons. Therefore, in an atmosphere containing nitrogen, hydrogen evolution occurs simultaneously with ammonia formation. This evolution of hydrogen diverts 25–35 per cent of the total reductants available for the nitrogenase reaction, which is regarded as an intracellular wastage of energy in the overall process of nitrogen fixation, and the reactions can be summarized as follows:

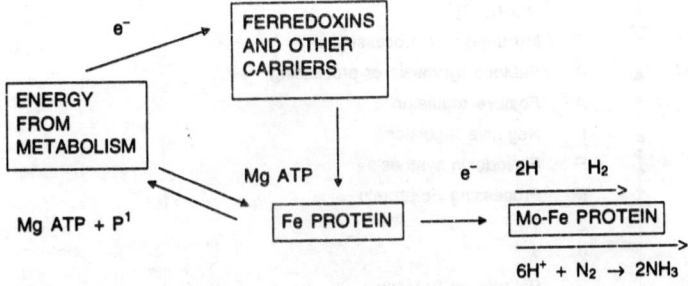

In *Azotobacter vinelandii*, two additional nitrogenases have been recognized, in addition to the nitrogenase described earlier. These nitrogenases have been studied in mutants of *A. vinelandii*. One of these contains vanadium instead of molybdenum and the other has neither molybdenum nor vanadium. The characterization of these nitrogenases, has posed fresh problems in pinpointing evidences to demonstrate the essentiality of molybdenum for nitrogen fixation and characterization of the site at which nitrogen binds to nitrogenase.

Genetics

Our knowledge of the genetic analysis of the facultative anaerobe *Klebsiella pneumoniae* has served as a model system for the analysis of *nif* genes of other nitrogen-fixing microorganisms. In *K. pneumoniae*, the *nif* region constitutes a cluster of chromosomal genes next to the genes regulating the biosynthesis of histidine. By genetic procedures involving the isolation of mutants lacking *nif* genes (*nif*), complementation analysis, cloning of *nif* genes, identification of *nif*-coded polypeptides and DNA sequencing, the genes involved in nitrogen fixation and their organization in the chromosome have been deciphered in *K. pneumoniae*. Barring *nif J*, the entire *nif* cluster of *K. pneumoniae* has now been sequenced. The complete *nif* cluster constitutes 21 genes *nif JCHDKTYENXUSVWZMFLABQ*, of which T, W, and Z are the three potential new genes (Fig. 38). Thus the physical map of *nif* cluster in *K. pneumoniae* is well understood and all the genes have been cloned into various vectors, which has facilitated the screening for *nif* homology in other nitrogen-fixing microorganisms. This has resulted in the analysis of *nif* genes in *Azotobacter*, *Azospirillum*, *Rhizobium*, *Enterobacter*, cyanobacteria, *Frankia* and other species.

The properties and functions of some of the *nif* gene products of *K. pneumoniae* have been fairly well understood. These relate to the nitrogenase enzyme, the electron transfer system and the regulatory func-

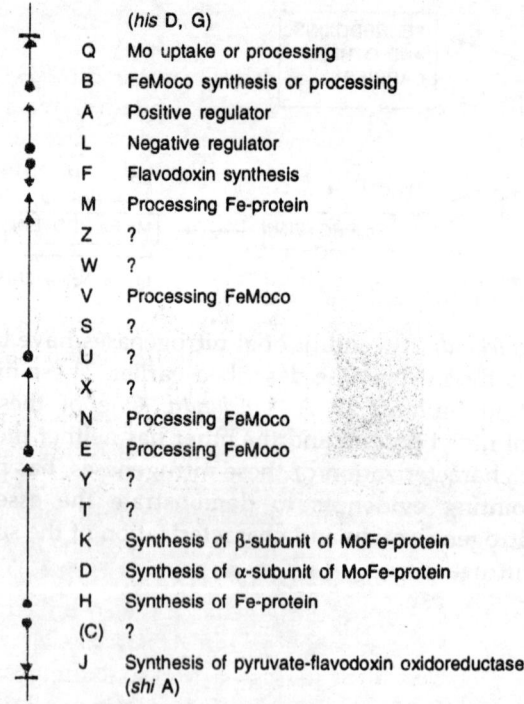

	(*his* D, G)
Q	Mo uptake or processing
B	FeMoco synthesis or processing
A	Positive regulator
L	Negative regulator
F	Flavodoxin synthesis
M	Processing Fe-protein
Z	?
W	?
V	Processing FeMoco
S	?
U	?
X	?
N	Processing FeMoco
E	Processing FeMoco
Y	?
T	?
K	Synthesis of β-subunit of MoFe-protein
D	Synthesis of α-subunit of MoFe-protein
H	Synthesis of Fe-protein
(C)	?
J	Synthesis of pyruvate-flavodoxin oxidoreductase (*shi* A)

Fig. 38 The *nif* regulon of *Klebsiella pneumoniae*. The gene cluster is preceded by the *his* operon and followed by *shi A* on the *K. pneumoniae* chromosome. The arrows signify directions of transcription of the component operons; the biological functions of the encoded peptides are as known in late 1988. A twenty-first (C) has been proposed but is probably part of J (from Postgate, 1989).

tions. There are many other gene products where properties and functions are not clear. The Mo-Fe protein moiety of nitrogenase is encoded by *nif H*. The nucleotide sequence of *nif H* and *nif D* is not well established. Five genes (*nif BNEVQ*) are involved in the formation of functional Mo-Fe protein. The genes, *nif BNE* are involved either in synthesis or insertion of FeMoCO, a catalytic cofactor of nitrogenase. The gene *nif V* plays a role in substrate specificity while *nif Q* operates under conditions of Mo deficiency. The genes *nif MS* are believed to be necessary in the processing of the Fe protein. The roles of six genes *nif TYXUSW* are presently unknown. Components of the specific electron transfer are the products of *nif F* and *nif J* which contain acid labile sulphur. In the electron chain to nitrogenase, the physiological donor is pyruvate. Electrons are carried from the *nif J* protein having pyruvate flavodoxin oxydoreductase activity to the *nif F* protein which is the substrate of the Fe protein.

Ammonia (NH_4) totally represses *nif* gene product biosynthesis. The genes involved in glutamine synthetase, an enzyme which regulates NH_4 assimilation are referred to as *Gln* while *ntr* denotes genes whose products regulate nitrogen assimilation. The genes which determine uptake hydrogenase activity are known as *hup* genes. In *K. pneumoniae*, *ntrBC* are linked to *glnA*, the structural gene for glutamine synthetase whereas *ntrA* is unlinked. *GlnA* and *ntrBC* are organized in one or two operons transcribed from two promoters in the order P1glyAP2ntr BC. P1 promotes transcription under conditions of nitrogen limitation whereas P2 acts under nitrogen enriched situations. In this way, when nitrogen is the limiting factor, the biosynthesis of glutamine synthetase is derepressed including those operons under the control of *nif* which includes the *nif* regulator also. The product of *ntrC* acts as a general activator of all these operons which is again dependent on *ntrA* product. The organization of the *nif* cluster of *K. pneumoniae* is shown in Fig. 38.

From the current information on the organization and functioning of the *nif* gene is some N_2 fixing species, it can be reasonably concluded that a basic group of *nif* genes is perhaps common to all diazotrophs including *nif* HDK genes which are highly conserved. These are the genes concerned with the processing of the metal clusters of the nitrogenase plus the regulatory genes.

Genetic analysis of *Azotobacter vinelandii* and *A. chroococcum* have revealed that a major *nif* cluster comparable to that of one in *K. pneumoniae* and comprising of genes *nif HDKTY*, *nif ENX*, *nif USV*, *nif WZM*, *nif F*, also occur in *Azotobacter*, The *nif ABQ* genes are also linked to each in a separate cluster and have been sequenced in *Azotobacter*. *A. chroococcum* is capable of producing two nitrogenases—a Mo containing one in N-free molybdenum added medium, a Va containing one in the presence of vanadium and absence of molybdenum. *A. vinelandii* can produce three enzymes—a Mo nitrogenase, a Va nitrogenase and one without either of these two metals in a medium devoid of those metals. Even though *nif* regulation in *Azotobacter* appears to be considerably similar to *K. pneumoniae*, the situation is complicated because of the unknown mechanisms through which the three types of nitrogenases are synthesised by the bacteria. Nevertheless, the *nif* regulation in *Azotobacter* is believed to be similar with regard to *nir/nif* system of *K. pneumoniae*.

Azospirillum spp. contain plasmids ranging from 90 to 120 megadaltons (Mda). They are difficult to cure, non-conjugative and not easy to purify. Different species of *Azospirillum* cannot be differentiated on the basis of plasmid content. There are no indications that plasmids of the same molecular weight correspond to the same molecular species. Spontaneous loss of plasmids has been reported in strain sp7 of *A. brasilense*. No phenotype has been associated with the plasmids. *Nif* genes are likely to be chromosomal in nature.

A small number of mutants impaired in N_2 fixation and metabolism (ex: impaired in glutamine synthetase activity) have been obtained using chemical mutagens. Hybridization of DNA of *Azospirillum* with *Klebsiella pneumoniae nif* probes revealed homology with *nif* HDK and *nif* A. The transcriptional organization of the *nif* HDK cluster of *A. brasilense* sp7 was examined by Tn5 site directed mutagenesis, genetic complementation and analysis of products. Similarly, the presence of *nif* loci in the 20 KB region *nif* HDK was also examined by carrying out over 50 Tn5-induced mutations. The results showed the presence of *nif* loci about 5 and 12 Kb downstream from *nif* K. Using heterologous *nif* probes from *K. pneumoniae* and *Azorhizobium caulinodans* new *nif* regions were characterized as *nif* E proximal to *nif* K and *nif* US distal to *nif* K. Eventhough the corresponding gene(s) were not identified, an additional locus was detected between *nif* E and *nif* US. Nitrogen-fixing genes (fix) have been identified in rhizobia but not in *K. pneumoniae*. These genes were, however, characterized as *fix* ABC in *A. brasilense* by hybridization experiment with *fix* ABC genes of *Bradyrhizobium japonicum* and *fix* A of *Azorhizobium caulinodans*. Many of these results have been obtained by Claudine Elmerich and her colleagues at the Institute of Pasteur, Paris and summarized in Fig. 39.

Enterobacter agglomerans, a nitrogen fixing bacterium isolated from the innermost rhizosphere of wheat contains all genes essential for nitrogen fixation on large (100–150 Kb) self-transmissible indigenous plasmids in the form of a continuous package.

nif	H	D	K	E	?	US	fix ABC
			→				
R	X	P	PBGR	R R	H	P	PXBgP B

Fig. 39 *A. brasilense nif* cluster. The arrow indicates the direction of transcription and the question mark an unidentified *nif* locus. Restriction sites: B: Bam H1, Bg: Bg1 II, H: Hind III, P: *Pst* I, R: *ECOR1* X:*Xhol* (from Elmerich *et al.*, 1988).

When 22 strains of *Bacillus azotofixans*, a new nitrogen-fixing species were tested for DNA homology with *Klebsiella pneumoniae nif* genes, it was found to contain only sequences which are homologous to structural *nif* genes of *K. pneumoniae*.

Nitrogen fixation by methanogenic bacteria such as *Methanobacterium* and *Methanosarcina* have been recently discovered. By using *Klebsiella pneumoniae* and *Anabaena* probes, 14 methanogenic species were analysed and found to contain homology to *nif H*, suggesting the very ancient origin of *nif* genes since archaebacteria to which methanogens belong occupy the lowest position in the evolution of microorganisms.

The availability of sophisticated DNA recombination techniques has made it possible to think in terms of transfer of *nif* genes to a larger variety of prokaryotic organisms and even to eukaryotic cells. This kind of exercise may ultimately lead to the introduction of *nif* genes into other eukaryotic systems. The main emphasis behind all these recombinant DNA research in relation to biological nitrogen fixation is to render crop plants self-sufficient with regard to nitrogen, one of the key elements in crop growth and grain production.

The first step in the genetic engineering strategy has been to construct an appropriate vehicle for transferring nitrogen-fixing (*nif*) genes. The vehicles are plasmids (extra-chromosomal circular DNA molecules) which are small amplifiable self-replicating units. The most useful plasmid has been RDI which has been used to transfer *K. pneumoniae nif* genes to *Agrobacterium tumefaciens*, *Rhizobium meliloti* and *Azotobacter vinelandii*.

The second step is to construct nitrogen-fixing cereal plants by transferring *nif* genes into eukaryotic plants. In this approach, the success depends on the construction of plasmid vehicles which overcome the barriers of DNA uptake, DNA replication and gene expression in the higher plants. One of the promising methods which has been proposed is to introduce *nif* genes into '*Agrobacterium tumefaciens*' which produces a tumour called "crown gall" in a large number of dicotyledonous plants upon wounding and subsequent infection. The bacterium has a plasmid that causes crown gall tumour and induces the plant to synthesize opines, which are nitrogen-rich compounds. The underlying mechanism is the insertion of T-DNA (transfer DNA, a segment of the plasmid) into a chromosome of the plant cells which are infected. (For more information, see Chapter 16.) Through this genetic modification, other daughter plant cells acquire the property to regenerate crown-gall tumours without the need for fresh *A. tumefaciens* infection. The altered plant cells synthesize their own hormones and opines necessary for tumour growth. The procedure envisaged is to cut open the plasmid at a site within the T-DNA region and the foreign gene (in this instance, *nif* genes) can be spliced into it. This reconstituted DNA is replicated when the tumour cells of a crown gall are grown in tissue culture and the dividing cells continue to carry

T-DNA. The transmission may continue through successive progeny. This approach provides a model for transferring genes from a prokaryote to a eukaryote. Other two possible approaches for molecular cloning of *nif* genes involve plant viruses like cauliflower mosaic virus and chloroplast DNA.

Yeast is an example of simple eukaryote. Genes for nitrogen fixation have been inserted in yeast (*Saccharomyces cerevisiae*). The first step is aimed at producing a hybrid plasmid from the yeast cell and *E. coli* and in the second step *nif* genes from *K. pneumoniae* are cleaved into another *E. coli* plasmid to produce another hybrid plasmid. The two types of yeast cell genomes are then integrated to form a novel yeast genome carrying *nif* genes. Although integration of *nif* genes into the yeast genome has been achieved, the expression of *nif*, i.e., the ability of altered yeast cells to fix N_2 is yet to be achieved.

Mechanism of Nitrogen Fixation

The nitrogenase reaction has two essential steps: (1) electron activation by a suitable donor or adenosine di-phosphate (ADP) and (2) substrate reduction. These two steps of the reaction take place at different sites on the nitrogenase molecule but are interdependent. Purified preparations of nitrogenases are highly sensitive to oxygen, specially the Fe protein part of the enzyme. However, it is believed that an undefined respiratory system exists in *Azotobacter* near the site of nitrogen fixation which actively 'scavenges' oxygen so as to prevent the inactivation of nitrogenase.

Energy requirements for nitrogenase reaction come from the cellular metabolic cycles in the form of adenosine triphosphate or ATP (roughly 12 to 20 moles of ATP per mole of molecular nitrogen reduced). Pyruvate functions both as an electron donor and an energy source. In the phosphoroclastic reaction, pyruvate forms acetyl phosphate which in the presence of adenosine diphosphate or ADP gives rise to ATP. The reductants are the strongly reducing naturally occurring electron carrier proteins, ferredoxin and flavodoxin. Dithionite ($Na_2S_2O_4$) and certain dyes such as methyl viologen and benzyl viologen can also serve as artificial extracellular sources of electron donors. Since all nitrogen-fixing microorganisms contain hydrogenase, this enzyme system in cells catalyzes the transfer of electrons from pyruvate or hydrogen to ferredoxin or flavodoxin.

Ferredoxins are naturally occurring iron-sulphur (Fe-S) electron carrier proteins capable of undergoing reversible oxidation and reduction. They have been isolated from plants, blue-green algae and bacteria such as *Clostridium pasteurianum*, *Azotobacter vinelandii*, *Rhizobium japonicum*, *Anabaena cylindrica*, *Bacillus polymyxa*, *Chromatium* sp. and *Desulfovibrio gigas*. Ferredoxins differ in molecular weight, iron and sulphide contents and biological activity. Such electron carrier proteins isolated from several nitrogen-fixing organisms can react not only with the nitrogenase of specific microorganisms but also be effectively interchanged with other microorganisms.

Flavodoxin is a flavoprotein first isolated from *Clostridium pasteurianum* in media containing low concentrations of iron and was found to replace ferredoxin as a electron carrier in a large number of reactions. Most of the nitrogen-fixing microorganisms are now known to possess flavodoxins. Subsequently, such electron carriers have been isolated from other anaerobic bacteria like *Peptostreptococcus elsdenii* and *Desulfovibrio* spp. An electron carrier named azotoflavin has been isolated from *Azotobacter vinelandii* possessing biological activity similar to ferredoxins.

The role of pyruvate and ferredoxin in nitrogenase reactions can be illustrated as follows (from Fottrell, 1968):

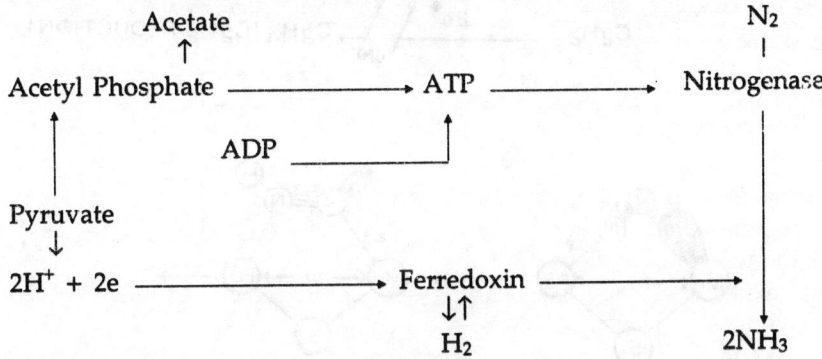

Nitrogenase, in addition to reducing atmospheric N_2 to NH_3, can also reduce certain other compounds, as follows:

$C_2H_2 \rightarrow C_2H_4$; $HCN \rightarrow CH_4 + HN_3$; $H^+ \rightarrow H_2$;
$HN_3 \rightarrow N_2 + HN_3$; $N_2O \rightarrow N_2 + H_2O$.

According to Hardy and his associates of the Du Pont Laboratory, U.S.A., the active site of the enzyme for substrate reduction is believed to be composed of an Mo-Fe dinuclear site bridged by sulphur, having the proper size and electron characteristics to provide Mo-Fe distance of about 3.8 Å (Fig. 40). This distance is specific so as to accommodate various nitrogenase substrates including N_2 and to exclude others. The first reaction in nitrogen reduction is the formation of a linear complex of N_2 with the Fe of nitrogenase. This is followed by transfer of electron from Mo which is the end point of the electrons activating system, resulting in the formation of diimide which is stabilized by hydrogen bonding from the protein as well as the metal-nitrogen bonds. Successive addition of

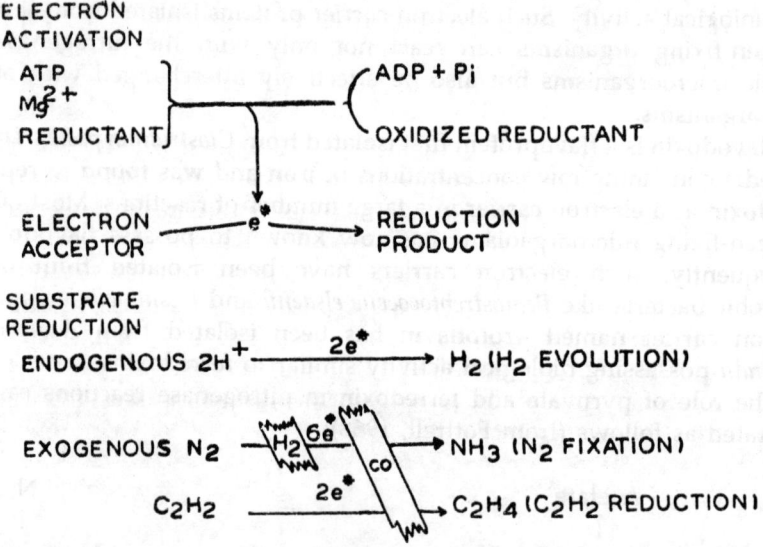

ELECTRON
ACTIVATION

ATP
Mg^{2+} ⎱   ⎰ ADP + P$_i$
REDUCTANT ⎰   ⎱ OXIDIZED REDUCTANT

ELECTRON ————— e* —————→ REDUCTION
ACCEPTOR   PRODUCT

SUBSTRATE
REDUCTION

ENDOGENOUS 2H$^+$ ———— 2e* ————→ H$_2$ (H$_2$ EVOLUTION)

EXOGENOUS N$_2$ —H$_2$— 6e* ——CO——→ NH$_3$ (N$_2$ FIXATION)

C$_2$H$_2$ ———— 2e* ——CO——→ C$_2$H$_4$ (C$_2$H$_2$ REDUCTION)

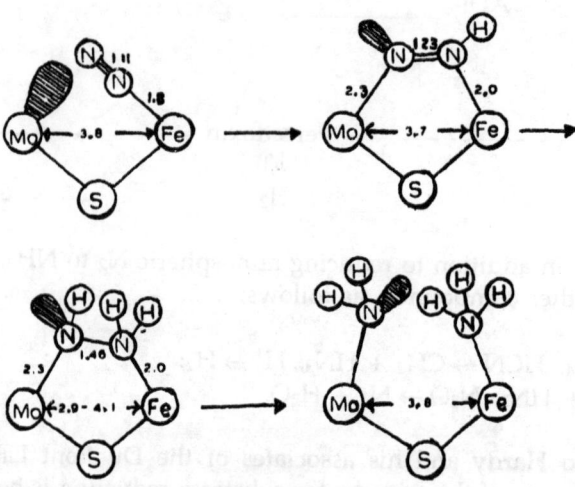

Fig. 40. *Top*—Two step reaction mediated by the enzyme nitrogenase; *Bottom*—proposed intermediates and dinuclear active site for N$_2$ reduction by nitrogenase with enzyme-bound diimide and hydrazine as intermediates. Suggested bond lengths are indicated in Å and shaded areas represent electron pairs which may be protonated; for details see text (from Hardy *et al.*, 1971).

electrons produce hydrazine followed by cleavage of NN bond to yield 2 moles of NH$_3$. The increase in the NN bond length during reduction is accompanied by compensating changes in MNN angles so that Mo-Fe distance remains constant.

Response of Plants to *Azotobacter* Inoculation

Inoculation of soil or seed with *Azotobacter* is effective in increasing yields of crops in well-manured soil with high organic matter content. Besides the ability to fix atmospheric nitrogen, *Azotobacter* is also known to synthesize biologically active substances such as B-vitamins, indole acetic acid and gibberellins in pure cultures (Table 12). The organism possesses fungistatic properties even on certain pathogenic species such as *Alternaria* and *Fusarium*. These attributes of *Azotobacter* explain the observed beneficial effects of the bacteria in improving seed germination, plant growth, plant stands and vegetative growth.

Table 12 Abilities of seven strains of *Azotobacter chroococuum* isolated from the rhizosphere of sugarcane to fix nitrogen and produce indole compounds and gibberellin-like substances (GLS) in pure cultures (from D.L.N. Rao, 1975)

Strains	N_2 fixed mg/g sucrose utilised	µg IAA equivalents/ml culture filtrate	GLS (intensity of spots on paper chromatograms)
S1	10.64	2.5	+ + +
S2	6.16	4.0	+ +
S3	6.51	4.0	+ +
S4	8.50	5.0	+ + +
S5	6.72	3.0	+
S6	6.96	2.5	+ + +
S7	7.70	1.5	+ +

Several experiments conducted in temperate regions of the world show that nitrogen fixation in *Azotobacter* inoculated soils is not more than 10 to 15 kg of N/ha/year, depending on the availability of carbon sources. Bacterial preparations containing *Azotobacter* cells under the name 'azotobacterin' were being produced and used in the erstwhile USSR and East European countries such as Czechoslovakia, Rumania, Poland, GDR, Bulgaria and Hungary where bacterization of seeds with azotobacterin has proved beneficial in increasing yields of crops such as wheat, barely, maize, sugarbeet, carrot, cabbage and potato. The increase in yield of field crops was not more than 12% over corresponding uninoculated controls (Table 13). Experiments with *Azotobacter* cultures and crop plants at the Indian Agricultural Research Institute, New Delhi, lead us to believe that significant increases in growth (Fig. 41) and yield of wheat, rice and vegetable crops could be obtained in pot trials. However, under field conditions, such uniform trends towards increases in yield are not always reproducible. Apart form *Azotobacter*, bacteria like *Beijerinckia* and *Derxia* are also found to occur in tropical soils which are capable of fixing nitrogen but field experiments are needed before any conclusion could be made regarding their utility and benefits to crop production.

Table 13 Effect of Azotobacterin on yield of field crops (from Mishustin and Shilnikova, 1969)

Crop	Average yield in control (metric cwt/ha)	Increase from Azotobacterin (%)
Spring wheat	15.8	8.2
Winter wheat	21.3	9.8
Oats	17.1	12.0
Barley	21.0	9.0
Maize	36.2	8.0
Sugarbeet	283.1	7.0
Potatoes	178.0	8.0
Average		8.86

The population of *Azotobacter* in the rhizosphere of crop plants and in uncultivated soil is generally low. Often inoculation of soil or seed does not improve the situation. To overcome this limitation, repeated application of *Azotobacter* during different stages of growth of a crop is now being recommended with the object of increasing the number of bacteria in soil. Some experiments on inoculation of soil with *Azotobacter* with different doses on inorganic N fertilizer have demonstrated the possibility of saving considerable amount of N fertilizer while still attaining desired yields of rice. Field trials with new and efficient cultures of *Azotobacter* have shown that the yields of sorghum, maize and cotton can be substantially increased by *Azotobacter* inoculation (Table 14).

Nitrogen Fixation in the Root Zone of Rice

Using acetylene reduction method, a group of scientists at the International Rice Research Institute, Manila have demonstrated significant amount of N_2 fixation in the root soil system of rice plants (Table 15).

The studies done at IRRI further point out that submerged soils have a greater capacity to fix atmospheric nitrogen than non-submerged soil. The amounts of nitrogen fixed in 1 ha estimated by using theoretical conversion factor ($C_2H_2 : N_2 = 3 : 1$) during the dry season were 79.8 kg/ha in planted, submerged fields; 42.5 kg/ha in unplanted, submerged fields; 5.4 kg/ha in the planted, upland fields and 2.7 kg/ha in the unplanted,

Fig. 41. Response of plants to inoculation with *Azospirillum brasilense* in unsterilized soil. A—Pellicle of *Azospirillum* formed two mm below the surface of semi-solid sodium malate agar medium indicating the microaerophilic nature of the bacterium; B—*Azosprillum brasilense* as seen under a phase contrast microscope showing the spirillar morphology of the bacterium; C, D—*Chrysopogon fulvus* and *Cenchrus ciliaris*—left to right—uninoculated control; carrier alone; carrier inoculated with *A. brasilense*; 20 kg N/ha (urea); 20 kg N/ha + *A. brasilense*; E—*Hordeum vulgare* (barley)—left to right—uninoculated control; inoculated with *A. brasilense*; 40 kg N/ha (urea).

Table 14 The effect of *Azotobacter chroococcum* inoculation in field on the yield of three crops, 1978–79 (experimental results of S.T. Shende)

Crop	Location of field trials in India	Without Azotobacter	With Azotobacter	C.D. at 5%	% increase due to Azotobacter
Sorghum (kg/ha)	Pali	1280	1400	122	9.3
	Dharwar	2360	3260	1056	38.1
Maize (kg/ha)	I.A.R.I.	780	1340	480	71.1
	Dharwar	320	4370	990	36.5
Cotton (kg/ha)	Surat	1254	1339	241	6.7
	Indore	366	401	104	9.5
	Khandwa	559	708	165	20.6

Table 15 Acetylene reducing activities of intact soil plant system, plant removed and remaining soil measured by anaerobic $C_2H_2C_2H_4$ assay (from IRRI Annual Report, 1972)

Incubation time (h)	Cumulative ethylene formed (n mol/tube)		
	Intact soil plant system	Plant removed	Remaining soil
3	8	29	11
5	29	42	13
10	406	70	19
24	2360	375	27

upland fields. In the field, most nitrogen is fixed in soils during the reproductive and ripening phases of the growth of rice plants. The absolute figures reported by IRRI scientists may not be as important as the indication that large quantities of atmospheric nitrogen are being continuously fixed in the rhizosphere of rice plants.

The aerenchyma present in the rice plant transfers air from the atmosphere to the rhizosphere. The root system of rice in a submerged field is located in the anaerobic soil zone. The air transferred by the rice plant to the root zone may contain enough nitrogen for the N_2 fixing activity of bacteria associated in the rhizosphere which belong to the genera, *Beijerinckia, Azotomonas, Pseudomonas, Flavobacterium, Azospirillum* and *Azotobacter*. Gas chromatographic analysis of the gases in IRRI rice soils under flooded conditions (sampled from several experimental plots which did not receive fertilizer N) showed that submerged soils planted to rice variety IR-20 contained more nitrogen gas than unplanted soil at all locations, at tillering (34 days after transplanting), during panicle initiation (48 days after transplanting) and heading (77 days after transplanting).

Another important factor to be taken into consideration is the amount of carbohydrates available in the root zone for nitrogen fixation. The results obtained at IRRI further point out that at the tillering stage, 96%

of ^{14}C fed to experimental plants was located in stem and leaves and 4% in roots. At maturity, 66% of ^{14}C was in the straw, 18% in panicles, and 15% in roots. The cumulative amount of ^{14}C in the water culture solution was less than 1% of the total ^{14}C which remained in plant tissues. These results point out that organic carbon provided by rice roots is an important factor contributing to N_2 fixation but the amount of carbon exuded in the root medium is comparatively smaller than that held by the plant tissue.

Availability of cellulosic substrates in flooded soil is yet another factor to be reckoned with in bacteria-mediated nitrogen fixation in rice fields. Application of rice straw enhances nitrogen fixation whereas addition of ammonium sulphate (combined nitrogen) retard nitrogen fixation (Table 16). Interestingly, the fluctuations in nitrogen fixation are related to the number of *Azotobacter* colonies present in rice soil under different treatments. The waterlogged soil planted to rice provides an ideal habitat for facultative anaerobic bacteria other than *Azotobacter* to function both in presence and absence of oxygen and fix nitrogen (Table 17).

Table 16 Nitrogen-fixing activity of paddy soil in pots under various treatments (Kalininskyaya *et al.*, 1973)

Treatment	Atom % excess ^{15}N in soil sample after incubation*	Mean nitrogen fixed in mg/kg soil/month
1. Control	0.183	8.65
2. 50 kg ((NH₄)₂SO₄/ha	0.095	4.22
3. Rice straw (5 t/ha)	0.221	11.92
4. Rice straw (5 t/ha) + 50 kg (NH₄)₂SO₄/ha	0.226	11.73
5. Rice straw (10 t/ha)	0.770	39.34
6. Rice straw (10 t/ha) + 50 kg (NH₄)₂SO₄/ha	0.780	50.00

*Atom % excess ^{15}N in the gas mixture = 51.10.

From green house pot-culture experiments, using rice plants grown under flooded conditions the International Rice Research Institute, Manila (IRRI Annual Report, 1978) has brought out a nitrogen balance sheet which indicates that photosynthetic nitrogen-fixing microorganisms increase the accretion of nitrogen to rice plants in flooded soil. In terms of N in four crops (mg/pot), the amount of nitrogen removed by the crop in standard pots where blue-green algae were spontaneously growing was 967. When pots were covered with black cloth so as to exclude blue-green algae, the amount of nitrogen removed by the crop was reduced to 915 whereas in fallow-flooded pots without plants no significant amount of nitrogen was fixed. In pots containing blue-green algae plus P and Fe, the amount of

Table 17 C_2H_2 reducing activity of bacteria isolated from waterlogged paddy soil (from V.R. Rao, 1973)

Bacteria	$N_2[C_2H_2C_2H_4]$ fixed in* µg/ml medium/day	
	Aerobic system	Anaerobic system
Mycobacterim flavum	5.72	7.70
Pseudomonas sp.	3.79	5.32
Bacillus sp.	0.084	0.140
Bacillus polymyxa	7.88	7.88
Bacillus polymyxa	3.65	5.36
Mycobacterium sp.	5.30	13.65
Pseudomonas sp.	0.14	1.00
Bacillus polymyxa	11.38	11.38
Bacillus polymyxa	6.83	6.83
Bacillus polymyxa	5.32	9.10
Bacillus polymyxa	3.50	3.15
Mycobacterium sp.	15.17	15.93

*All cultures were grown in glucose medium; measurements were made after 7 days incubation.

nitrogen recovered from the crop increased to 1093 and this was further increased to 1316 in the presence of *Azolla*.

From the foregoing account, it is clear that the waterlogged rice ecosystem provides a congenial microhabitat for nitrogen-fixing bacteria and blue-green algae to thrive. The current knowledge indicated that photosynthetic blue-green algae are the dominant nitrogen fixers in rice fields.

Nitrogen Fixation in the Rhizosphere of Grasses and Weeds (Associative Symbiosis)

During the last decade, a group of Brazilian workers observed specific and abundant colonization of the rhizosphere of the genus *Paspalum* (grasses) with *Azotobacter paspali*, a powerful nitrogen fixer. Sizeable amounts of acetylene were reduced (nitrogenase activity) by the rhizosphere soil samples of *Paspalum notatum* indicating the role of *A. paspali* in fixing nitrogen *in situ* in the vicinity of an actively growing plant. Quite recently, nitrogen fixation in the rhizospheres of many forage grasses such as *Pennisetum purpureum, Brachiaria mutica, Brugulosa, Digitaria decumbens, Panicum maximum, Melinis minutiflora* and *Hyparrhenia rufa* and weeds such as *Heracleum sphondylium, Anthriscus sylvestris, Mercurialis perennis, Rumex acetosa, Convolvulus arvensis, Viola caniana* and *Stachys sylvatica* have been reported.

Although *Spirillum lipoferum* as a nitrogen-fixer was known since 1963, it was Dobereiner and associates in Brazil who in 1975 highlighted and attributed the nitrogen-fixation potential of some tropical forage grasses such as *Digitaria, Panicum, Brachiaria*, maize, sorghum, wheat and rye to the activity of *S. lipoferum* in their roots. Subsequently, the bacterium has been isolated from many tropical countries and found to reside inside the roots and aerial parts of plants. Dobereiner coined the name 'Associate Symbiosis' to denote the occurrence of nitrogen-fixing *Spirillum* in plants which has now been recently enlarged to encompass other possible bacterial associations by adopting the terminology 'Diazotrophic biocoenocis'. The taxonomy of *Sprillum* has been re-examined and the genus has been named as *Azospirillum* with four species—*A. lipoferum, A. brasilense, A. amazonense* and *A. seropedica*.

Due to its microaerophilic nature, *Azospirillum* can be isolated on a semi-solid malate containing medium by enrichment procedures. The development of white, dense and undulating fine pellicles (Fig. 41A) on a semi-solid malate medium is a characteristic feature of *Azospirillum*. During enrichment, the dominant organisms on a sodium malate medium are characteristically curved rods of various sizes with prominent refractive fat droplets (Fig. 41B). This organism is Gram-negative and contains poly-β-hydroxy butyrate granules. Microscopic examination reveals polymorphism and spirillar movement. For fixation of molecular nitrogen, the bacterium needs microaerophilic surroundings (low O_2 conditions) but can grow profusely on ammonium containing medium without fixing nitrogen. The ability of the organism to fix nitrogen has been verified by the acetylene reduction test and uptake of $^{15}N_2$ gas. In terms of mg N_2 fixed/g of substrate added, the range of fixation, as measured by Kjeldahl method, varies from 12 to 36 among different isolates of *A. brasilense*. The organism is also known to produce growth substances such as IAA, kinetins and gibberellins.

Some New Nitrogen-fixing Bacterial Associations

As stated earlier, the genus *Azospirillum* appears to be widespread in the roots of many graminaceous species. *A. amazonens* is an acid tolerant and sucrose utilizing species isolated from the root system of sugarcane and sweet sorghum. In Pakistan, salt affected soils abound in Kallar grass (*Leptochloa fusca*) on whose root surfaces *Azospirillum halopraeferans* has adapted its existence to the harsh ecological surroundings with optimum requirement of 41°C and 0.25 per cent salt concentrations. A bacterium named *Herbaspirillum seropedicae* has been isolated from washed and surface sterilized roots of maize, rice and sorghum in Rio de Janeiro. The bacterium has bipolar flagella and N_2 fixation is more O_2 and pH tolerant than that of azospirilla. At the International Rice Research Institute, Philippines, a bacterium named as *Pseudomonas diazotrophicus* has been isolated from roots of rice and weeds in low land rice growing tracts but not from upland rice regions. A new species of *Campylobacter* (McClung and Patriquin, 1980) has been reported to occur in the roots of a water weed (*Spartina alterniflora*), which appears to be very sensitive to oxygen. A strain of *Bacillus azotofixans* capable of fixing nitrogen efficiently was isolated from surface sterilized roots of grasses, wheat and sugarcane which, like *Azotobacter paspali* isolated from the roots of *paspalum* grass, is also nitrate dependent for growth and capable of fixing nitrogen in the presence of combined nitrogen. The residual sugar in sugarcane trash and roots of weeds associated with several cultivars of sugarcane in many parts of Brazil contain *Azotobacter nitrocaptans* in the range of 10^3 to 10^7 per g wet weight basis. This nitrogen-fixing species appears to grow well in 10 per cent cane sugar or 1 per cent cane juice, acidified with acetic acid to pH 4.5. It is a small Gram-negative aerobic bacillus and can use ethyl alcohol for growth. Some plant associated nitrogen-fixing bacteria which have been investigated since 1974 are shown in Table 18.

Table 18 Several nitrogen-fixing bacteria (from Boddey and Dobereiner, 1988)

Nitrogen-fixing bacteria	In association with
Azospirillum lipoferum	Many grasses and cereals such as maize, wheat, sorghum, rice
Azospirillum brasilense	Many grasses and cereals such as maize, wheat, sorghum, rice
Pseudomonas sp.	Rice
Pseudomonas sp.	*Deschampsia caespitosa*
Campylobacter nitrofigilis	*Spartina alterniflora* (Salt march grass)
Azospirillum amazonense	Many grasses in Amazon and Brazil
Bacillus azotofixans	Several grasses, wheat, sugarcane
Herbaspirillum seropedicae	Many grasses and cereals (Maize)
Azospirillum halopraeferans	Kallar grass
Azotobacter nitrocaptans	Sugarcane

The question of colonization and mode of entry of these diazotrophs into the root interior remains to be fully answered. The 2,3,5-triphenyl terrazolium dichloride (TTC) mediated reduction staining of cell contents including live bacteria has been advocated as one way of substantiating the presence of bacteria but this has obvious limitations in pinpointing the authenticity of particular species of a microorganism. Electron micrographs showing the presence of the bacteria of the genus *Campylobacter* in the aerenchyma of field grown *Spartina alterniflora* have been cited as proof of bacterial colonization. More recently, through microscopic evidence aided by the use of the protein A-gold technique and silver amplification, diazotrophs have been shown to predominate the roots of kallar grass (*Leptochloa fusca*) and wheat. The technique has revealed that *A. brasilense* cccupy the intercellular spaces of the root cortex. However, the bacterial cells were not detected either in the endodermis or the vascular bundle.

Response of Plants to *Azospirillum* Inoculation

Field experiments carried out by the Indian Agricultural Research Institute in different parts of India have revealed that seed inoculation of sorghum (*Sorghum bicolor*), bajra (*Pennisetum americanum*) and ragi (*Eleusine corocana*) increased grain and fodder yields in different agro-climatic conditions of India (Table 19). Similar results have been obtained by scientists in Israel who also find responses of millets to *Azospirillum* inoculation. Currently, *A. amazonense* has been isolated from the root zone of coconut in Kerala, India (Fig. 42).

The Concept of C4-dicarboxylic Acid Pathway

As mentioned earlier, the availability of easily assimilable carbon substrates in the root region by way of root exudates is the crucial factor in nitrogen fixation by non-symbiotic bacteria. Tropical grasses of agricultural value such as sugarcane, maize, sorghum, *Paspalum*, *Pennisetum*, *Panicum*, *Digitaria* and *Melinis* which possess C4-dicarboxylic acid pathway of photosynthesis (Hatch and Slack pathway) and are capable of making efficient use of high light intensities and temperatures prevalent in tropics might harbour unknown nitrogen fixers. Some of these plants are already

Table 19 Effect of *Azospirillum brasilense* inoculation on sorghum (*Sorghum bicolor*) (Var. CSH 5) (Field trial 1978–79)

Location	Nitrogen level (kg N/ha)	Grain yield (q/ha)**		Per cent increase in yield
		Uninoculated	Inoculated	
Udaipur	0	30.1	39.3	35.6*
	40	51.4	53.3	3.7*
Delhi	0	16.5	20.2	23.4
	40	20.5	26.0	26.8*
Pantnagar	0	11.7	17.8	65.6
	40	15.6	28.1	61.3*
Coimbatore	0	31.3	37.0	15.2
	40	49.8	51.4	3.2
Dharwar	0	23.6	31.6	33.9*
	40	29.1	36.7	26.1*
Hyderabad	0	30.3	41.7	37.6
	40	45.6	43.1	–

*Significant increase over corresponding control. **1q = 100 kg.

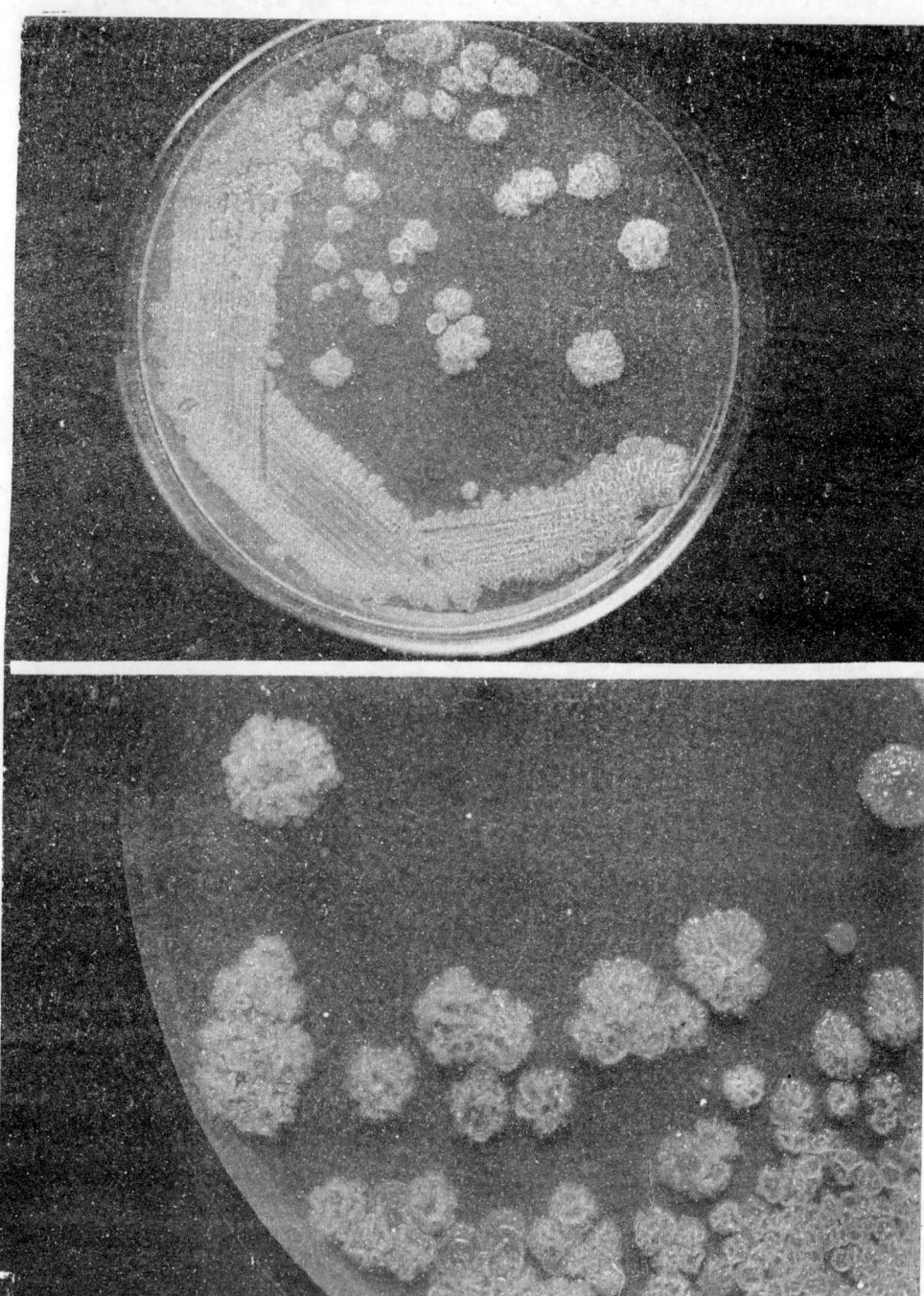

Fig. 42 Massive white colonies of *Azospirillum amazonense* isolated from the roots of coconut palm (*Cocos nucifera*) on potato infusion agar lacking malate (Courtesy S.R. Prabhu, CPCRI, Kasargod, Kerala).

known to harbour abundant population of *Beijerinckia* and *Azotobacter* in their rhizosphere and hence detailed investigations on the C/N status of the rhizosphere of these plants under varying environmental stresses may have to be undertaken.

Nitrogen Fixation by *Rhizobium* in a Free-living State

It was generally agreed upon that *Rhizobium* can fix atmospheric nitrogen only in the root nodules of legumes and that too when it is in the bacteroid stage of its life cycle. All attempts to obtain nitrogen fixation by pure cultures of *Rhizobium* with or without extracts of host plants failed in the past and it was postulated that some of the genes determining *Rhizobium's* ability to fix atmospheric nitrogen (*nif* genes) reside in the host plant and hence the need for symbiosis between the host and the bacterium in the bacteroid tissue of the root nodule.

Interesting experimental findings have emerged some years ago which show that *Rhizobium* possesses the entire complement of genes for nitrogen-fixation which is normally latent and become active only under special conditions. One of the first evidences came from the successful transfer of *nif* genes from *Rhizobium trifolii* to a non-nitrogen-fixing strain of *Klebsiella aerogenes*. Secondly, a strain of *Rhizobium* sp. cowpea group was induced to fix atmospheric nitrogen in the presence of diffusible substances from callus tissue of leguminous as well as non-leguminous plants, thereby indicating that plant tissues contain some substances which stimulate bacteria to fix nitrogen. The third set of evidences have come forth simultaneously from two groups of Australian and one group of Canadian workers. A simple synthetic medium was evolved which contains besides other ingredients, certain key metabolites such as a pentose sugar (arabinose or xylose), a dicarboxylic acid (such as succinate) and a relatively small amount of fixed nitrogen (such as glutamine, glutamate or nitrate) to induce nitrogen fixation by a strain of cowpea type of *Rhizobium* in a free-state on a solid agar medium. The ability of fix nitrogen has been verified by C_2H_2 reduction test as well as by ^{15}N enrichment procedures.

These results lead us to believe that we may have to reconsider the recognition of legume-*Rhizobium* association within the nodules as a true instance of symbiosis. In a more restricted way, the term symbiosis is valid since the specialised structure of nodule may be meant only to restrict the access of oxygen to rhizobia for proper functioning of the enzyme nitrogenase.

Selected References

Ausubel, F.M. 1979. Application of recombinant DNA technology to the study of nitrogen fixation. pp. 257–280. In *Recent Advances in Biological Nitrogen Fixation*, Ed. N.S. Subba Rao, Oxford and IBH Publishing Co., New Delhi.

Baldani, J.I., Baldani, V.L.D., Seldin, L. and Dobereiner, J. 1986. Characterization of *Herbaspirillum seropedicae* Gen. sp. Nov. A root-associated nitrogen fixing bacterium. *Int. J. Syst. Bacteriol.*, 36, 86–93.

Becking, J.H. 1963. Fixation of molecular nitrogen by an aerobic *vibrio* or *Spirillum* sp. *Antonie Van Leeuwenhoeck, J. Microbiol.*, 29, 326.

Bergersen, F.J. Ed. 1980. *Method for Evaluating Biological Nitrogen Fixation*, John Wiley & Sons, New York.

Bishop, P.E., Jarlenski, D.M.L. and Hetherington, D.R. 1980. Evidence of alternative nitrogen fixation system in *Azotobacter vinelandii*, *Proc. Natl. Acad. Sci.* 77, 7342–7346.

Boddey, R.M. and Dobereiner, J. 1988. Nitrogen fixation associated with grasses and cereals: Recent results and perspectives for future research. *Pl. Soil*, 108, 53–65.

Burns, R.C. and Hardy, R.W.F. Eds. 1975. *Nitrogen Fixation in Bacteria and Higher Plants*. Vols. I and II. Springer Verlag, Berlin.

Cavalcante, V. and Dobereiner, J. 1988. A new acid tolerant nitrogen-fixing bacterium associated with sugarcane. *Pl. Soil*, 108, 23–31.

Dobereiner, J., Day, J.M. and Dart, P.J. 1972. Nitrogenase activity and oxygen sensitivity of the *Paspalum notatum-Azotobacter paspali* association. *J. Gen. Microbiol.*, 71, 103–116.

Dobereiner, J. 1974. Nitrogen fixing bacteria in the rhizosphere. In *The Biology of Nitrogen Fixation*, pp. 86–120. Ed. A. Quispel, North Holland Publishing Co., Amsterdam.

Dobereiner, J., Burris, R.H. and Hollaender, A. Eds. 1978. *Limitations and Potentials for Biological Nitrogen Fixation*, Plenum Press, New York.

Dobereiner, J., Reis, V.M. and Lazarini, A.C. 1988. New N_2 fixing bacteria in association with cereals and sugarcane. pp. 717–722. In *Nitrogen Fixation: Hundred Years After*. Eds. H. Bothe, F.D. de Bruijn and W.E. Newton, Gustav Fischer, Stuttgart, New York.

Elmerich, C., Gallimand, M., Vieille, C., Delorme, F. and deZamaroczy. 1988. Nitrogen fixation genes of *Azospirillum*, pp. 327–331. In *Nitrogen Fixation: Hundred Years After*. Eds. H. Bothe, F.J. de Bruijn and W.E. Newton, Gustav Fischer, Stuttgart.

FAO 1982. FAO Soils Bulletin 49. *Application of nitrogen-fixing systems in soil management*. FAO, Rome.

Gibson, A.H. and Newton, W.E. Eds. 1981. *Current Perspectives in Nitrogen Fixation*. Aust Acad. of Science.

Hardy, R.W.F., Burns, R.C., Herbert, R.R., Holsten, R.D. and Jackson, E.K. 1971. *Biological nitrogen fixation: A key to world protein. Pl. Soil.* sp. vol., 561–590.

Hardy, R.W.F., Burns, R.C. and Parshall, G.W. 1971. The biochemistry of nitrogen fixation. In *Advances in Chemistry*, Series No. 100. *Bioinorganic Chemistry*. pp. 218–247, American Chemical Society.

Hardy, R.W.F. and Gibson, A.H. Eds. 1977. *A Treatise on Dinitrogen Fixation*, Section IV, *Agronomy and Ecology.*, Wiley, New York.

Hollaender, A. Ed. 1977. *Genetic Engineering for Nitrogen Fixation*. Plenum Press, New York.

International Rice Research Institute, Annual Report, 1972.

Kalininskaya, T.A., Rao V.R., Volkova T.N. and Ippolitcv. 1973. Nitrogen fixing activity of soil under rice crop studied by acetylene reduction assay. *Mikrobiologiya*, 42, 481–485.

Krieg, N.R. and Tarrand, J.J. 1978. Taxonomy of the root associated nitrogen-fixing bacterium *Spirillum lipoferum*. In *Limitations and Potentials for Biological Nitrogen Fixation in the Tropics*. Eds. J. Dobereiner et al., pp. 317–333, Plenum Press, New York.

Kurz, N.G.W. and La Rue, T.A. 1975. Nitrogenase activity in rhizobia in the absence of host plant, *Nature*, Lond., 256, 407–409.

Levanony, H., Bashan, Y., Romano, B. and Klein, E. 1989. Ultrastructural localisation and identification of *Azospirillum brasilense* cd on and within wheat root by immuno-gold labelling. *Pl. Soil*, 117, 207–218.

McComb, J.A., Elliot, J. and Dilworth, J.M. 1975. Acetylene reduction by *Rhizobium* in pure culture *Nature*, Lond., 256, 409–410.

McClung, C.R. and Patriquin, D.G. 1980. Isolation of a nitrogen-fixing *Campylobacter* species from the roots of *Spartina alterniflora* Loisel, *Can. J. Microbiol.*, 26, 881–886.

Merrick, M.J. 1988. Organization and regulation of nitrogen fixation genes in *Klebsiella* and *Azotobacter*. pp. 283–302. In *Nitrogen Fixation: Hundred Years After*, Eds. H. Bothe, F.J. de Bruijn and W.E. Newton, Gustav Fischer, Stuttgart.

Mishustin, E.N. and Shilnikova, V.K. 1969. The biological fixation of atmospheric nitrogen by free-living bacteria. In *Soil Biology. Reviews of Research*, pp. 72–109, UNESCO, Paris.

Mulder, E.G. and Brotonegoro, S. 1974. Free-living heterotrophic nitrogen fixing bacteria. In *The Biology of Nitrogen Fixation*, pp. 37–85, Ed. A. Quispel. North Holland Publishing Co., Amsterdam.

Murray, P.A. and Zinder, S. 1984. Nitrogen fixation by a methanogenic archibacterium. *Nature*, 312, 284–286.

Newton, W.E., Postgate, J.R. and Rodriguez-Barreuco. Eds. 1977. *Recent Developments in Nitrogen Fixation*. Academic Press, New York.

Newton, W.E. and Orme-Johnson, W.H. Eds. 1980. *Nitrogen Fixation*, University Park Press, Baltimore, U.S.A.

Pagan, J.D., Child, J.J., Scowcroft, W.R. and Gibson, A.H. 1975. Nitrogen fixation by *Rhizobium* cultured on a defined medium, *Nature*, Lond., 156, 406–407.

Postgate, J.R. 1971. Ed. *The Chemistry and Biochemistry of Nitrogen Fixation*. Plenum Press, London, New York.

Postgate, J.R. 1974. Prerequisites for biological nitrogen fixation in free-living heterotrophic bacteria. In *The Biology of Nitrogen Fixation*. pp. 663–683. Ed. A. Quispel, North Holland Publishing Co., Amsterdam.

Postgate, J.R. 1975. *Rhizobium* as a free-living nitrogen fixer. *Nature*, Lond., 256, 363.

Postgate, J.R. 1989. Trends and perspectives in nitrogen fixation research. *Advances in Microbial Physiology*, 30, 1–22.

Rao, V.R. 1973. *Non-symbiotic nitrogen fixation on paddy fields.* Doctoral Thesis, Academy of Sciences, U.S.S.R., Moscow.

Renihold B., Hurek, T., Fendrik, I., Pot, B., Gillis M., Kersters, K., Thielemans, S. and DeLey, J. 1987. *Azospirillum halopraeferans* sp. nov., a nitrogen fixing organism associated with roots of Kallar grass *Leptochloa fusca* (L) (Kunth.). *Int. J. Syst. Bacteriol.,* 37, 43–51.

Renihold, B. and Hurek, T. 1989. Localization of diazotrophs in the root interior with special attention to the kallar grass association. pp. 209–218. In *Nitrogen fixation with non-legumes.* Eds. F.A. Skinner, R.M. Boddey and I. Fendrik, Kluwer Academic Publications, London.

Seldin, L., VanElsas, J.D. and Penido, E.G.C. 1984. *Bacillus azotofixans* sp. nov., a nitrogen fixing species from Brazilian soils and grassroots. *Int. J. Syst. Bacteriol.* 34, 451–456.

Seldin, L., Bastos, M.C.F. and Penido, E.G.C. 1989. Identification of *Bacillus azotofixans* nitrogen fixing genes using heterogenous *nif* probes. pp. 179–187. In *Nitrogen Fixation with Non-legumes.* Eds. F.A. Skinner, R.M. Boddey and I. Fendrik, Kluwer Academic Publishers, London.

Sibold, L. and Souillard, N. 1988. Genetic analysis of nitrogen fixation in methanogenic archaebacteria, pp. 705–710. In *Nitrogen Fixation: Hundred Years After.* Eds. H. Bothe, F.J. de Bruijn and W.E. Newton, Gustav Fischer, Stuttgart.

Stewart, W.D.P. 1977. Blue-green algae. In *A Treatise on Dinitrogen Fixation,* Section III. *Biology.* pp. 63–123. Eds. R.W.F. Hardy and W.S. Silver, John Wiley & Sons, New York.

Stewart, W.D.P. and Gallon, J.R. Eds. 1980. *Nitrogen Fixation.* Academic Press.

Subba Rao, N.S., Tilak, K.V.B.R., Singh, C.S. and Lakshmi Kumari, M. 1979a. Response of a few economic species of graminaceous plants to inoculation with *Azospirillum brasilense, Curr. Sci.* 48, 133–134.

Subba Rao, N.S., Tilak, K.V.B.R., Lakshmi Kumari, M. and Singh, C.S. 1979b. *Azospirillum*—a new bacterial fertilizer for tropical crops, *Sci. Reporter,* C.S.I.R. (India), pp. 690–692.

Subba Rao, N.S. Ed. 1982. *Recent Advances in Biological Nitrogen Fixation,* Oxford and IBH Publishing Co., New Delhi.

Subba Rao, N.S. Ed. 1982. *Advances in Agricultural Microbiology,* Oxford and IBH Publishing Co., New Delhi.

Subba Rao, N.S. 1993. *Biofertilizers in Agriculture and Forestry,* 3rd Ed., Oxford and IBH Publishing Co., New Delhi.

Tien, T.M., Gaskins, M.H. and Hubell, D.H. 1979. Plant growth substances produced by *Azospirillum brasilense* and their effect on the growth of pearl millet (*Pennisetum americanum* L.), *Appl. Environ. Microbiol.,* 37, 1016–1024.

Watanable, I., So., R., Ladha, J.K., Katayama-Fujimura, Y. and Kuraishi, H. 1987. A new nitrogen-fixing species of pseudomonad: *Pseudomonas diazotrophicus* sp. nov. isolated from the root of wetland rice. *Can. J. Microbiol.,* 33, 670–678.

6. Nitrogen Fixation by Free-living Blue-Green Algae

General Aspects

Blue-green algae constitute an important group of microorganisms capable of fixing atmospheric nitrogen. They comprise unicellular, colonial and filamentous types (Fig. 43, Table 20). Some fossil forms have also been discovered which date back to pre-cambrian periods. Most of the nitrogen-fixing blue-green algae belong to the orders *Nostocales* and *Stigonematales* under the genera *Anabaena, Anabaenopsis, Aulosira, Chlorogloea, Cylindrospermum, Nostoc, Calothrix, Scytonema, Tolypothrix, Fischerella, Hapalosiphon, Mastigocladus, Stigonema* and *Westiellopsis*. In pure cultures, blue-green algae fix varying amounts of nitrogen ranging from 5.2 to 14.48 mg/100 ml of the medium, depending upon the incubation time.

In general, nitrogen fixation is associated with forms possessing heterocysts, although there are reports of fixation by unicellular and filamentous non-heterocystous strains. The plankton of lakes contains species of nitrogen-fixing algae which are invariably heterocystous forms such as *Anabaena*. The number of heterocysts could be taken as a rough parameter to indicate the nitrogen-fixing capacity of blue-green algae in spite of the fact that the amount of nitrogen fixed is dependent on physiological and environmental factors such as intensity of light, concentration of inorganic nitrogen sources, concentration of dissolved organic nitrogen compounds, temperature and aeration of the substrate. There is also a diurnal fluctuation in the quantity of nitrogen fixed by a given species of blue-green alga.

Heterocysts

Heterocysts are large, thick-walled, apparently empty cells appearing amidst normal pigmented cells (Fig. 43). However, when viewed through an electron microscope they seem to have a complex lamellar network. A large body of evidence has accumulated showing that heterocysts are the sites of nitrogen fixation and recent reports provide further direct evidence of nitrogenase activity in preparations made from isolated heterocysts. A mature heterocyst is surrounded by a multilayered envelope and shows

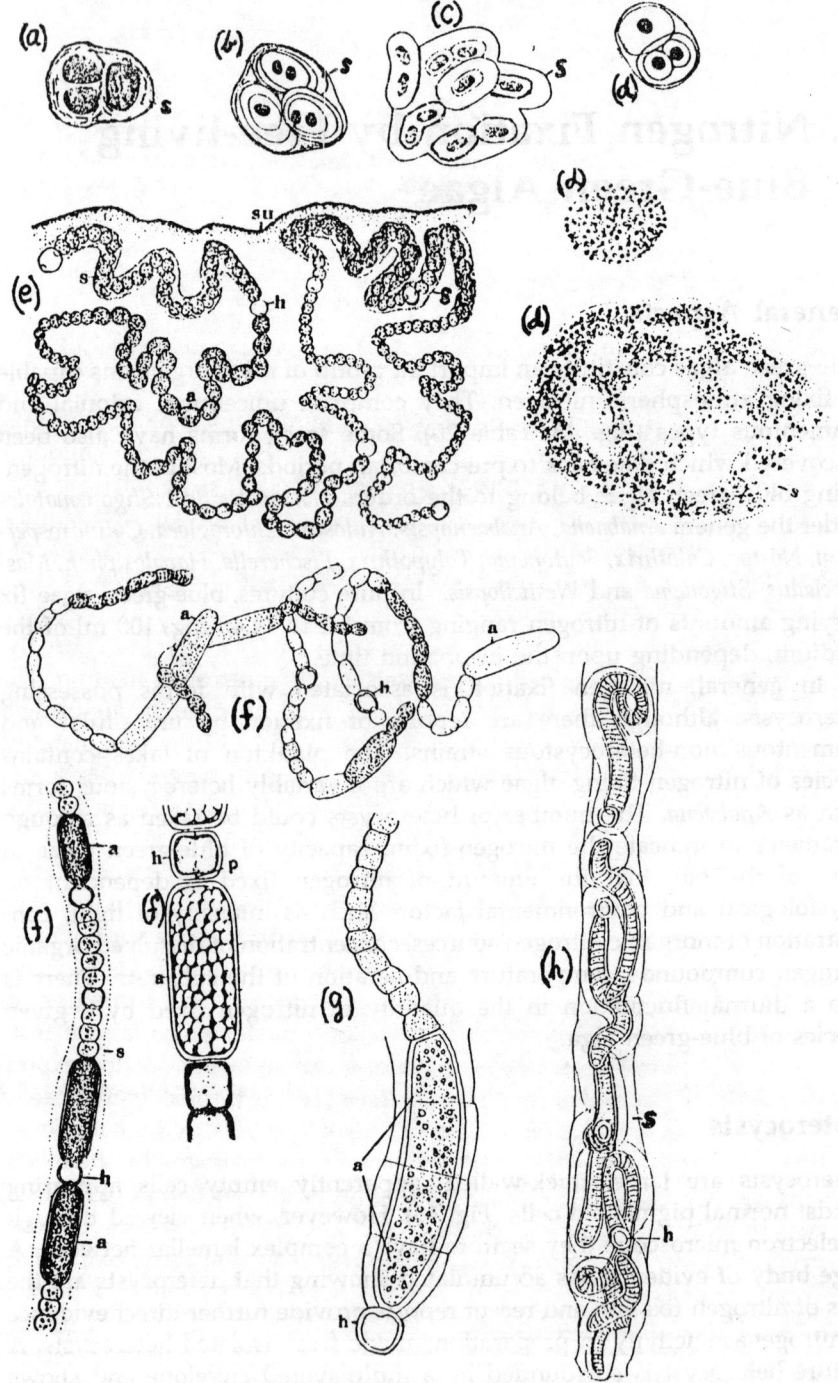

Fig. 43 Some representative genera of blue-green algae:
(a) *Chroococcus* sp.; (b) *Gloeocapsa* sp.; (c) *Gloeothece* sp.; (d) *Microcystis* spp. ; (e) *Nostoc* sp.; (f) *Anabaena* spp.; (g) *Cylindrospermum* sp.; (h) *Scytonema* sp. s—slime; h—heterocysts; a—akinete; p—pore of heterocyst; su—surface (from Fritsch, 1979).

Table 20 Nitrogen-fixing genera of blue-green algae

Unicellular—*Synechococcus, Gloeocapsa (Gloeothece), Aphanothece, Dermocarpa,*
    Xenococcus, Myxosarcina, Chroococcidiopsis, Pleurocapsa group

Filamentous, non-heterocystous—*Plectonema boryanum, Lyngbya, Trichodesmium,*
    Oscillatoria, Pseudoanabaena, Microcoleus, Schizothrix,
    LPP group

Filamentous, heterocystous—*Anabaena, Nostoc, Nodularia, Cylindrospermum, Scytonema,*
    Calothrix, Anabaenopsis, Mastigocladus, Fischerella, Tolypothrix,
    Aulosira, Stigonema, Hapalosiphon, Chlorogloeopsis,
    Camptylonema, Gloeotrichia, Nostochopsis, Rivularia,
    Scytonematopsis, Westiella, Westiellopsis, Wollea, Chlorogloea

an elaborate cytoplasmic membrane system devoid of granular inclusions. Heterocysts provide a congenial environment for the effective functioning of nitrogenase, to generate energy and reductant required for nitrogen fixation, to bring the nitrogen fixed into organic combination and to maintain a dual transport system for getting carbon and sending out nitrogen into the vegetative cells.

The transition or metamorphosis of a vegetative cell of a blue-green alga into a heterocyst is a gradual process from a CO_2 fixing and O_2 evolving cell into an anaerobic cell conducive for active nitrogen fixation. The heterocyst has a multilayered cell wall connected by cytoplasmic bridges to neighbouring photosynthetic vegetative cells. These bridges regulate the flow of molecules between the two types of cells and therefore a series of regulated physiological and biochemical changes lead to nitrogen fixation and assimilation.

Experiments carried out with *Anabaena variabilis* have shown that under light, nitrogenase activity in isolated heterocysts is dependent on a supply of H_2 whereas in dark the activity is dependent on a supply of O_2 and H_2. The principal product of fixation of nitrogen which is translocated from heterocysts to vegetative cells is glutamine. The formation of glutamine in heterocysts is dependent on the transfer of glutamate from vegetative cells. These conclusions have been made by conducting experiments using N^{13} labelled nitrogen. A disaccharide probably maltose, a product of photosynthesis, moves from vegetative cells to the heterocysts where it is metabolized to glucose 6 phosphate and oxidized by the oxidative pentose pathway. It is believed that pyridine nucleotide (NADPH) reduced by this pathway can combine with O_2 and thus provide conducive environment for the reduction of electron carrier ferredoxin (Fd). Heterocysts lack photosystem II activity but photosystem I which is present can also reduce ferredoxin. The reduced ferredoxin can donate electrons to the nitrogenase which reduces N_2 to NH_4^+ as well as release H_2. Uptake hydrogenase, capable of recycling H_2 is present only in heterocysts. Heterocysts have high levels of glutamine synthetase (GS) and low levels of glutamine oxoglutarate amido transferase (GOGAT). Glutamate formed in vegetative cells and which gets transferred to the heterocysts reacts with NH_4^+ to form glutamine. The latter moves into the vegetative cells where it reacts with alphaketoglutarate (alpha KG) to provide 2 molecules of glutamine. It is believed that glutamine or a metabolite of glutamine other than NH_4^+ is a component of the system that represses heterocysts differentiation. These results have come from the work of Wolk and colleagues of the Michigan State University, East Lansing, Michigan and Haselkorn of the Chicago University, USA and schematically reproduced in Fig. 44.

Algal Nitrogenase

Considerable progress has been made in studies on the nitrogenase enzyme obtained from cell-free extracts of nitrogen-fixing algae. The first report of nitrogenase activity came from preparations of *Anabaena cylindrica*. Subsequently, nitrogenase activity was shown in isolated heterocysts and also in extracts from non-heterocystous *Plectonema*. The soluble nitrogenase from both heterocystous and non-heterocystous algae appear very similar. One distinctive feature of algal nitrogenase is its high oxygen sensitivity, which can be overcome by exclusion of oxygen from the medium containing cell-free extracts. Blue-green algae evolve oxygen during photosynthesis and obviously, a built-in mechanism must exist at the cellular level to protect the highly sensitive nitrogenase from oxygen. Nitrogenase enzyme from filamentous forms such as *Plectonema* is more sensitive to oxygen than that from heterocystous algae indicating the protective mechanism afforded by the heterocyst which prevents the oxygen inactivation of the nitrogen-fixing system under aerobic conditions. However, it is not yet clear whether the nitrogenase enzyme is exclusively located in the heterocyst, since many workers have also detected nitrogenase activity in normal vegetative cells of heterocyst bearing algae.

Nitrogenase of BGA is very similar to that of other N_2 fixing bacteria. It catalyses the reduction of protons, cyanide and C_2H_2 besides N_2. The enzyme requires ATP, reductant and Mg^{2+} and is oxygen sensitive. As in other bacteria, the Fe-protein component binds to MgATP and transfers electrons to the MoFe protein which binds to the substrate and reduces it. The MoFe and Fe proteins of *Anabaena cylindrica* and *Plectonema boryanum* are known to cross react with the Fe proteins of *Azotobacter* and *Clostridum*. The O_2 sensitivity of nitrogenase is overcome in serval ways, the chief among which is the structural modification of some of the vegetative cells into thick walled heterocysts. Further the physiological environment in heterocysts is conducive for N_2 fixation, due to the absence of photosystem

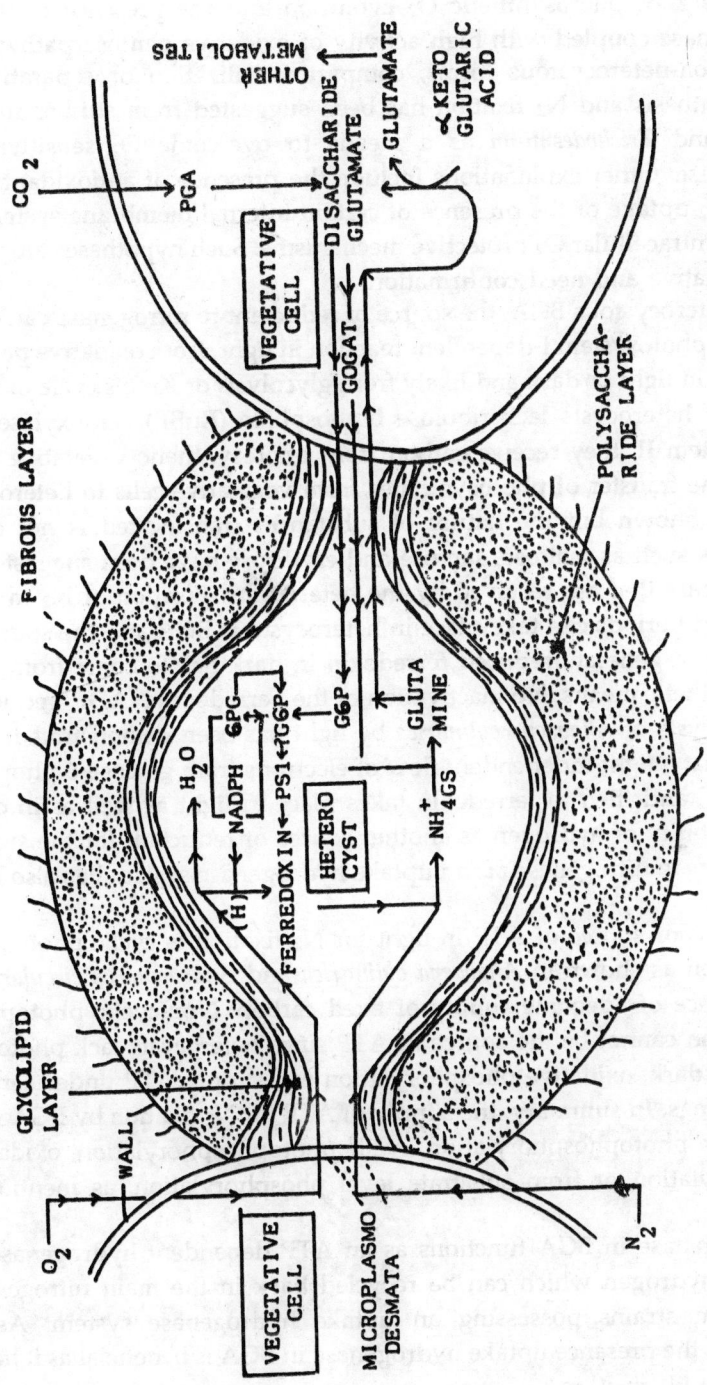

Fig. 44 The principal interactions between a heterocyst and vegetative cells of a blue-green algal filament. (Unified version of Haselkorn et al., 1980 and Wolk, 1980).

GS—glutamine synthetase; GOGAT—glutamine oxoglutarate amidotransferase; NADPH—pyridine nucleotide; G6P—glucose-6-phosphate dehydrogenase; 6PG—6-phosphogluconate dehydrogenase; PGA—phosphoglyceric acid.

II, the lack of photosynthetic O_2 evolution and the presence of uptake hydrogenase coupled with high activity of oxidative pentose pathway.

In non-heterocystous forms, compartmentalization or separation of photosynthesis and N_2 fixation has been suggested from studies in *Plectonema* and *Trichodesmium* as a means to overcome O_2 sensitivity of nitrogenase. Other explanations include the presence of antioxidants, O_2-linked H_2 uptake or the presence of certain internal membrane systems to serve as intracellular O_2 protective mechanisms. Such hypotheses are merely speculative and need confirmation.

In heterocystous BGA, the source of reductant to nitrogenase can come from the photosystem I-dependent reaction in light, from oxidative pentose pathway in light or dark and likely from glycolysis or Kreb's cycle or both. Since the heterocysts lack ribulose biphosphate (RuBP) carboxylase and Photosystem II, they receive carbon from photosynthetic vegetative cells. Indeed the transfer of photosynthates from vegetative cells to heterocysts has been shown but the nature of substances translocated is not clear. Substrates such as maltose, sucrose and erythrose have been suggested as likely sugars that are received by the heterocysts. Fixation of N_2 in dark for limited periods can take place in heterocystous BGA and the source of reductant to generate reduced ferredoxin in dark has to come from non-photosynthetic means. No inactivation of the Ferredoxin-$NADP^+$ reductase in heterocysts of *Anabaena cylindrica* by light has been demonstrated, suggesting that a light-independent flow of electrons from glucose-6-phophate to $NADP^+$ and then to ferredoxin takes place in light as well as in dark. The possibility of hydrogen as another source of reductant for the supply of electrons to nitrogenase *via* an uptake hydrogenase system has also been suggested.

The major source of ATP in light for N_2 fixation is cyclic photophosphorylation as shown in *Anabaena cylindrica* and *Anabaenopsis circularis* in the presence of adequate supply of fixed carbon. Non-cyclic photophosphorylation cannot be the source of ATP since heterocysts lack photosystem-II. In dark, oxidative phosphorylation can supply ATP under aerobic surroundings. In summary, the supply of ATP for N_2 fixation by BGA may come from photophosphorylation, cyclic photophosphorylation, oxidative phosphorylation or from substrate level phosphorylation, as mentioned above.

Nitrogenase in BGA functions as an ATP dependent hydrogenase to generate hydrogen which can be recycled back in the main nitrogenase reaction in strains possessing an uptake hydrogenase system. As in *Rhizobium*, the presence uptake hydrogenase in BGA is beneficial as it helps to enhance N_2 fixation.

Ammonia Assimilation

As mentioned earlier, N_2 fixation results in NH_4 formation in heterocysts which reacts with glutamate translocated from vegetative cells to form glutamine with the aid of glutamine synthetase. Most of this glutamine gets back into vegetative cells where it is metabolized into glutamate and other amino acids by glutamate synthase or glutamine oxoglutarate amidotransferase (GOGAT) located in vegetative cells. The glutamate formed may eventually form a substrate for more glutamine synthesis or else be transported back into heterocysts. In this way, the first stable nitrogen fixation is NH_4^+ and the first organic product of assimilation of NH_4^+ is glutamine followed by glutamate. Alanine may be formed through alanine dehydrogenase or transamination reaction in some species of BGA and aspartate in other species by transamination reaction probably from glutamate.

Genetics

While it is easy to obtain mutants of bacteria, it is relatively difficult to obtain mutants of blue-green algae, particularly those of filamentous forms. The work done in India by the late R.N. Singh and co-workers reveal that it is possible to obtain large number of mutants of nitrogen-fixing blue-green algae such as *Nostoc calcicola, N. linckia, Cylindrospermum maius* and *Gloeotrichia ghosei*. Genetic recombinations and transformations among strains of *Anacystis nidulans* have also been achieved. In general, mutants of blue-green algae have been obtained which neither possess heterocysts nor exhibit nitrogenase activity. On the contrary, mutants capable of fixing nitrogen have not been obtained from wild strains of non-nitrogen fixing blue-green algae. Three classes of genes important in heterocyst differentiation and nitrogen fixation have been cloned. They are *nif* genes coding for structural components of nitrogenase, glutamine synthetase and ribulose biphosphate carboxylase. Differences in the organization of *nif* genes between *Anabaena* and *Klebsiella* have been demonstrated.

Importance in Rice Cultivation

In India and South East Asia, rice is the most important staple food and its nitrogen nutrition is associated with nitrogen-fixing blue-green algae. The soil conditions in rice fields provide a congenial environment for the growth of nitrogen-fixing blue-green algae. According to some earlier estimates, fixation of nitrogen by algae in rice fields amounts to approximately 49 kg/ha under normal conditions which can be doubled if optimum amounts of phosphate and molybdenum are available in the soil. Some of the nitrogen fixing algae isolated from rice fields belong to the genera *Aulosira, Anabaena, Anabaenopsis, Calothrix, Campylonema, Cylindrospermum, Fischerella, Hapalosiphon, Michrochaete, Nostoc, Westiella, Westiellopsis* and *Tolypothrix.* Besides fixing atmospheric nitrogen blue-green algae synthesize and excrete several vitamins and growth substances (Vitamin B_{12}, auxins and ascorbic acid) which contribute towards better growth of rice plants (Fig. 45).

A survey conducted in rice growing areas in India has revealed that only about 33 per cent of the 2213 rice field soil samples harboured nitrogen fixing BGA species. Obviously management practices that are location specific may have influenced BGA occurrence. The situation is also true with regard to other Asian rice growing regions. Predators of BGA such as chytridious fungi, myxobacteria, cyanophages, copepods, snails and mosquito larvae graze algal growth. Phosphatic fertilizers stimulate the growth of BGA and similarly organic matter encourages the development of algal growth in general. Algal growth by itself improves the organic matter status of soils.

The flooded rice plant ecosystem is extremely complex, physically, chemically as well as microbiologically (see Fig. 31 in Chapter 4). One of the effects of flooding in uncropped rice field is a fall in O_2 content. However, in rice cropped soil, due to aeremyma in the rice plant, O_2 is capable of moving

Fig. 45 Growth response to blue-green algal inoculation to rice plants grown in pots.

from the leaf blade to the root cortex. This results in the oxidation of soil around the actively growing root system. Flooding of soil results in ammonium accumulation and nitrate instability. Ammoniacal nitrogen, the dominant form of mineral nitrogen in low-land rice soil is liable to fixation by clay, loss by volatilization, nitrification, denitrification, leaching, run off and seepage. About 60–80% of N absorbed by crops (40–50 kg N/ha/Yr) can be attributed to the native nitrogen pool. Approximately 60% of the rice yields (2–4 t/ha) can be obtained without the application of N fertilzer. The soil N does not show decreasing trends by rice plantings and harvest of grains indicating the existence of biological mechanisms to renew the depleted N from the soil N pool. Legumes, *Azolla* and N-fixing bacteria and blue-green algae take part in biological fixation of N. The fixed N is mostly mineralised to NH_4^+ which is the key process of N nutrition in waterlogged soil is subjected to environmental stresses.

Field trials conducted in different parts of India have shown significant increases in grain yields of many rice varieties by inoculation of rice fields with blue-green algae (Table 21). The high yielding varieties of rice such as I.R. series demand heavy doses of nitrogenous fertilizers up to 120 kg N/ha. It is now well known that any form of combined nitrogen is detrimental to the process of nitrogen fixation by microorganisms. The observed increased yields of rice due to algal inoculation, even under heavy doses of nitrogenous fertilizers, could be attributed to the combined effect of biologically fixed nitrogen and the growth substances secreted by blue-green algae.

Depending upon the requirement, algal biofertilizer can be produced by individual farmers and also on commercial basis, using one of the four methods, viz. (i) Trough method, (ii) Pit method, (iii) Field scale production, and (iv) Nursery-cum-algal biofertilizer production.

The Trough Method of Algal Production

1. Prepare shallow troughs (2 m × 1 m × 23 cm) of galvanised iron sheet or permanent tanks. The size can be increased if more material is to be produced.
2. Spread 8–10 kg of soil (4 kg/m^2) and mix well with 200 g of superphosphate.
3. Add water (5–15 cm) to the troughs and adjust pH to neutral (pH 7).
4. To prevent insects, add carbofuron (3% granules, 20 g. per tray) or malathion or BHC or any other suitable insecticide.
5. Add the starter culture of BGA (can be obtained from nearby University of Agriculture or Research Centre) to the trough. Within 10 days, a mat of BGA appears which when dried results in flakes.
6. The trough can be used again by using the flakes from the previous batch.

Table 21 Increase in the yield of rice crop (kg/ha) grown without chemical N fertilizer to algal inoculation at the rate of 10 kg/ha of dried soil-based BGA under field conditions in different parts of India (from Venkataraman, 1979)

Chemical Nitrogen	Orissa	Bihar	Madhya Pradesh	Maharashtra	Uttar Pradesh	Andhra Pradesh	Average
O Nitrogen	2979	2305	2416	2066	3525	3636	2079
ON + BGA	3710	3062	2820	2555	4356	4434	2541

The Pit Method of Algal Production

Polythene lined shallow pits can serve as containers instead of troughs, which makes it less expensive.

The Field Method of Algal Production

Near the rice cultivation area, 40 m^2 plots are bunded and flooded with water (2.5 cm) with application of superphosphate at 12 kg/40 m^2. Carbofuran, BHC or any other insecticide is used to avoid insect predators and mosquitoes. A starter culture of BGA is sprinkled over the body of water. Mats of BGA growth appears under hot humid sunny surroundings. When the plot dries up, the flakes of BGA are collected and used as inoculum at the rate of 10 kg/ha.

Limitations

The entire procedure is labour intensive and frequent monitoring is required to find out if the selected strains of nitrogen fixing blue-green algae have really established in the field without admixture of other unwanted microorganisms.

Selected References

De, P.K. 1939. The role of blue-green algae in nitrogen fixation in rice fields. *Proc. R. Soc.*, 127B, 121–139.

De Datta, S.K. 1987. Nitrogen transformation processes in relation to improved cultural practices for lowland rice. *Pl. Soil*, 100, 47–69.

Desikachary, T.V. 1972. Taxonomy and Biology of Blue-green Algae, Univ. of Madras, India.

Fogg, G.E. 1939. Nitrogen fixation. In *Physiology and Biochemistry of Algae*. pp. 161–170. Ed. R.A. Lewin, Academic Press.

Fogg, G.E. 1971. Nitrogen fixation in lakes. *Pl. Soil*, sp. vol., 393–401.

Fritsch, F.E. 1979. *The Structure and Reproduction of the Algae* Vol. I and II. CUP-Vikas Students edition.

Haselkorn, R., Mazur, B., Orr, J., Rice, D., Wood, N. and Rippka, R. 1980. Heterocyst differentiation and nitrogen fixation in cyano bacteria (blue-green algae). In *Nitrogen Fixation*, Vol. 2, pp. 259–278. Ed. W.E. Newton and W.H. Orme-Johnson, University Park Press, Baltimore, U.S.A.

Rice, D., Mazur, B.J. and Haselkorn, R. 1982. Isolation and physical mapping of nitrogen fixation genes from the cyanobacterium *Anabaena* 7120, *J. Biol. Chem.* 257, 1315–63.

Singh, R.N. 1961. *The Role of Blue-Green Algae in Nitrogen Economy of Indian Agriculture*, Indian Council of Agricultural Research, New Delhi.

Stewart, W.D.P. 1974. Blue-green Algae. In *The Biology of Nitrogen Fixation*. pp. 202–287. Ed. A. Quispel, North Holland Co., Amsterdam.

Thomas, J. 1970. Absence of the pigments of photosystem II of photosynthesis in heterocysts of a blue-green alga. *Nature*, Lond., 228, 181–183.

Venkataraman, G.S. 1972. *Algal Biofertilizers and Rice Cultivation*. Today and Tomorrow Printers and Publishers, New Delhi.

Wolk, C.P. 1980. Heterocysts, ^{13}N and N$_2$-fixing plants. In *Nitrogen Fixation*. Vol. 2, pp. 279–292. Ed. W.E. Newton and W.H. Orme-Johnson, University Park Press, Baltimore, U.S.A.

7. Nitrogen Fixation by Symbiotic Blue-Green Algae

Lichens

There are some blue-green algae which exist in association with fungi, liverworts, ferns and flowering plants. Some of them fix atmospheric nitrogen. The alga-fungus association to form lichens (Fig. 46) occurring on soils, rocks and tree-tops is yet another instance of symbiosis wherein the genus *Nostoc*, *Calothrix* or other unidentified blue-green algae fix nitrogen and, in turn, obtain protection and space from the fungal partner. The ability of lichens to fix nitrogen has been proved by the use of ^{15}N in genera of lichens such as *Collema*, *Stereocaulon*, *Leptogium*, *Lichina* and *Peltigera*. Recently, the list of lichens capable of fixing nitrogen has been expanded to include *Lobaria*, *Massalongia*, *Nephroma*, *Pannaria*, *Parmeliella*, *Placopsis*, *Placynthium*, *Polychidium* and *Sticta* which reduce acetylene to ethylene and thus demonstrate nitrogenase activity. Apart from releasing the nitrogen fixed, the alga *Nostoc* is known to provide biotin, riboflavine, thiamine, nicotinic acid and pantothenic acid to support the growth of the fungal partner. Fixation of nitrogen by lichens is dependent on the intensity of light and the moisture content of the thallus although it is well known that lichens can withstand very severe environmental stress.

Lichens appear as crusts, foliose or shrubby fruiticose growths on rocks, walls, trees or on ground. There are about 13,500 species of lichens. Lichen fungi are mostly ascomycetes and are known as mycobionts. Lichen growth is a slow process even with plentiful nutrients, the reasons for which are not clearly understood. The algal partner of a lichen is either a true alga also known as 'phycobiont' (85 per cent of lichens) or a cyanobacterium also known as 'cyanobiont' (15 per cent of lichens). Together, the cyanobiont and the phycobiont are called 'photobionts' and there are many unidentified photobionts. One particular genus of a lichen usually has one genus or species of a photobiont but in some instances two photobionts (one a true alga and the other a cyanobacterium) coexist in the same lichen species. Of the two photobionts, one situated in the thallus of the lichen happens to be a true green alga while the other is a blue-green alga situated in special nitrogen fixing structures known as cephalodia that are either within or on the surface of the thallus. The commonest photobionts found

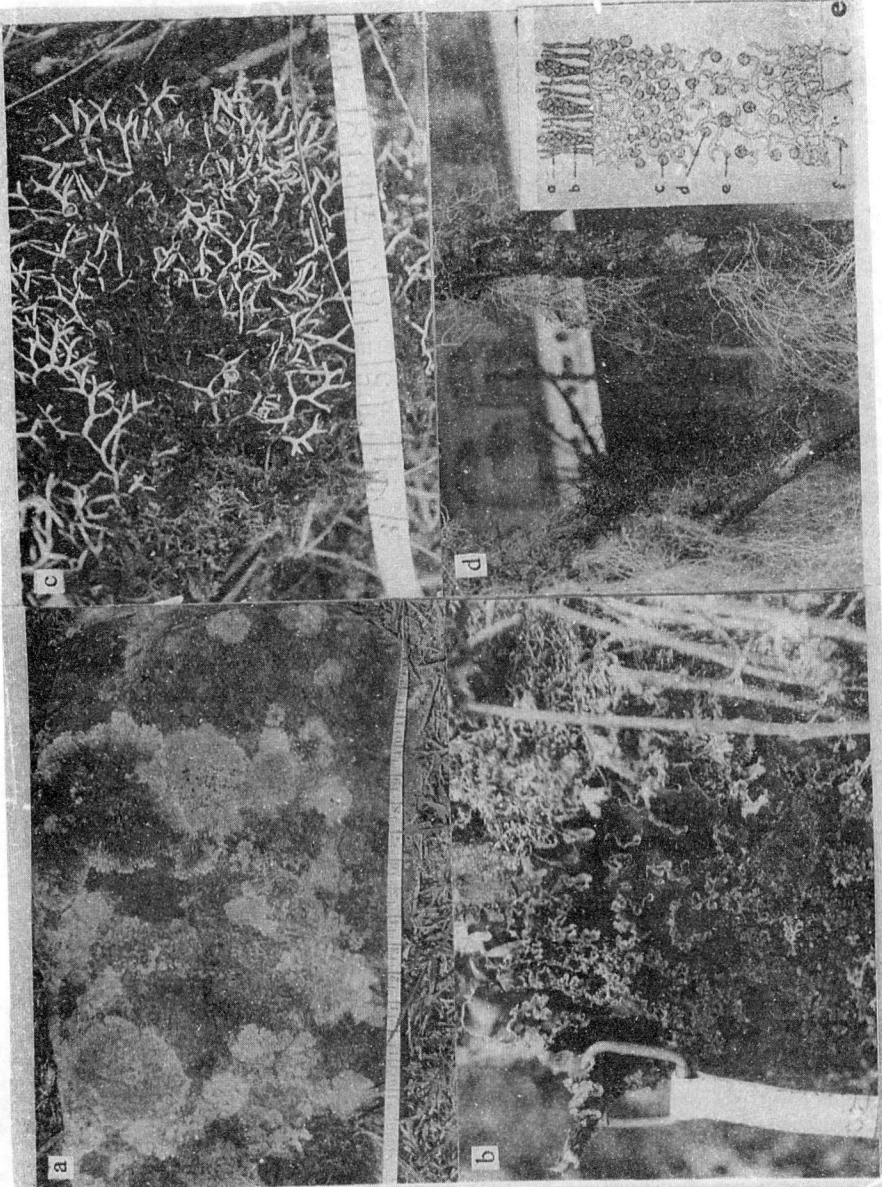

Fig. 46 Lichens (Courtesy, D.J. Hill, Bristol, U.K.)

a) Crust and foliose lichens on siliceous rock (New South Wales, Australia)

b) Foliose lichen with green alga photobiont. (*Anzia* sp. mountains in Papua, New Guinea)

c) Foliose lichen with cyanobacterial photobiont on tree trunks and ground (*Peltigera sp.* Solomon Islands)

d) Fruitcose lichen (*Usnea sp.* Solomon Islands)

e) Diagrammatic view of a section of the lichen showing (a) sterile hypha, (b) spore sac (ascus), (c) upper cortex, (d) unicellular alga, (e) mycelial filament, and (f) holdfast hypha (Courtesy: Dr. Awasthi, Lucknow).

in the majority of lichen species (70 per cent) are *Trebouxia* and *Pseudo-trebouxia*. The taxonomy of the lichen photobionts is somewhat controversial inspite of the fact that they have been cultured and many of them are difficult to identify *in situ*.

Beyond occupying space in a symbiotic association, the mycobiont does not seem to supply any essential nutrient to the photobiont in return to photosynthetic products received from the blue green alga in the form glucose as shown from ^{14}C tracer studies. The true green algal photobiont appears to transfer polyols, depending upon the genus. Polyols contained in the green alga enable the alga to absorb water from air and become physiologically active from a desiccated and dormant state without the aid of natural water. On the other hand the blue green alga does not contain polyols but has the capacity to saturate the organs with elevated CO_2 concentration. The consequence is the formation of high levels of carbohydrates in lichens that may serve as an adaptation to overcome dessication stress involved in the continuous natural process of drying and wetting with rain water.

Stages in Lichen Re-establishment

The mechanism by which a photobiont acquires a mycobiont or *vice-versa* has been an interesting point of investigation in microbial ecology. When ascospores are liberated, some of them germinate to form a mycelial mat, at which stage, contact with algal cells is established. An algal binding protein (lectin) which binds the photobiont cell wall to the fungal cell wall has been implicated in the process of recognition. The fungal mycelial mat eventually envelops the phycobiont cell. Experiments on resynthesis of lichens in culture have revealed that green alga photobionts which happen to be the incorrect symbionts turn yellow and die while appropriate symbiotic photobionts remain green and healthy thereby suggesting that some sort of recognition phenomenon is operating.

The establishment of symbiosis begins by carbohydrate transfer to the fungus and modification of the structure and physiology of symbiont cells. For example, cultured *Nostoc* cells are smaller than those in a symbiotic state. In another case, in the lichen *Nephroma laevigatum*, the cyanobiont (Nostoc) gets enlarged and shapeless remaining as single cells whereas the same cyanobiont assumes a filamentous form in culture medium. The next step is the formation of a primordial lobe leading to the development of a mature thallus in about 5–6 months followed by spore formation.

Nitrogen Fixation and Assimilation

Nitrogen fixation by cyanobacteria in lichen thalli is generally 2–3 times greater than in free-living state as revealed by studies carried out with the cephalodium (structure containing nitrogen fixing blue-green alga) of

Peltigera aphthosa and many other lichens. The cyanobiont (also called photobiont) of blue-green alga in cephalodia of *P. aphthosa* has low levels of glutamine synthetase (GS) and glutamate synthase (GOGAT) activity and hence unable to utilize all the ammonia derived from N_2 fixation. The excess ammonia diffuses out of the cyanobiont and is absorbed by the mycobiont which incorporates it into glutamate *via* glutamic dehydrogenase. Since the cyanobiont does not incorporate nitrogen into its own cells, the cells get N starved and this physiological state triggers the development of more heterocysts to fix nitrogen. The high frequency of heterocysts formation accompanied by the rapid transfer of ammonia out of the cells into the mycobiont can explain the augmented rate of N_2 fixation in symbiosis within the lichen thallus when compared to free blue-green alga in culture. The carbon and nitrogen interrelations in the cephalodium of *Peltigera aphthosa* has been depicted in Fig. 47.

There are various types of interactions in the organisation of the interface between mycobionts and photobionts in the thalli of lichen. Apart from mere juxtaposition of the two symbionts, in many crustose lichens the mycobiont actually penetrates the photobiont cell wall to form haustoria. On the contrary in other lichens the mycobiont remains non-invasive except that the cell walls very close to the photobiont are thin and invaginated followed by the cell contents exhibiting aggregation of mitochondria near the cell walls. All these manifestations are regarded as adaptations to perform the function of transferring nutrients, especially carbohydrates from photobiont cells.

Blue-green Algal Association with Bryophytes

Certain mosses and liverworts are also known to be inhabited by blue-green algae. The lower surface of the thallus of *Anthoceros* contains a species of *Nostoc* which stimulates the formation of papillae within the cavities of the host. The papillae provide space for the formation of compact colonies of the alga, which in turn, help in the nitrogen nutrition of the host plant. *Nostoc sphaericum* inhabits the cavities occurring on outgrowths near the edges of the lower surface of *Blasia* and *Cavicularia* (genera of liverworts). The dependence of the host on the lower symbiont for its nitrogen nutrition has been established in the case of *Blasia*, using [15]N. Another instance of an association between a species of *Hapalosiphon* and *Sphagnum* (a moss) has also been cited in literature together with ^{15}N data on translocation of fixed nitrogen.

Blue-green Algal Association with Higher Plants

The only angiosperm to develop a symbiotic association and fix nitrogen (Table 22) with a nitrogen-fixing blue-green alga is *Gunnera*. This genus

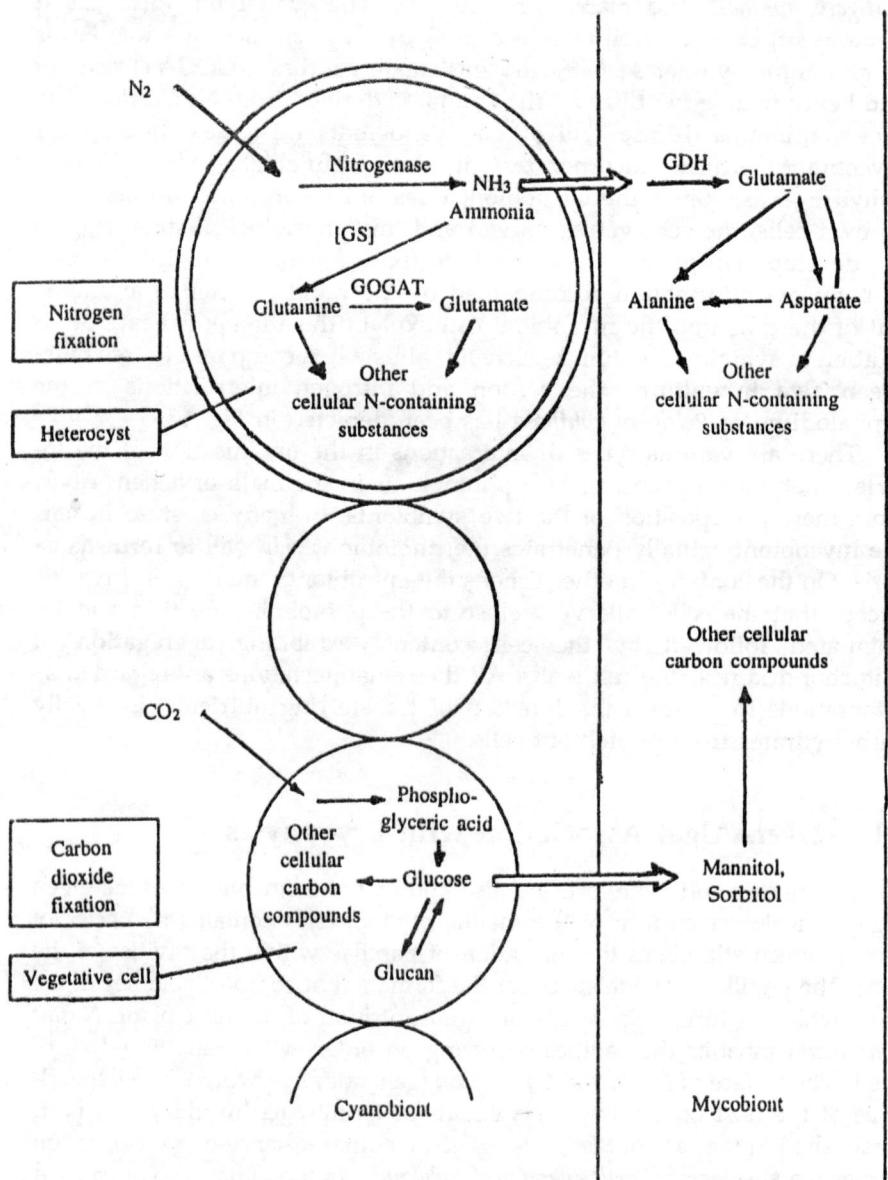

Fig. 47 Diagrammatic representation of interrelationships between a cyanobiont and a mycobiont in lichen (*P. apthosa*) symbiosis (After Hill, 1994).

has about 40 species which are herbaceous and widely distributed in the southern hemisphere. The plants have mucilage-filled cavities called glands near the bases of petioles. *Nostoc* cells penetrate the cells of the

Table 22 Acetylene reduction by *Gunnera macrophylla* tissue fragments containing *Nostoc* glands*

Per cent of C_2H_2 in air	Incubation time (min)	C_2H_4 produced (n moles/mg sample N/30 min)	C_2H_4 (nmoles/mg sample protein/min)
7.6	0–30	99.7	0.53
	30–60	115.0	0.60
	60–90	125.9	0.67
	90–120	142.1	0.76
	120–150	154.8	0.83
9.4	0–30	188.7	1.00
	30–60	169.5	0.90
	60–90	227.7	1.21
	90–120	184.8	0.99
	120–150	266.1	1.42

*Experiments were carried out between 08.30 and 11.00 h. In each experiment the total fresh weight of the excised tissue was about 9 g, the dry weight about 1 g. (From Becking, 1976.)

gland when infection takes place naturally. Two glands known as apical papillate glands, just below the point of cotyledonary attachment are formed which produce mucilage enabling the growth of *Nostoc*. From the mucilage, *Nostoc* cells penetrate the interior of the glands and then the host cells. Other blue-green algal species are also capable of forming associations with *Gunnera* under artificial inoculated conditions.

An excellent example of algal association with higher plants is the occurrence of endophytes *Anabaena* or *Nostoc* in the *coralloid* masses on the roots of *Cycadaceae* (Figs. 48 and 49). Cycads produce stubby apogeotropic profusely dichotomously branching tannin-rich coral-like (coralloid) root masses which become infected with cyanobacteria. These roots appear in addition to normal ordinary roots which are often very tuberous.

The functions of coralloid roots are not clearly understood. It has been suggested that lenticels present on these roots facilitate gas exchange between the plant and the atmosphere. Cyanobacteria may enter through lenticels or through breaks in the root's dermal layers. The cyanobacteria are restricted to intercellular spaces of the cyanobacterial zone although occasional penetration of the cortex has been reported in *Cycas revoluta*. Invariably the cyanobacteria are heterocystous belonging to the genus *Nostoc*. In some cases *Anabaena* and *Calothrix* have been encountered. Endophytes have been found in distinct zones in the cortex of the coralloid nodules on roots of genera such as *Cycas, Encephalartos, Zamia, Ceratozamia, Macrozamia* and *Stangeria*. The blue-green alga, *Nostoc cycadeae* isolated from *Cycas, Encephalartos* and *Macrozamia* have been shown to fix atmospheric nitrogen in the free state as well as in association with their host plants, as revealed by experiments using [15]N. When washed coralloid and normal roots were exposed to ^{15}N labelled gas, coralloid roots showed enrichment with ^{15}N while normal roots showed no such en-

Fig. 48 A—*Cycas* sp. in natural habitat.; B—A mass of coralloid roots taken out of soil beneath the plant; C—Transverse section of coralloid root showing the dark algal region.

Fig. 49   Blue-green algal association with roots of *Encephalartos* spp.
A—*Encephalartos middelburgensis* in natural habitat in South Africa; B—Transversely
sectioned coralloid root of *Encephalartos transvenosus* showing the dark cyanobacterial zone; C—Transverse section of uninfected coralloid root of *E. transvenosus*
showing tannin-rich cells; D—Transverse section of infected coralloid root of *E.
transvenosus*; CZ—cyanobacterial zone, P—periderm; E—cyanobiont trichome with
a terminal heterocyst (H) in the coralloid root of *E. paucidentatus* (from Grobbelaar,
1993).

richment. However, coralloid roots of two species of *Encephalartos* have been
shown to reduce acetylene to ethylene rather weakly. The speed with which
the transfer of fixed nitrogen takes place from the site of fixation into different

parts of the hosts varies from 1½ hours (in *Gunnera* symbiosis) to 48 hours (in cycad symbiosis).

Azolla-Anabaena Associaton

A species of *Anabaena* (*A. azollae*) is associated with the aquatic fern *Azolla* occurring in a ventral pore in the dorsal lobe of each vegetative leaf. The endophyte fixes atmospheric nitrogen and resides inside the tissues of the water fern. *Azolla* is being used as green compost for rice cultivation in North Vietnam. An added advantage is that the plant multiplies fast and provides higher yields of green compost (200–300 t ha/yr) than conventional green manure plants such as *Sesbania*, *Crotalaria*, and *Tephrosia* which are known to yield 30–50 t ha/yr. The disadvantages are that the plant is susceptible to parasites and sensitive to fluctuations in temperature, the most favourable range of temperature being 20–28°C. It is also necessary to prevent algal growth in rice fields for utilization of *Azolla* since algae tend to overgrow and inhibit the proliferation of the water weed. It is reported from Vietnam that a 10-ton layer of *Azolla* enables the rice yield to increase by 10–25% over corresponding *Azolla*-free rice fields. The benefit from *Azolla* growth to the associated rice crop has been variously estimated. It ranges from 95 kg N/ha/yr to 670 kg N/ha/yr, depending on the method used in determining the amount of nitrogen fixed.

The common species of *Azolla* in India is *A. pinnata* (Fig. 50A). It is recommended that *Azolla* nurseries are raised in small plots (50–100 sq m) or in concrete tanks with 5–10 cm deep water (pH 7–8) containing superphosphate at 4–8 kg P_2O_5/ha after seeding the plots with *Azolla* innoculum at the rate of 0.1 to 0.4 kg per sq m. These nurseries have to be planned several weeks ahead of the date set for transplanting rice seedlings (Fig. 50B). At the end of 2–3 weeks, when full growth of *Azolla* takes place, the water is drained and the *Azolla* growth is incorporated into the rice fields by ploughing the mass (10–20 t/ha) into the puddled rice field. This is followed by transplanting of rice seedlings within 7 days. Alternatively, *Azolla* is grown as a dual crop with the main crop of rice. As and when the *Azolla* mat formation takes place, it is ploughed into the field, a process which can be repeated depending on the growth of *Azolla*. Field experiments in India have demonstrated that 10 t/ha of *Azolla* is equivalent to 25 to 30 kg N/ha and similarly, an application of 20 kg N/ha as ammonium sulphate with *Azolla* is equivalent to 40 kg N/ha as ammonium sulphate (Table 23).

Survival of *Azolla*

Survival of *Azolla* after a drought period is associated with the production of sexual organs since plants die due to desiccation. The plant produces

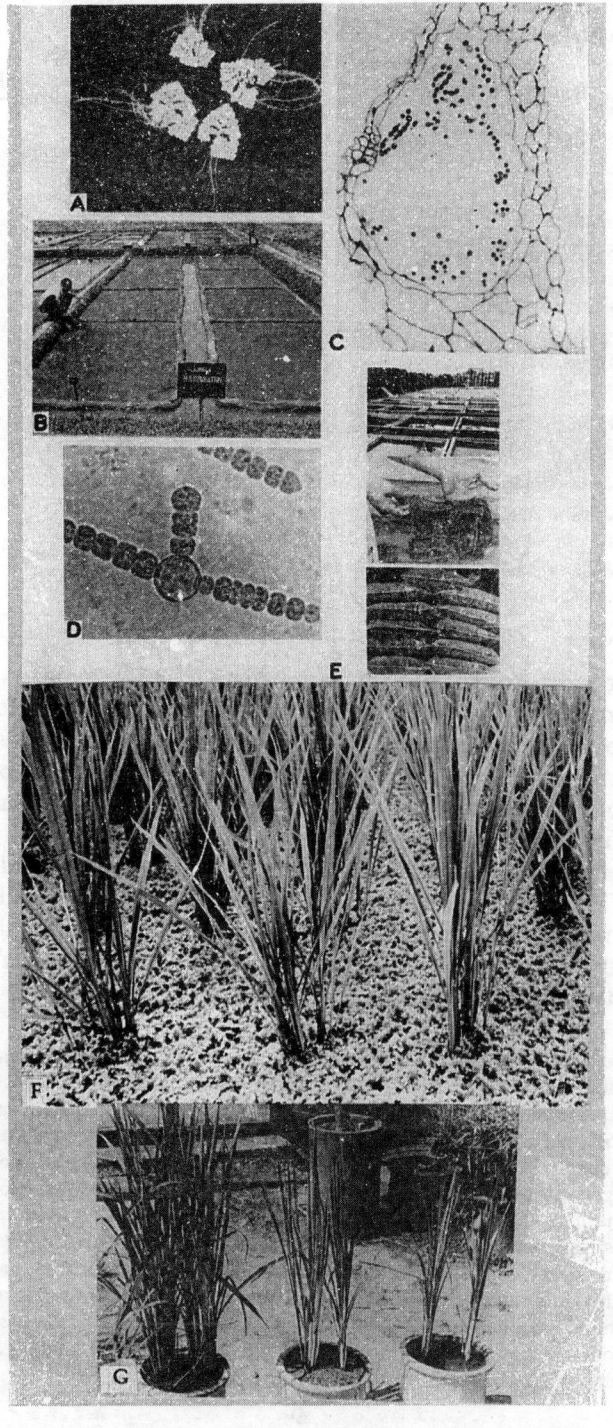

Table 23 Effect of *Azolla* and N fertilizer on the yield of short duration rice variety Kalinga-2 during Rabi season (Singh, 1977)

Treatments	Grain yield		Straw yield	
	kg/ha*	% increase over control	kg/ha*	% increase over control
1. Control (no *Azolla*)	1722	–	1325	–
2. Incorporated *Azolla* (10 tons fresh wt/ha)	2423	41	2087	58
3. Incorporated *Azolla* (20 tons fresh wt/ha)	2623	53	2587	95
4. Unincorporated *Azolla* (10 tons fresh wt/ha)	2400	39	2025	53
5. 20 kg N/ha (basal)	2208	28	2100	58
6. 40 kg N/ha (basal)	3187	85	3437	159
7. 60 kg N/ha (basal)	3518	104	3737	182
8. 80 kg N/ha (basal)	3894	126	4650	251
9. 30 kg N/ha + *Azolla* (10 tons fresh wt/ha)	3461	101	2837	114
10. 50 kg N/ha + *Azolla* (10 tons fresh wt/ha)	3576	108	3032	129
*Mean value	C.D. (1%)		355.78 kg/ha	526 kg/ha

microsporocarps (male) as well as macrosporocarps (female). *Azolla* plants have the main rhizome bearing alternate secondary and tertiary branches. Sporocarps are initiated at the place of attachment of branches and replace the lower lobe of the first leaf of a branch. The dorsal lobe covers the sporocarp. The sporocarps occur in pairs, both male and female or male plus female side by side. Macrosporocarps are really small oval-shaped with pointed ends measuring 0.75–1.0 mm in length and 0.5 mm in breadth. They are yellowish green in colour especially at the tip which becomes brown and hard due to lignification and tannin deposits. It is through the macrocarp tip that vegetative cells of *Anabaena* get into the plant and establish symbiosis. A mature microsporocarp is really larger

---

Fig. 50 *Azolla* and blue-green algae: (A) *Azolla pinnata* fronds showing rhizoids floating in water; (B) Nursery of *A. pinnata* at the Central Rice Research Institute (Courtesy: Dr. P.K. Singh); (C) L.S. of *Azolla* showing dotted algal cells in one of the dorsal lobes of *Azolla* frond representing *Anabaena azollae* cells (Courtesy: Dr. G.A. Peters of Beattle-Kettering Research Laboratory); (D) A filmentous blue-green alga showing vegetative cells surrounding a heterocyst, the site of N2 fixation (Courtesy Dr. Bansi Dhar); (E) Mass multiplication of blue-green algae (BGA) in rectangular concrete vessels where a mat of the algae develops which upon sun-drying can be collected and packed in plastic packets to serve as seed material for BGA propagation at the cultivators' fields (Courtesy: Dr. G.S. Venkataraman); (F) Fronds of *Azolla* floating around rice plants in the field; (G) Response of rice plants to inoculation with *Azolla*: From left to right *Azolla* incorporated into soil, *Azolla* left floating and control pots without *Azolla* or any other manure (from P.K. Singh).

than a macrosporocarp, globular in shape and measures about 2 mm in length and 1.5 mm in breadth. As many as 120 or more microsporangia may develop inside a single microsporocarp. Within each microsporangium 32–64 microspores develop. On the other hand, a mature macrocarp contains a single functional macrospore.

The micro- and macrosporocarps which are seasonally produced decay and sink to the bottom of a pond or any body of water in which *Azolla* grows. The decay and rupture of these sporocarps results in a meshed consortium described as macrosporangium complex. A mature microspore does not get freed from this macrosporangium complex and therefore development of micro- and macrogametophyte takes place in close proximity which makes matters easy for fertilization. The macrogametophyte (female) develops an archegonium (female apparatus) in which the egg cell (macrospore) is situated. The microgametophyte formed from a microspore also remains entangled with its appendages and develops an antherdium (the male apparatus). Fertilization of the egg cell with a spermatozoid (antherozoid) from the antherdium results in the formation of a zygote or embryo.

The development of the embryo takes place in the macrosporic complex and only when the seedling emerges with its cotyledons, the unaided eye can recognize the juvenile *Azolla* seedling with its rhizoidal development. The bilobed leaf formation takes place and the mature plant develops in about 21 days. The life cycle of *Azolla* is shown in Figs. 51 and 52.

Limitations

While the application of *Azolla* as an organic manure in rice cultivation has potentialities, its adoption by farmers as a regular practice depends on how certain major limiting factors are overcome. The phosphate requirement for the optimum growth of *Azolla* biomass has to be determined together with pesticide schedules to minimize the incidence of pests and diseases. In dual cultivation of *Azolla* with rice, the water requirements of the fern and the rice have to be borne in mind in situations where water is scarce. If the practice of incorporation of *Azolla* has to be advocated in a given situation, maintenance of separate ponds to raise the *Azolla* biomass is necessary. This obviously brings in the question of land allocation which otherwise could be used for crop cultivation. The requirement of biomass to the tune of 10 t/ha of fresh weight with the attendant problem of autolysis and the labour intensive nature of the entire operation renders the practice only suitable for developing nations where cost of labour is relatively cheap. Multiplication of *Azolla* depends on vegetative propagation and spore production by sexual reproduction is more an exception than a rule. Research in understanding factors promoting spore production may perhaps make the practice a lot easier by use of a small mass of spores to develop a nursery of *Azolla* instead of resorting to use of large quantities of fresh biomass. Transportation adds

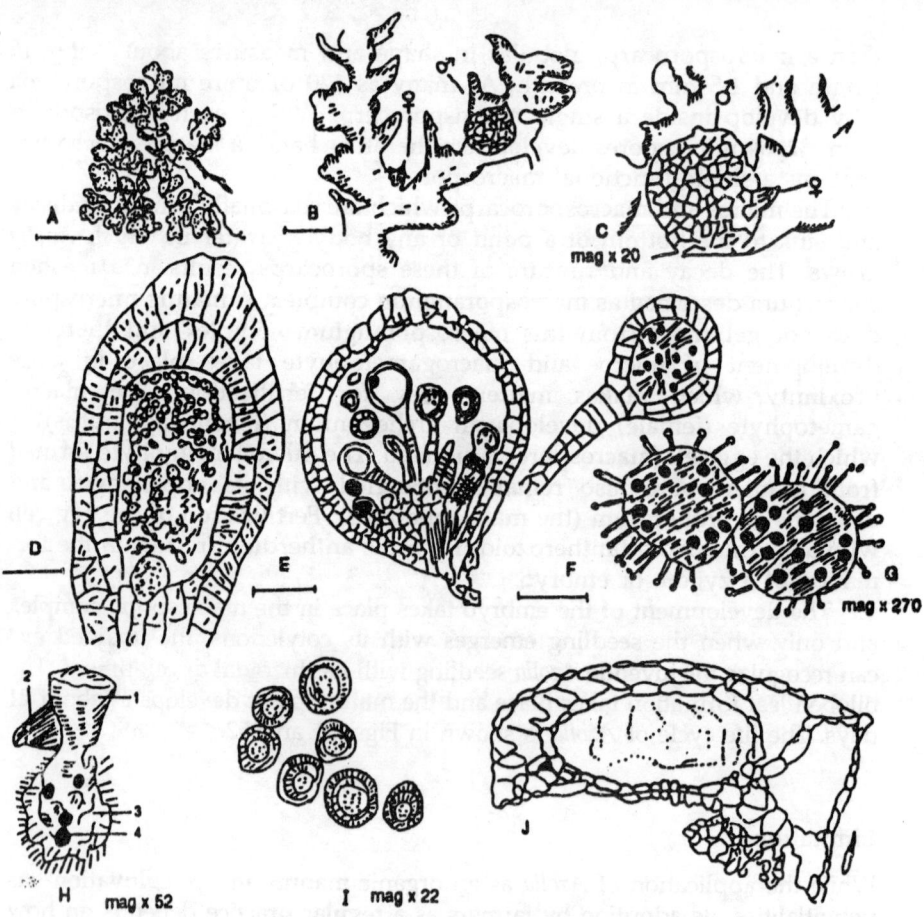

Fig. 51 A—*Azolla pinnata* showing the main rhizome; alternate branching, and the leaves, bar = 0.5 cm. B—Macrosporocarp (female) and microsporocarp (male) of *A. filiculoides* on two different branches, bar = 1 mm. C—A pair of sporocarps, macro and micro, in the axil of the lower lobe of *A. filiculoides*:. × 20. D—Young developing macrosporocarp where the *Anabaena* filaments fill the cavity with a basal developing megaspore; bar = 40 μ. E—Longitudinal section through a young microsporocarp of *A. filiculoides* showing number of stalked microsporangia on a basal placenta; bar = 160 μ. F—A microsporangium with microspores amidst developing massulae: G— Free massulae of *A. filiculoides* with anchor-like projections called 'glochidia' which help in attachment to macrospore facilitating fertilization. The microspores are embedded in the massulae; × 270. H—An *Azolla* seedling (1) emerging from a fertilized megaspore (2) of the megasporic complex which includes adhering massulae (3) with microspores (4) × 52. I—Very young seedlings of *Azolla* freely floating on water, × 22. J—A section through an older dorsal left lobe of *Azolla* showing *Anabaena azollae* filaments in the leaf cavity.(After Becking, 1987.)

to the expense of the operation in situations where nurseries of *Azolla* are located away from cultivators fields.

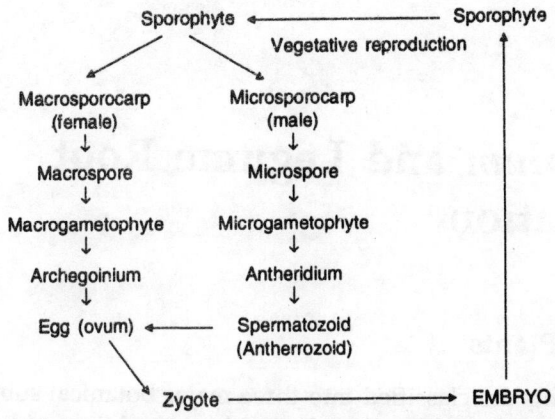

Fig. 52 Life cylce of *Azoila*. (After Becking, 1987.)

Selected References

Becking, J.H. 1976. Nitrogen fixation in some natural ecosystems in Indonesia, pp. 539–550. In *Symbiotic nitrogen fixation in plants*. Ed. P.S. Nutman, Cambridge Univ. Press, London.

FAO. 1977. Recycling of organic wastes in agriculture. Food and Agricultural Organisation of the U.N., Rome.

Grobbelaar, N. 1994. The cycad-cyanobacterium symbiosis pp. 96–140. In Symbioses in Nitrogen-fixing Trees. Eds. N.S. Subba Rao and C. Rodriguez-Barrueco. Oxford and IBH Publishing Co. Ltd., New Delhi.

Hill, D.J. 1994. The nature of the symbiotic relationship in lichens, *Endeavour* 13, 96–103, Elsevier Science Ltd., Great Britain.

Millbank, J.W. 1974. Associations with blue-green algae. In *The Biology of Nitrogen Fixation*, pp. 238–264, Ed. A. Quispel, North Holland Publishing Co., Amsterdam.

Silvester, W.B. and Smith, D.R. 1969. Nitrogen Fixation by *Gunnera-Nostoc* Symbiosis. *Nature*, 244, 1231.

Silvester, W.B. and McNamara, P.J. 1976. The infection process and ultrastructure of the *Gunnera-Nostoc* symbiosis. *New Phytol.*, 77, 135–141.

Singh, P.K. 1977a. *Azolla* plants as fertilizer and feed. *Indian Fmg.* 27, 19–21.

Singh, P.K. 1977b. Effects of *Azolla* on the yield of paddy with and without application of N-fertilizer. *Curr. Sci.*, 46, 622–644.

Singh, P.K. 1980. Introduction of "Green Azolla" biofertilizer in India. *Curr. Sci.*, 49, 155–156.

8. *Rhizobium* and Legume Root Nodulation

Leguminous Plants

Leguminous plants are classified into three major botanical subfamilies of the family Leguminoseae—the Ceasalpinioideae, the Mimosoideae and the Papilionoideae. There are nearly 750 genera and 18,000–19,000 species of leguminous plants of which 500 genera and approximately 10,000 species belong to the subfamily Papilionoideae. Not all legumes bear nodules on their root system and it is known that certain tree forms do not possess them at all. Hardly 16% of Leguminoseae have so far been examined for nodulation of which 95% of Mimosoideae, 26% of Ceasalpinioideae and 90% of Papilionoideae possess root nodules.

The origin of leguminous plants and the evolution of bacterial symbiosis are largely speculative. Evidence from fossil legumes does not provide much help in judging the exact time of origin of Leguminoseae. These plants are likely to have originated in sub-humid tropical, subtropical or temperate regions. The type of soil in which the first symbiotic legumes developed is also conjectural. It might have been in acid, neutral, alkaline or calcareous soil. Nodule bacteria were probably free-living nitrogen fixers before they became symbiotic under conditions of low availability of essential soil nutrients. It is likely that the slow growing, symbiotically promiscuous cowpea-type *Rhizobium* was the ancestral type of nodule bacterium which has persisted till today in association with many modern genera of Leguminoseae.

The Discovery of Symbiotic Nitrogen Fixation

In the 19th century, even though scientists had understood the value of mineral nutrition of plants some suspected that plants could obtain nitrogen from the atmosphere. However, it was fortuitous that two German chemists, Herman Hellriegel and Herman Wilfarth, presented very convincing experimental reports in 1886 and 1888 to distinguish the innate ability of legumes to fix elemental nitrogen in the atmosphere from the inability of cereal plants to perform the same function. Earlier, Boussingault in 1883 had done somewhat similar experiments but was not able to clearly point out the special symbiotic ability of legumes. The results of these discoveries are presented in Tables 24 and 25.

Table 24 Results of pot experiment conducted by Boussingault (from Fred *et al.*, 1932)

Plant cultivated	Duration of culture (months)	Weight (g)		Nitrogen (g)		Gain or loss in nitrogen
		Seed	Crop	Seed	Crop	
Clover	2	1.576	3.220	0.110	0.120	10.010
	3	1.632	6.288	0.114	0.156	0.042
Wheat	2	1.526	2.300	0.043	0.040	– 0.003
	3	2.018	4.260	0.057	0.060	0.003
Pea	3	1.211	4.990	0.047	0.100	0.053

Table 25 An important result obtained by Hellriegel and Wilfarth (from Fred *et al.*, 1932)

No.	Nitrogen in calcium nitrate per pot (g)	Oats	Peas
		Average weight of grains and straw (g)	Average weight of seed and vines (g)
1	none	0.390	4.380
2	0.056	5.876	4.128
3	0.112	10.961	9.132
4	0.168	15.997	–
5	0.224	21.357	9.725
6	0.336	30.175	11.352

In 1679, Malpighi provided the first diagrams and descriptions of root nodules of plants but considered them as insect galls. In 1866 Woronin saw fungal hyphae in root nodules which looked like threads, probably the 'infection threads' as we now know them. Therefore, subsequent observations led to the belief that root nodules were caused by fungi. Ward in 1889 studied the development of root nodules from the root hair to the origin of nodule protrusions.

More than one hundred years have elapsed since Hellriegel and Wilfarth (Fig. 53) presented their findings in several lectures delivered in 1886 to learned societies. Their conclusions were: (1) The Leguminoseae behave with regard to their nitrogen nutrition principally different from the Gramineae, (2) The Gramineae can rely for their nitrogen nutrition solely on the nitrogenous substances which are present for assimilation in the soil and their development is directly related to the available nitrogen supplies, (3) Besides the soil nitrogen, the Leguminoseae can use nitrogen from a second source available in the form of free elementary nitrogen of the atmosphere, (4) The Leguminoseae by themselves cannot assimilate atmospheric nitrogen but need the cooperation of vital microorganisms, (5) It is

Fig. 53 The two German chemists Herman Hellriegel (left) and Herman Wilfarth (right) who, in 1888, discovered that leguminous plants such as peas and clover possess the unique ability to fix elemental nitrogen from the atmosphere as opposed to cereals such as wheat and oats. (Courtesy: Prof. Dr. Sc. G. Schilling, Martin-Luther University, Halle-Wittenberg, Germany).

essential that certain species of microorganisms enter into symbiosis with the legume, (6) The root nodule is not merely a storehouse of protein but has a causal relation with the assimilation of free nitrogen. Such startling observations made after careful experimentation, have no doubt stood the test of time as we now look back more than one hundred years later.

Rhizobium Classification

Beijerinck in Holland was the first to isolate and cultivate a microorganism from the nodules of legumes in 1888. He named it *Bacillus radicicola* which is now placed in Bergey's *Manual of Determinative Bacteriology* under the genus *Rhizobium*.

Bacteria belonging to the genus *Rhizobium* live freely in soil and in the root region of both leguminous and non-leguminous plants. However, they can enter into symbiosis only with leguminous plants, by infecting their roots and forming nodules on them, the only exception being root nodulation in *Trema* (*Parasponia*) by a *Rhizobium* sp. The term symbiosis generally denotes a mutual beneficial partnership between two organisms. In legume root nodule symbiosis, the legume is the bigger partner while the *Rhizobium* is the smaller partner, often referred to as the 'microsymbiont'. When a nodule becomes senescent after a period of nitrogen fixation, decay of tissue sets in liberating motile forms of *Rhizobium* into soil which normally serve as a source of inoculum for the succeeding crop of a given species of legume. The genus *Rhizobium* has been placed in Bergey's *Manual of Determinative Bacteriology* in such diverse families as Azotobacteriaceae, Mycobacteriaceae, Myxobacteriaceae and Pseudomonadaceae. Speciation of *Rhizobium* based on the Linnaean concept has proved difficult and therefore, the cross-inoculation grouping based on the classical studies of Fred, Baldwin and McCoy is being generally followed. The principle of cross-inoculation grouping is based on the ability of an isolate of *Rhizobium* to form nodules in a limited number of species of legumes related to one another. All rhizobia that could form nodules on roots of certain legume types have been collectively taken as a species. This system of classification has provided a workable basis for the agricultural practice of legume inoculation. Under this scheme, seven species are generally recognized (Table 26).

The system of cross-inoculation grouping of rhizobia is not perfect since bacteria have been found to cross-infect or interchange between groups. However, until a better system of classification has been perfected, it appears as if we have to be content with the cross-inoculation grouping as a convenient and workable method of classifying root nodule bacteria into species.

Table 26 Cross-inoculation groups of *Rhizobium*

Rhizobium spp.	*Cross-inoculation grouping*	*Legume types*
R. *leguminosarum*	Pea group	*Pisum, Vicia, Lens*
R. *phaseoli*	Bean group	*Phaseolus*
R. *trifolii*	Clover group	*Trifolium*
R. *meliloti*	Alfalfa group	*Melilotus, Medicago, Trigonella*
R. *lupini*	Lupini group	*Lupinus, Orinthopus*
R. *japonicum*	Soybean group	*Glycine*
R. sp.	Cowpea group	*Vigna, Arachis*

Root nodule bacteria have been differentiated, on the basis of growth on a defined substrate, as fast growers and slow growers. Computer analysis of one hundred morphological and physiological characters (colonial character, vitamin, carbohydrate and nitrogen nutrition, antibiotic sensitivities and infective attributes) of nodule bacteria has revealed a new trend in their classification. Fast-growing pea and bean rhizobia (*R. trifolii, R. leguminosarum* and *R. phaseoli*) have been united under a common species name, *R. leguminosarum* Frank (the type species in Bergey's classification) while the fast-growing *R. meliloti* Dangeard retains a distinctive status as a species. The agrobacteria (*A. radiobacter* and *A. tumefaciens*) have been clubbed to a species, *Rhizobium radiobacter* (Beijerinck and Van Delden Lohnis). The slow-growing lupin, soybean and cowpea rhizobia, *R. lupini, R. japonicum* and *R. sp.* (cowpea group) have been assigned to a separate genus *Phytomyxa* and given a new name *Phytomyxa japonicum* Kirchner. In this way, root nodule bacteria could be delimited into two genera, *Rhizobium* and *Phytomyxa*, the former having three species, *R. leguminosarum, R. meliloti* and *R. radiobacter* and the latter with only one species, *R. japonicum*.

After re-examination of the above proposal, other workers have suggested that the genus *Rhizobium* need not be split at the generic level and that *Agrobacterium* need not necessarily be merged with *Rhizobium*. In other words, a *status quo* has been proposed and we are back to the cross-inoculation grouping of rhizobia as the only practical approach to the problem for the present.

Based on the ability of rhizobia to produce acid or alkali on yeast extract mannitol agar medium, the fast-growing *R. phaseoli, R. trifolii, R. leguminosarum* and *R. meliloti* have been grouped as acid producers whereas the slow-growing *R. japonicum, R. lupini* and *R. sp.* (cowpea) have been grouped as non-acid producers. The slow-growing non-acid producing rhizobia have been considered to be the ancestral forms of rhizobia since they are associated with primitive tropical legumes growing in acidic environment.

Attempts have been made to find the base composition of pure DNA (expressed as molar percentage of guanine and cytosine) of several rhizobia. Based on such studies, a regrouping of species has been suggested in which the fast-growing peritrichous strains having a low % (G + C) composition in the range 58.6–63.1% belong to *R. leguminosarum* and *R. meliloti*. On the other hand, the subpolarly flagellated, slow-growing strains having a somewhat higher % (G + C), mostly in the range of 62.8–65.5% correspond to *R. japonicum*. Since it is now believed that a knowledge of the base composition of DNA would clarify the present classification of bacteria in general, more investigations on these lines intended to clarify the ticklish problems in the taxonomy of *Rhizobium* are bound to be useful in future.

The following revision has been proposed by the international committee on systematic bacteriology and included in the ninth edition of Bergey's *Manual of Determinative Bacteriology*.

The genus *Rhizobium* will consist of three reorganized species: *R. leguminosarum*, which will contain three biovars (biovar *trifolii*, biovar *phaseoli*, and biovar *viceae*); *R. meliloti*; and *R. loti*. The reorganization combines into one the former species of *R. leguminosarum*, *R. trifolii*, and *R. phaseoli*. The fast-growing members of the cowpea rhizobia and the former species *R. lupinus* have been included in the species *R. loti*. The new genus, *Bradyrhizobium*, is made up of one species, *B. japonicum*, which consists of the former species *R. japonicum*, plus the slow-growing members of the cowpea rhizobia. The newly proposed classification of *Rhizobium* is as follows (from Elkan, 1984).

GENUS I: *Rhizobium*

R. *leguminosarum*
  biovar *trifolii*
  biovar *phaseoli*
  biovar *viceae*
R. *meliloti*
R. *loti*—fast-growing, sub-polar flagellated strains from *Lotus* and *Lupinus* with strong affinity for *L. corniculatus*, *L. densiflorus*, and *Anthyllis vul neraria* (but also nodulates *Ornithopus sativum*). Includes the fast-growing strains nodulating *Cicer*, *Sesbania*, *Leucaena*, *Mimosa*, and *Lablab*.

GENUS II: *Bradyrhizobium*

Slow-growing, polar or sub-polar flagellated strains nodulating soybean, *Lotus uliginosus*, *L. pendutulatus*, and *Vigna*. Includes those slow-growing strains nodulating *Cicer*, *Sesbania*, *Leucaena*, *Mimosa*, *Lablab*, and *Acacia*. The possibility exists that other species will eventually be defined within this genus, but for the present it is suggested that, other than *B.*

japonicum (the type species), the various cultures be designated ex. *Brady-rhizobium* sp. (*Vigna*), *Bradyrhizobium* sp. (*Cicer*), etc.

Recently, two more genera have been added to the family Rhizobiaceae. They are *Sinorhizobium* and *Azorhizobium*, nodulating soybean and *Sesbania*, respectively (1993).

The application of serological methods has helped in the delineation of strains among different rhizobia. There are at least two distinct kinds of antigens associated with the rhizobial cell—on the main body of the cell (somatic) and on the flagella. Agglutinations of a suspension of specific *Rhizobium* due to flagellar antigens can be distinguished from those due to somatic ones by the nature of its reaction. By this means, it has been shown that rhizobia are serologically heterogenous. The combined results of both somatic and flagellar reactions have served to distinguish strains within a cross-inoculation group. Serological methods can be used as a means of obtaining information on the distribution of strains that can be recognised within an area, on widely separated areas, on the plant or within a nodule. Serologically, it is known that a single nodule contains a homogeneous population of a single strain of *Rhizobium*, although it is not uncommon to find more than one strain on the same plant. Serological heterogeneity of isolates from the same locality has also been recorded. Serological methods have helped to distinguish nodules formed by inoculum applied to seeds from those formed by strains of *Rhizobium* already present in soil.

Cultural Characteristics

Rhizobium can live on relatively simple synthetic media. It has been found that glutamate is much superior to nitrate or the ammonium ion as nitrogen source. It is incapable of fixing atmospheric nitrogen on ordinary media (for exception, see Chapter 5) but can only do so in nodules on roots of a legume partner. The bacterium is unable to utilize cell wall materials such as cellulose, lignin or pectin. This fact has to be reckoned within any attempt to explain the mode of entry of *Rhizobium* into legume roots. In general, fast-growing rhizobia such as *R. trifolii*, *R. leguminosarum* and *R. phaseoli* grow vigorously with most sources of carbohydrates. On the other hand, slow-growing rhizobia are more specific in their requirements. They utilize sodium citrate, xylose, mannitol, arabinose, galactose, fructose and rarely dextran.

Vitamin requirements of rhizobia differ among species and strains. Many workers have observed thiamine, biotin and calcium pantothenate requirements for rhizobial growth. However, no growth responses have been observed to nicotinic acid, piridoxin, folic acid, p-aminobenzoic acid, inositol, vitamin B_{12} and riboflavin. On the other hand, inhibition of growth of *R. japonicum* by biotin and nicotinic acid has been reported.

The calcium requirement of *Rhizobium* varies with species. The essentiality of this element for the growth of *Rhizobium* has been debated. However, it has been concluded that deficiency of calcium in the presence of magnesium reduces the growth of *R. trifolii*. The calcium deficient cells become swollen and vacuolated. Low concentrations of vanadium and cobalt have been shown to stimulate the growth of some species of rhizobia. Iron at 10 ppm has also been reported to be stimulatory for the growth of *Rhizobium*.

Rhizobium cells in clover rhizosphere are small to medium sized (0.5–0.9 × 1.2–3.0 μ), Gram-negative rods. They are motile when young and have bi-polar, sub-polar or peritrichous flagella (Fig. 54). Cells contain characteristic granules of polymerised B-hydroxybutyrate (40–50% of cell dry weight) which stain with Sudan Black and appear as highly refractive bodies under phase contrast illumination. Most strains produce gum (extracellular polysaccharide slime) of varying composition. Cell wall of *R. trifolii* contains glucosamine, muramic acid, glutamic acid, diaminopimelic acid and a wide range of amino acids, characteristic of Gram-negative bacteria.

Under electron microscope, two membranes averaging 7–9 μ in diameter are apparent in *R. trifolii*. The outer membrane is the rigid cell wall and the inner membrane is the protoplasmic lining. In between the two membranes is a regular non-rigid inter-membrane area (50 μ). The *Rhizobium* cell has a large irregularly shaped nuclear region in the centre surrounded by a narrow region of denser cytoplasm.

Genetics

Genetic studies on *Rhizobium* began by obtaining records of simple mutations, the action of phage on *Rhizobium* and transformation attempts with different strains of nodule bacteria. Markers generally employed are resistance to antibiotics and biochemical requirements of an amino acid or a simple sugar. X-rays have been used to obtain a range of biochemical mutants. *Rhizobium* has been successfully used as an acceptor bacterium for transformation of tumour-inducing capacity from *Agrobacterium tumefaciens*, an instance of inter-generic transformation. Transformation of nodulating capacity between *Rhizobium* species has been reported. In this way, *R. japonicum* has been made to nodulate lucerne and *R. meliloti* to nodulate lupin. In *R. trifolii*, avirulent strains have been made virulent by transformation experiments. Mutations to ineffectiveness are common in soil. It has been shown that phage resistant mutants of *R. trifolii* tend to be ineffective. When wild-type ineffective strains are treated with phage,

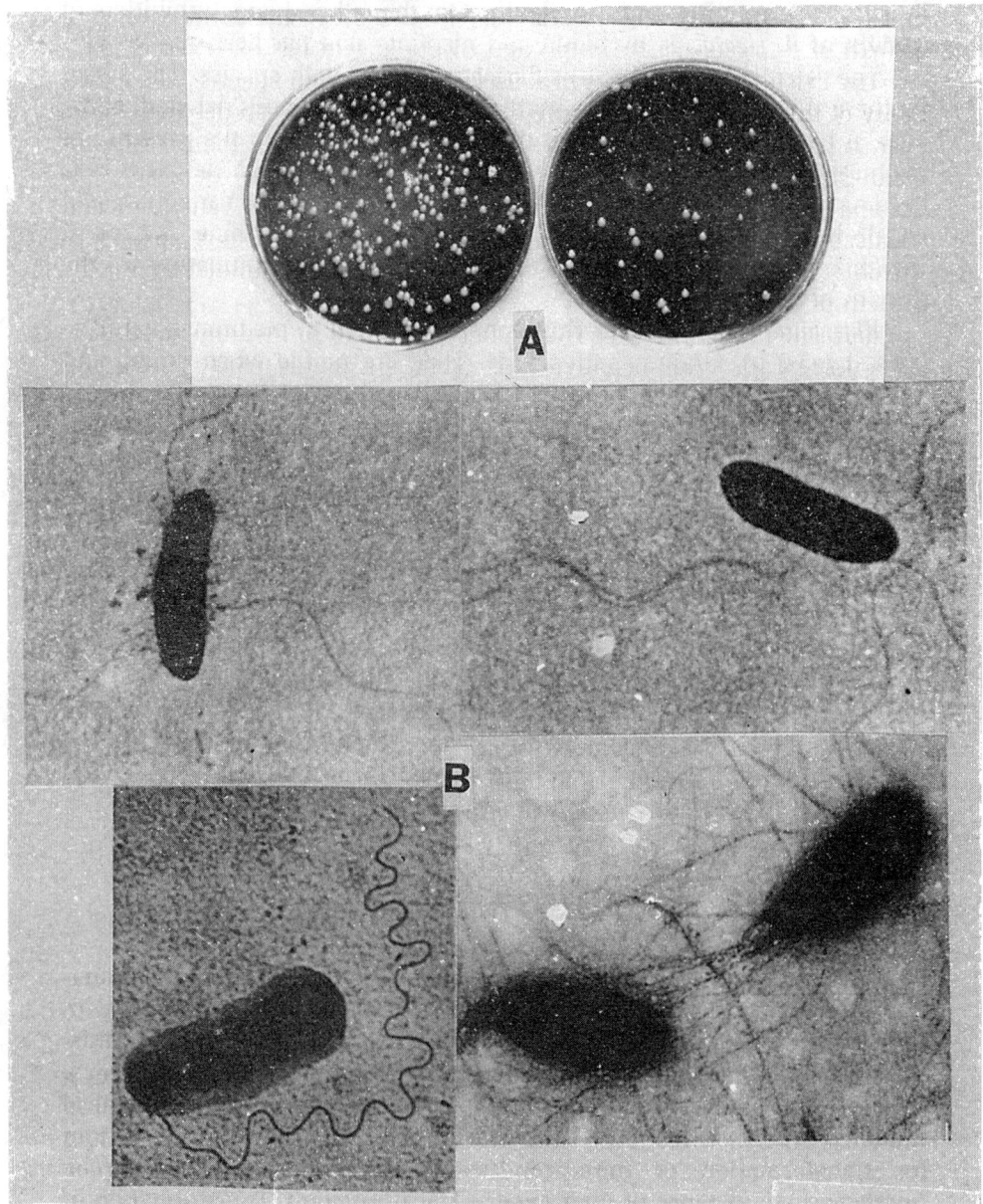

Fig. 54 A—Rhizobial colonies on YEMA medium having Congo red showing glistening colonies; B—Electron photomicrographs of rhizobia showing polar, sub-polar and peritrichous flagella. (Courtesy: Prof. J. De Ley, Belgium).

effective variants do not arise. On the other hand, ineffective mutants occasionally revert to effectiveness when treated with phage. Loss of effectiveness is found to be closely associated with mutation to viomycin and neomycin resistance in strains of *R. leguminosarum, R. trifolii* and *R. meliloti*. Both viomycin resistance and ineffectiveness remain unchanged in clones re-isolated from nodules. Ability to form nodules (ineffectiveness) on the homologous host is, however, retained in all antibiotic resistant strains. Genetic circular linkage maps have been constructed for *R. meliloti, R. leguminosarum* and *R. lupini* and studies are in progress to locate more genes on the nuclear material.

Conjugation of *R. phaseoli* with *R. trifolii* as donor has been accomplished which extended the progeny to nodulate *T. repens*. This acquired ability was got rid of by treatment of bacteria with acridine orange. Other experiments with *R. meliloti* have also shown the loss of ability to nodulate upon treatment with acridine orange. These results suggest that infectivity is a plasmid controlled property, thereby indicating the possibility of obtaining strains with a wide host range. Over the past ten years, the availability of methods to analyze DNA, more particularly from plasmids, has made it possible to identify high molecular weight plasmids (megaplasmids with mol. wt. more than 100×10^6) in fast-growing rhizobia. These plasmids carry heritable factors for nodulation and nitrogen-fixation.

The use of transposon-induced mutation technology coupled with other novel methods of *in vivo* strain constructions using R plasmids has improved our knowledge concerning the genes controlling symbiotic functions. Mutants of *Bradyrhizobium japonicum* requiring tryptophan or unable to nodulate (Nod⁻), fix nitrogen (fix⁻) or nodulate but not fix nitrogen (Nod⁺, fix⁻) have been obtained by physical and chemical mutagenesis. Similarly, mutants with more hydrogenase (hup⁺) or ones lacking in this enzyme (hup⁻) and those unable to curl root hairs (Hac⁻) have also been obtained. These studies have helped in understanding the genetic control of symbiosis.

The question whether all or part of the genes responsible for legume symbiosis are located on the rhizobial chromosome or on extrachromosomal elements is being increasingly examined with the discovery of extrachromosomal megaplasmids controlling the functioning and expression of the nitrogen-fixation process (*nod* and *nif* functions) in legumes. However, it is not easy to say how many are partitioned to the chromosome even though genetic mapping data have shown that *fix⁻* mutants of *R. meliloti* can be traced to the bacterial chromosome. Plasmids are known to carry *nod* and *nif* function in fast-growing *Rhizobium* species (*R. meliloti, R. leguminosarum, R. trifolii*) whereas evidence for the location of these genes in the slow-growing *Bradyrhizobium japonicum* is gradually emerging.

The current knowledge on the genetics of legume root nodulation is complicated owing to the capacity of rhizobia to enter plant roots, induce nitrogen-fixing nodules and regulate the translocation of ammonium ion through a series of events which are controlled by genes located in large megaplasmids, as stated earlier. These genes include the nitrogen fixation genes (*nif*), the common nodulation genes (*nod* ABC) and the host range genes (host specific *nod* H and Q). Several *nod* genes (*nod* ABC), *nod* FE and *nod* H have been identified in *R. meliloti* on a 'Sym' plasmid, by means of genetic analysis and DNA sequencing. The common *nod* ABC seem to be required in all *Rhizobium* systems for eliciting root hair deformation and root cell division. The root hair deformation and cell division processes appear to be caused by the products of (or) signals from *Rhizobium nod* ABC genes. The *nod* H and Q genes which are *Rhizobium* species dependent are necessary for deciding host specificity. If *nod* ABC genes are mutated, then live *Rhizobium* cells cannot deform root hairs whereas if *nod* H is mutated, specificity may be lost but not root hair deformation. The root hair deformation activity is known as 'Had' activity. The signal product of *nod* ABC genes from *R. meliloti* causing Had activity in alfalfa root hairs has been enriched, purified and named as Nod Rm1 factor which is a sulfated and acylated glucosamine oligosaccharide.

Flavonoids Induce Nodulation

In the last decade, the following root secreted flavonoids from leguminous plants have been known to induce transcription of nodulation (*nod*) genes:

Plant source	Compounds
Alfalfa seed (*Medicago sativa*)	Luteolin
Clover seedling (*Trifolium* spp.)	Geraldone
Soybean seedling (*Glycine max*)	Genistein, Daidzein
Bean seed (*Phaseolus vulgaris*)	Delphinidin, Petunidin
Bean root (*P. vulgaris*)	Naringenin
Vetch root exudate	Naringenin, Liquiritigenin
(*Vicia sativa* sub sp. *nigra*)	

In some strains of *R. meliloti*, as many as 15 nodulation genes have been identified. Some of them are barely expressed in *R. meliloti* in a free-living state but are induced to express themselves manifold in the presence of root exudates. In alfalfa (lucerne, *Medicago sativa*), the induction factor in the root exudate is a flavone known as luteolin (3', 4', 5', 7'-tetrahydroxy flavanone). During the initiation of symbiosis, the two symbionts (the *Rhizobium* and the legume) interact at the molecular level of gene expression by signalling to each other. The product of *nod* D gene in *Rhizobium* interacts with the flavonoid compounds secreted by the roots and this follows the activation of other *nod* genes in a host dependent manner (Fig. 55). Several flavonoid molecules produced by legumes have been found to induce or block the transcription of *nod* genes in *Rhizobium* and *Bradyrhizobium*. The active inducers in root exudates of clover are 7,4-dihydroxyflavone, umbelliferone and formononetin, and soybean roots

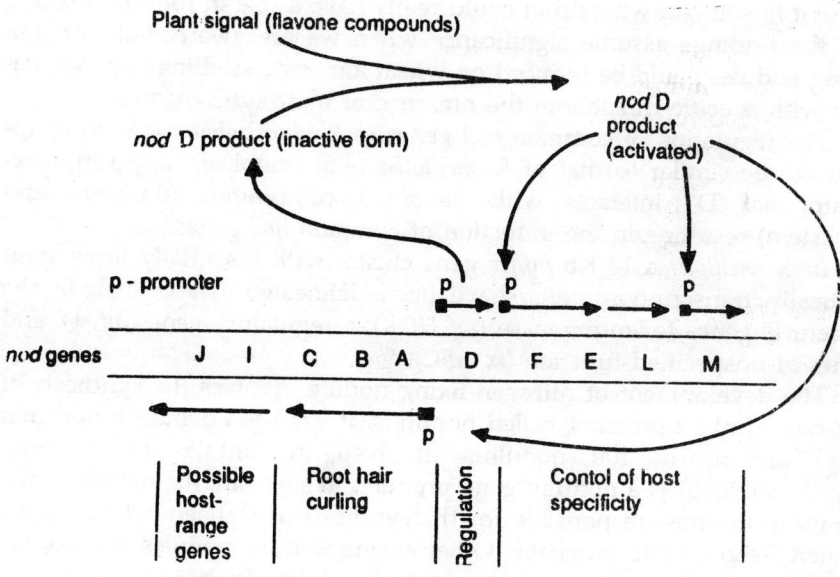

Fig. 55 A schematic diagram of events at the molecular level in the root region in *Rhizobium*-legume interaction. The protein product of *nod D* gene in *Rhizobium* is activated by contact with plant signal by way of flavonoid compounds. This then interacts on the promoter gene which controls *nod ABC* genes action to induce nodule formation.

release naringenin, genistein and diadzein (isoflavones). The positive action of *nod* genes in *R. leguminosarum* biovar *trifolii* is mediated by *nod* D gene and the interaction of *nod* D gene product with the plant secreted inducer and anti-inducer compounds. There are nodule inducer as well as anti-inducer regions on the clover root, the plant cells behind the root tip zone·being the major site for the secretion of the inducer compounds. This region is followed upwards with regions on the root alternatively producing stimulatory or inhibitory substances. Apparently, the primary determining factor for sites of nodule initiation in clover and other legumes could be the ratio of stimulator (inducer) : inhibitor (anti-inducer) in the vicinity of potential infection sites or the infection thread during infection. The anti-inducers are coumarins and isoflavones. Three components are necessary for induction of *nod* genes—1) the *nod* promoter, (2) the inducing substance produced by the plant which is of flavonoid nature, and 3) the *nod* D gene product.

Some reports indicate that adding flavonoids (10 μm of luteolin or naringenin) to the root region of certain varieties of alfalfa (*Medicago sativa*) increased nodulation and N_2 fixation by *R. melitoli* under controlled defined experimental conditions. Experiments have also shown that certain flavonoids could be extracted from soils and are also present in roots of non-legumes such as wheat. It is too early to say whether these flavonoids present in soil and wheat root could really have a role in root nodulation, but the findings assume significance when we are aware that nitrogen fixing nodules could be induced on wheat and rape seedlings by inoculation with specific rhizobia in the presence of hydrolytic enzymes.

The regulation of common *nod* genes in *Bradyrhizobium japonicum* appears to be similar to that of *R. meliloti*-alfalfa symbiosis. A positive activator *nod* D1 interacts with flavonoid compounds (daidzein and genestein) resulting in the induction of common *nod* genes.

In *R. meliloti* a 14 Kb *nif/fix* gene cluster with essentially three symbiotically transcription units, have been delineated. These include the structural genes for nitrogenase (*nif HDK*), a regulatory gene (*nif A*), and genes of unspecified function (*fx ABC*).

The development of nitrogen-fixing nodule involves the synthesis of a group of plant proteins, called nodulins. It has been demonstrated that there are nearly 100 nodulins involved in nodule development. Leghaemoglobin is a nodulin gene product. In soybean, the nodulin gene family in the host responsible for the synthesis of leghaemoglobin is activated 7–8 days after infection. Other enzymes in the nodules, the uricase and glutamine synthetase are also controlled by nodulins.

Novel Approaches to Extend Root Nodulation

Recently, novel approaches to break the cell wall barrier in root hairs of white clover (*Trifolium repens*) by treating seedlings with cellulase and pectolyase followed by treatment with polyethylene glycol (PEG) and calcium chloride have been successful. By these treatments, the protoplasts of root hairs were exposed for reception to rhizobia which normally do not invade white clover root hairs due to the cell wall barrier. For instance, *Rhizobium loti* which can only infect species of *Lotus* was able to enter the root hairs of white clover and cause nodules under axenic experimental conditions. The success of this experiment depended on the critical levels and combinations of cellulase, pectinase and PEG. In fact, nodules on white clover roots induced by heterologous *R. loti* fixed nitrogen, were pink in colour and comparable to nodules normally induced by *R. meliloti*, a homologous *Rhizobium*.

Similarly, pre-treatment of seedlings of rice (*Oryza sativa*) and wheat (*Triticum aestivum*), both monocots, also resulted in root nodulation when inoculated with *R. trifolii* or *R. loti* or their mixture. The nodules were structurally sparsely infected and exhibited feeble nitrogenase activity.

Nodulation of rape seedlings (*Brassica napus*) by *R. leguminosarum* and *Bradyrhizobium* has also been observed by following similar procedures adopted for rice. Interestingly, Nod+ strain of R. leguminosarum induced nodules on a variety of rape but no such response was seen when inoculated with Nod– strain. The nodules had ultrastructural details similar to normal nodules and exhibited respectable amount of nitrogenase activity.

What appears to be intriguing was the formation of nodules on the roots of rape seedlings even without enzyme and PEG treatment when inoculated with *Rhizobium parasponium*, a strain which produces nodules on roots of a non-leguminous tree genus, Parasponia. Unlike temperate legumes such as clovers which are normally infected by rhizobia through root hairs, infection in tropical *Parasponia* and other genera such as *Aeschynomene, Sesbania* and *Stylosanthes* is through wounds and points of lateral root emergence. While explaining these observations, due cognizance of different modes of entry of rhizobia have to be taken into consideration. Needless to say, as Cocking *et al.* of the Nottingham University, U.K. say, these researches "may have important consequences for both basic and applied aspects of nitrogen fixation."

Ecology

As stated earlier, nodule forming bacteria occur in soil and in the root region of legumes as well as non-legumes. No selective medium has yet been formulated for isolating rhizobia from soils. The method followed for estimating rhizobia in soil is by the most probable number method where aseptically grown seedlings are inoculated with dilute suspensions of soil samples and the extent of nodulation is then observed, followed by statistical analysis of results. In the absence of legumes, the soil population of rhizobia declines. However, rhizobia are known to survive for 19 to 45 years despite the fact that they are non-spore formers. *Rhizobium lupini* and *R. japonicum* have been found to be comparatively resistant to high soil temperatures unlike *R. trifolii* and *R. meliloti*. It is remarkable that in spite of high temperature, tropical rhizobia nodulate *Acacia, Lotus* and *Psorales*, probably by adaptation to such temperature regimes.

Rhizobium is preferentially stimulated in the rhizosphere of legumes than in that of non-legumes. A given legume tends to promote the multiplication of bacteria able to infect it more than others, although evidence is available in literature to show that strains of different cross-inoculation groups are stimulated to the same degree by one and the same legume. Legumes excrete a large number of substances into the rhizosphere, principally sugars, amino acids and vitamins such as biotin and pantothenic acid although rarely thiamine. Whether legumes excrete specific substances which stimulate rhizobia more than other microorganisms is still an open question. Seeds of legumes produce diffusible antibiotics active against nodule bacteria. However, the antibiotic principle being water soluble could be got rid of by soaking seeds in water, followed by repeated washing. Certain non-legumes are also known to secrete substances from seed that are toxic to rhizobia.

The inhibitory or stimulatory effects of soil microorganisms such as bacteria, fungi and actinomycetes on *Rhizobium* are known. Culture filtrates of fungi isolated from soil and those isolated from washed nodules often inhibit the growth of rhizobia (Table 27). The failure of nodulation in certain parts of Western Australia has been attributed to the presence in soil of microorganisms antagonistic to rhizobia. Among the antibiotics, aureomycin, terramycin and ledermycin are most effective in inhibiting the growth of rhizobia the slow-growing ones less susceptible than fast-growing strains.

Table 27 The effect of soil fungi on rhizobia. The figures denote the number of isolates whose culture filtrates inhibited *Rhizobium* and those in parenthesis denote the total number of isolates tested (from Sethi and Subba Rao, 1968)

Fungi	*Rhizobium* spp.					
	A	B	C	D	E	F
Acrothecium (1)	0	0	0	0	0	0
Aspergillus* (10)	1	3	2	1	4	6
Cephalosporium (1)	1	0	0	1	0	1
Chaetomium (1)	0	1	0	0	0	0
Cladosporium (1)	0	1	0	0	0	0
Fusarium (9)	4	5	4	5	4	9
Mortierella (1)	0	1	0	0	0	1
Paecilomyces** (3)	0	0	0	0	1	1
Penicillium** (10)	0	0	0	1	1	3
Phoma** (1)	0	0	0	0	0	0
Rhizoctonia (1)	1	0	1	1	0	1
Rhizopus** (2)	0	0	0	0	0	0
Scolecobasidium (1)	0	0	0	1	18	1
Sordaria (1)	1	1	1	1	0	0
Thielavia* (3)	0	1	1	0	0	0
Trichoderma* (2)	0	0	0	0	0	0

Note: (1) The mycelial mats of some isolates at 10 μg/ml were inhibitory (*) or stimulatory (**) to *Rhizobium* spp.
(2) A: *R. trifolii*; B: *R.* sp. (cowpea group); C: *R. phaseoli*; D: *R. leguminosarum*; E: *R. japonicum*; F: *R. meliloti*.

Rhizobia may be eliminated from soil by bacteriophages. The alfalfa sickness resulting in poor crop of lucerne is attributed to the existence of bacteriophages of *Rhizobium*. However, contradictory evidences concerning bacteriophage activity in reducing the vigour of legumes and their nitrogen content, even under green house conditions, indicate the necessity for a greater understanding of the relationship between bacteriophages and legume-*Rhizobium* symbiosis.

Soil acidity could be one of the factors minimising the population of *Rhizobium* in soil. *R. meliloti* is very acid sensitive, while *R. japonicum* is able to tolerate pH as low as 3.5. Neutralization of soil with calcium hydroxide or calcium carbonate renders it favourable for rhizobial multiplication. The preponderance of ineffective strains of *R. trifolii* over effective ones has been attributed to soil acidity which is reversible by raising the pH.

Temperature affects growth as well as survival of *Rhizobium*. The alfalfa group of rhizobia are relatively more tolerant of higher temperatures than those of pea, clover and bean. The tropical cowpea rhizobia are more variable with regard to their susceptibility to higher temperatures.

Fungicides, herbicides and other plant protectants may prove toxic to rhizobia and reduce the inoculum in soil. The susceptibility of *Rhizobium* to these chemicals differs among different species. *R. lupini* and *R. trifolii* have been found to be most sensitive to 2,4D (2-4-dichlorophenoxy) butyric acid and MCPA (4-chloro-2-methyl phenoxyacetic acid) whereas *R. meliloti* is least sensitive. The minimum inhibitory concentration (MIC) of one and the same herbicide differs with different strains and species of *Rhizobium*. Stimulation of growth has also been reported for *R. trifolii* and *R. lupini* in the presence of 10 µg/ml 2,4-D. Pre-incubation of lotus rhizobia in a medium containing 2,4-D-B up to 100 µg/ml did not affect the capacity of the bacteria to effectively nodulate *Lotus corniculatus*.

Infection

Studies on clover and lucerne have revealed that the first reaction of the root system to the presence of rhizobia is the curling and deformation of root hairs. The formation of a typical "Shepherd's crook" on the root hair is generally considered as a necessary prelude to the formation of a thread-like structure visible inside the root hair called "infection thread" (Figs 56, 57). The curling effect has been attributed to indole acetic acid (IAA) produced in the root region by rhizobia. IAA is also known to be produced by seveal microorganisms other than rhizobia. In view of this, a specific root hair curling factor, believed to be a water soluble polysaccharide produced by rhizobia has been implicated in the typical curling of root hairs in which infection threads are formed.

There appears to be an intense interaction between the nucleus of root hair cell and the infection thread originating at the tip of the curled portion of the root hair. The nucleus guides the path of the infection thread in the

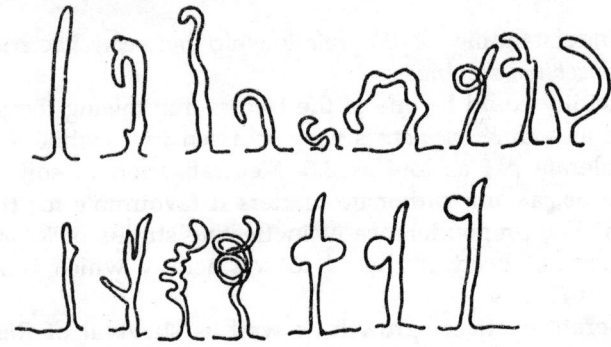

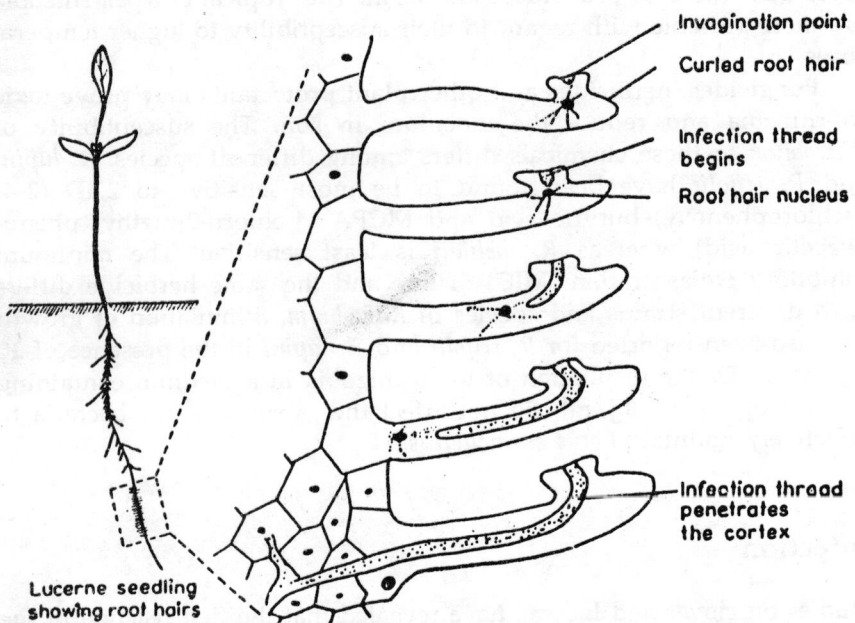

Fig. 56 Diagrams showing curling of root hairs due to the presence of *Rhizobium* in the root region and the formation of infection thread in root hairs of lucerne (*Medicago sativa*).

hair as is evident from the fact that in the event the nucleus gets disorganised, the growth of the thread also ceases. If the nucleus moves to the distal end of the hair and then wanders towards the proximal end near the cortex, the infection thread also traverses up and down before entering the cortex. Obviously, some kind of a message or impulse is transferred by the nucleus of the host to the contents of the infection thread.

Rhizobia are incapable of producing pectinase or cellulase in culture medium amended with pectin or cellulose. Recently, evidences have been

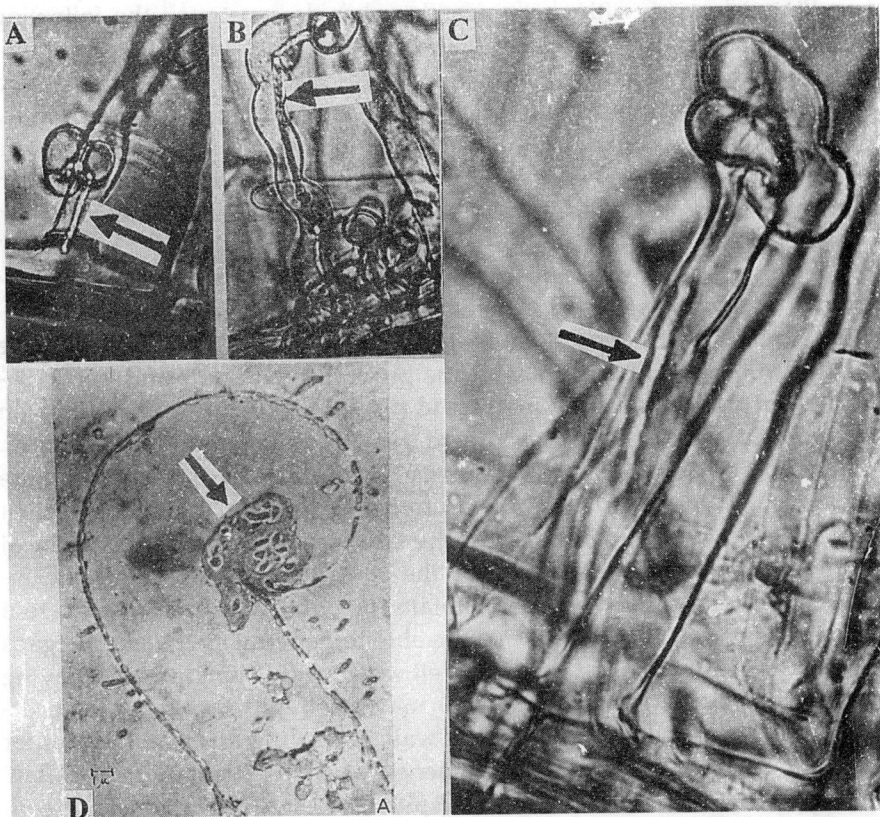

Fig. 57 *Rhizobium* entry through root hair: A—Light microscope view of a root hair of *Trifolium glomeratum* showing the typical 'shepherd's crook' in which the infection thread could be seen in the early stage of development; B—Light microscopic view of root hairs of *T. glomeratum*. Note the infection thread in one of them in an advanced stage of development; C—the same as in B where the thread is seen entering the cortex; D—Electron micrograph of longitudinal section through an infected root hair of *Trifolium parviflorum*. The section passes through a typical 'shepherd's crook' showing invagination. The infection thread appears as a rather thick lump which contains bacteria. Fragments of the thread can be seen inside the hairs. The arrows indicate the location of infection threads. (Courtesy Dr. K. Sahlman, Uppsala).

provided to show that legume roots liberate pectic enzymes, in response to the presence of homologous rhizobia in the root region. The validity of such induced production of pectinase by plant/homologous rhizobia associations has been discounted by other workers since results have not always been reproducible. However, detailed investigation on the role of enzymes in the mechanism of infection of legume roots by *Rhizobium* may throw more light on the problem.

Research work with clover seedlings has demonstrated the following important points with regard to root hair infection: (1) the infection of root hairs does not take place at random but takes place at a few well separated points, (2) these primary infection sites give rise to zones of infection by later infections of root hairs, (3) the number of infected root hairs increases exponentially until the first nodule is formed followed by a reduction in the number of infections thereafter, and (4) not all infections result in nodule formation.

Two modes of entry of rhizobia into the root hair have been suggested: (1) entry of small coccoid swarmers through the gaps in cellulose microfibrils, and (2) direct invagination of the root hair cell. The invagination hypothesis rests on the ground that auxins and pectic enzymes on the root surface interact to produce localized soft regions on the root hair facilitating the inward growth of the root hair cell wall. This inward growth has to take place against the hydrostatic pressure of the cell contents which appears to be an obvious argument against the hypothesis. A combination of the two hypotheses viz., the entry of swarmers through microfibrils of the root hair cell wall and the invagination of the cell wall may explain the entry of *Rhizobium* into root hairs. The coccoid swarmers may be able to penetrate between the gaps in the microfibrils of the root hair cell wall. At this juncture, overlaying of more primary or secondary wall material might result in the incorporation of the bacteria into the cell wall. Following this event, the outer wall of the root hair gets strengthened at a specified point to allow invagination of the inner layers. In spite of these explanations and conjectures, the *modus operandi* of infection is still not quite clear. Fine structure studies of infected root hairs showing the continuation of the wall of the infection thread with the cell wall of the root hair lend support to the invagination hypothesis.

As explained above, the infection thread appears to be of host origin. More than one infection thread is often seen in the same root hair. Upon its entry into the cortical cells of the root, the thread branches and then traverses intracellularly. It so happens that the contents of an infection thread (bacteria) are liberated into a tetraploid cell of the root cortex stimulating the cell to intense meristematic activity. A meristem is constituted in the differentiating cellular mass or the initial nodular tissue. Sooner or later, well differentiated areas are demarcated showing a diploid nodule cortex and a central tetraploid bacteroid zone, having vascular connections with the parent root system.

Not much work has been done on the mechanism of infection in plants where infection takes place directly through the cells of the root cortex. It is presumed that in such instances, rhizobia enter roots by mechanical injury caused to the roots (Fig. 58) or by some unexplained enzymatic process. However, based on the work done with clovers and lucerne, the physiological events leading to infection can be summarized as follows:

Normal root hair
↓
Exudation of organic substances by roots
↓
Accumulation of *Rhizobium* in the rhizosphere
↓
Tryptophane to indole acetic acid
↓
Root hair curling and deformation
↓
Involvement of lectins in rhizobial recognition
↓
Interaction of *nod* D gene product with plant root secreted flavonoids
↓
Incorporation of *Rhizobium* into cell wall and its participation in
'intussusception'
↓
Invagination of root hair cell to form an incipient infection thread
↓
Thread containing rod-shaped bacteria extending into root hair cell
guided by nucleus of the hair
↓
Entry of infection thread into root cortex and its branching.

Lectins

Several investigators are trying to unravel the factors behind the specific recognition between a *Rhizobium* strain and its homologous host. This recognition phenomenon has been implicated to plant lectins (proteins) which specifically bind to carbohydrate receptors on the rhizobial cell. The current evidence points out that the receptors for the specific binding of clover and soybean lectins to rhizobia are polysaccharides of the bacterial capsule. Most of the work on lectins have relied on seed lectin as source material but recent reports indicate that 'trifoliin', a specific plant protein has been isolated from the roots of clover and confirmed to exist on root surface by using immunofluorescence techniques. Some workers believe that the lectin hypothesis does not explain the enigmatic question of rhizobial invasion into leguminous roots. Although controversial, the lectin theory of *Rhizobium*-plant recognition did generate considerable investigation on specificity in *Rhizobium*-legume symbiosis.

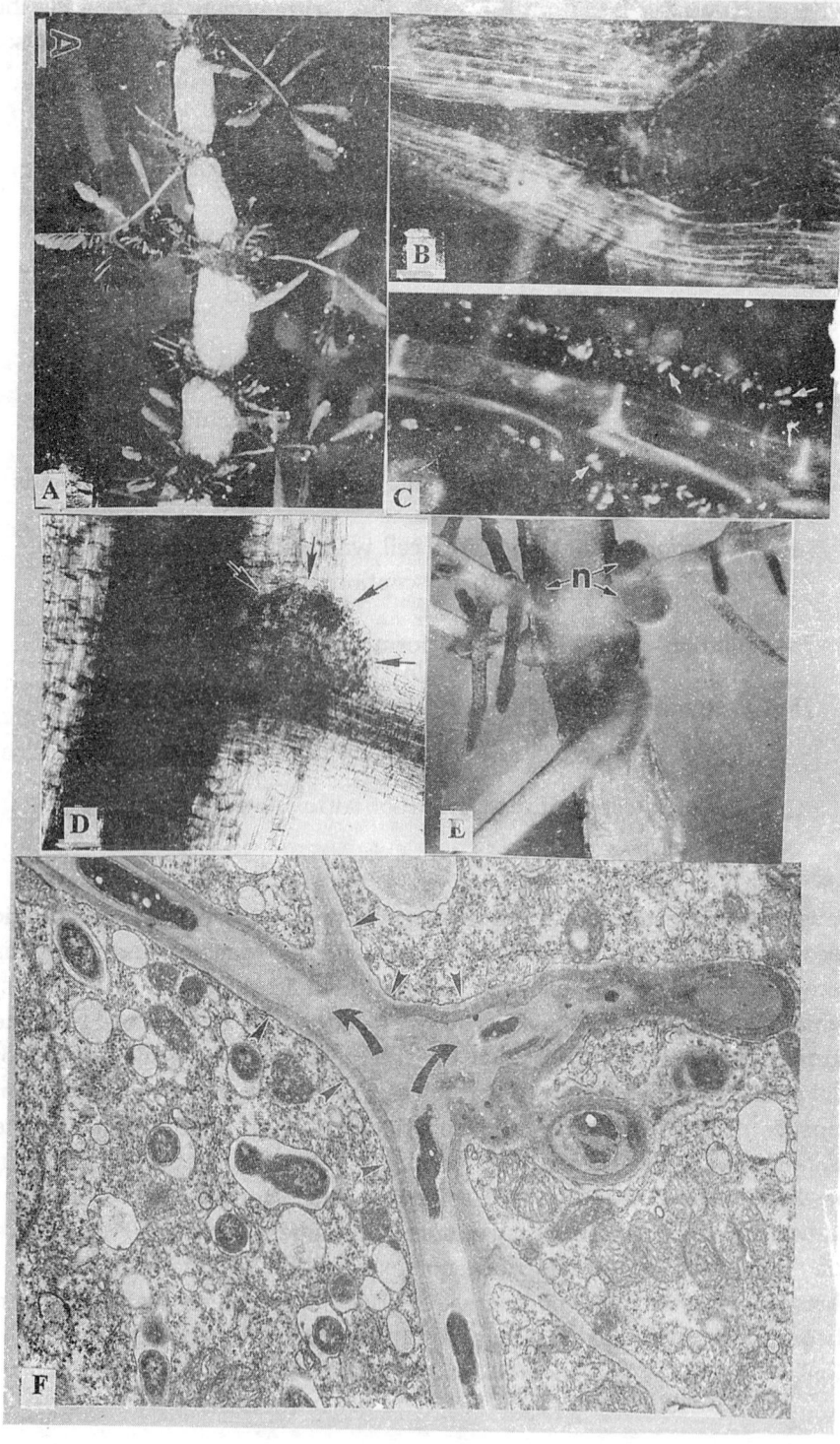

The synthesis of trifoliin A in white clover has been demonstrated by means of studies on the incorporation of labelled amino acids and tracing the lectin in the root exudate. *R. trifolii* grown in defined media produced polysaccharide receptors for trifoliin A which changed with the age of culture. These complementary protein-carbohydrate molecules (cross-reactive antigens) interact resulting in the recognition of the two symbionts with each other at the site of infection (Fig. 59). In soybean, it has been demonstrated that a fluorescein isothiocyanate (FITE) labelled soybean lectin preparation binds to cells of *Bradyrhizobium japonicum* which induced nodulation in plants. Exceptions to this rule, notably in tropical legumes nodulated by promiscuous cowpea miscellany rhizobia have been reported. For instance, concanavalin A from the tropical legume *Canavalia ensiformis* (Jack bean) can bind to many rhizobia unable to nodulate its roots.

There has been considerable amount of work on trifoliin. The surface of infective encapsulated *R. trifolii* contains an immunochemically unique polysaccharide that is antigenetically cross-reactive with a component on clover epidermal cells. Only those rhizobial strains which infect clover have this cross-reactive antigen (CRA) and clover roots preferentially absorb infective *R. trifolii* and its corresponding CRA. These sites on root hairs known as receptor sites accumulate at root hair tips and decrease towards the base of the root hair. On the contrary CRA is uniformly distributed along the root hair. Trifoliin specifically agglutinates *R. trifolii* and 2-deoxyglucose inhibits such agglutination. Similarly, 2-deoxyglucose inhibits the specific attachment mechanism of *R. trifolii* to clover root hairs. The nature of carbohydrate receptors for the lectin on *Rhizobium* cell has been the subject of several investigations. The structural differences in the lipopolysaccharide (LPS) and expolysaccharides (EPS) of *Rhizobium* are high and a search for relevant residues that function in recognition has become a major issue.

Within 4 hours, *R. trifolii* establishes a pattern of attachment on root hairs by clumping at the tip only by polar attachments along the site which can be inhibited by 2-deoxy-D-glucose. This initial reversible reaction becomes irreversible later with homologous *Rhizobium* and host combina-

Fig. 58 *Rhizobium* entry through cracks on the root surface in *Nepturia natans*, an aquatic legume: A—an adult plant in its natural habitat, floating in a pond of water; B—laser scanning confocal optisection of the open wound in the primary root where a lateral root has emerged; C—higher magnification of B showing fluorescent bacteria (white arrows) deep within the open cavity of the root wound; D—round nodule primordium (arrows) developing at the base of the root cortex; E—primary root showing nodules (n) attached to the base of lateral roots; F—transmission electron micrograph of intercellular infection and development of tubular intracellular infection threads which branches within the nodule (arrows) having bacteria in a single row. (From N.S. Subbarao *et al.*, 1995).

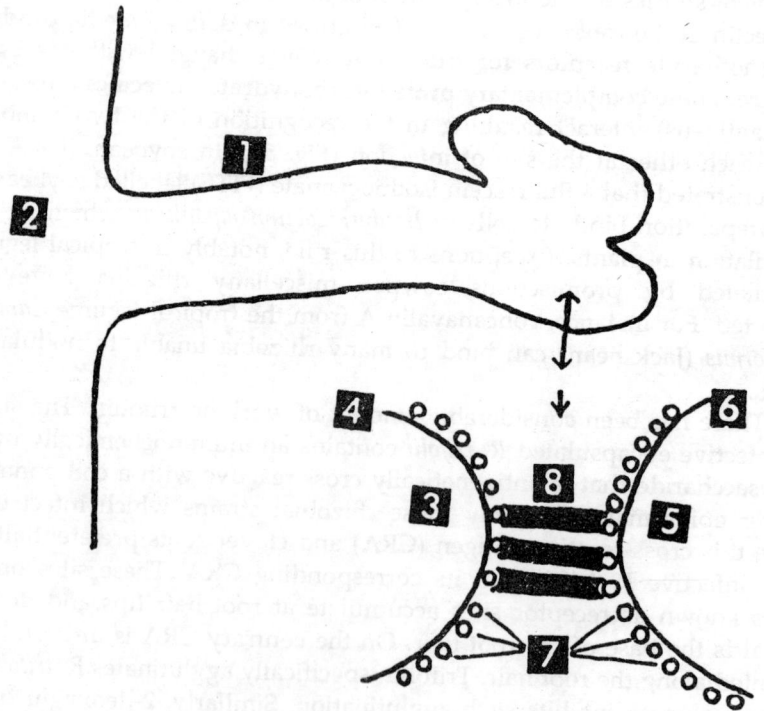

Fig. 59 Lectin (Trifoliin) mediated binding of *Rhizobium trifolii* cells to the wall of root hair of clover seedlings—1) enlarged diagram of a root hair of clover seedling; 2) root cortex; 3) root hair cell interior; 4) root hair cell wall; 5) interior of *R. trifolii* cell; 6) *Rhizobium* cell wall; 7) cross reactive antigens, one of which is carbohydrate receptor on bacterial cell wall; 8) Clover Lectin, Trifoliin. (Not to scale. Modified from Dazzo and Hubbell, 1975).

tions. These phases of attachment of rhizobia to root hairs appears to be controlled by *nod* genes.

Many genes are involved in the synthesis of *Rhizobium* polysaccharide (gum) which surrounds a colony of cells. Mutants defective in polysaccharide synthesis are often characterized by poor infectivity and nodule formation. The precise role of various polysaccharides such as capsular polysaccharides (CPS), exopolysaccharides (EPS) and lipopolysaccharides (LPS) and cyclic B-1, 2-glucans produced by rhizobia are not very clear. It has been demonstrated that addition of EPS from the parent strain to plants together with some acidic exopolysaccharide lacking (EPS⁻) mutants could restore the ability to induce nitrogen-fixing nodules which may point out the direct involvement of polysaccharides in the initiation of symbiosis.

Structure of the Nodule

The core of a mature nodule constitutes the 'bacteroid zone' which is surrounded by several layers of cortical cells. The relative volume of bacteroid tissue (16 to 50% of the dry weight of the nodules) is much greater in effective nodules than in ineffective ones. The volume of bacteroid tissue in effective nodules has a direct positive relationship with the amount of nitrogen fixed. Ineffective nodules produced by ineffective strains are generally small and contain poorly developed bacteroid tissue associated with structural abnormalities. In all ineffective associations, it has been shown that starch accumulates in the uninfected cell and dextran in the infected cell with glycogen in the bacteroid. Effective nodules are generally large and pink (due to leghaemoglobin) with well developed and organized bacteroid tissue.

A fully-developed bacteroid has no flagella and is surrounded by three unit membranes. There exists an intracytoplasmic membrane system in the bacteroids of nodules in subterranean clover. The nuclear region of bacteroids appears fragmented and is associated with granular cytoplasm. Bacteroids can be produced *in vitro* on a medium containing 3.5% of yeast extract. Caffeine and several other alkaloids also encourage bacteroid production on artificial media. Depending on the legume, each bacteroid or groups of bacteroids are surrounded by membrane envelopes whose identity has been variously interpreted, possibly because of the different techniques used in fine structure studies. The three hypotheses concerning them are: (1) they are formed *de novo* after the release of the bacteria from the infection thread, (2) they are extensions of the endoplasmic reticulum of the host cells, and (3) they are derived from plasmalemma by a process of phagocytosis. The number of bacteroids enclosed in membrane envelopes appears to vary from one to many depending on the species of legume.

The membrane envelope surrounding the bacteroids is also known as the peribacteroid membrane. The earlier concept that the peribacteroid membrane is only a slightly modified plasma membrane has to be viewed differently in the light of later findings. The multiplication of bacteroids and the formation of the peribacteroid membrane may not be synchronized cell events which result in the variability in the number of bacteroids enclosed by the membrane. The microsymbiont (*Rhizobium*) strain determines the particle density and the protein and fatty acid composition of the membrane. The peribacteroid membrane contains nodulins which may play special roles in the two-way transport of metabolites between the symbionts (Figs. 60, 61).

Function of the Nodule

Present evidences point out the fact that bacteroids are the sites of nitrogen fixation, although earlier, membrane envelopes surrounding a group of

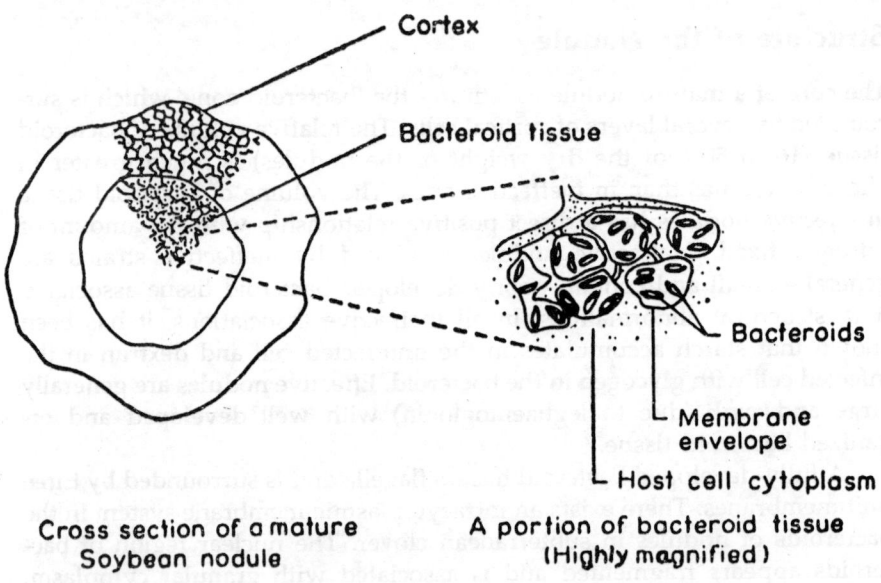

Cortex

Bacteroid tissue

Bacteroids

Membrane
envelope

Host cell cytoplasm

**Cross section of a mature
Soybean nodule**

**A portion of bacteroid tissue
(Highly magnified)**

Fig. 60 Diagrammatic sketch of a transverse section of a soybean nodule.

bacteroids were believed to be the sites of this reaction. Stable isotope ^{15}N has been used to trace the path of N_2 in nitrogen fixation. Earlier studies were done with intact or sliced nodules without adequate precautions to keep the experimental system under anaerobic conditions. In later studies, however, nodules were exposed to ^{15}N and crushed in an atmosphere of argon so as to prepare a brei under anaerobic conditions which retained the capacity to fix nitrogen. The brei was centrifuged, the bacteroids collected in the form of a pellet, washed and assayed for ^{15}N. The bulk of the activity was recorded in bacteroids and not in soluble fractions indicating that bacteroids are the primary sites of nitrogen fixation. The acetylene-ethylene reduction technique designed to determine the nitrogenase activity of different fractions of nodules has been of great help in these investigations.

Further confirmation on the role of bacteroids in N_2 fixation has come from studies on cell-free extracts of bacteroids which contain crude nitrogenase enzyme with sufficient activity to fix up to a maximum of 9 to 13 μ moles of N_2/min/mg protein. Using special purification procedures, nitrogenase has been purified and is found to consist of two protein fractions—Mo-Fe protein and Fe protein. The enzyme has been successfully extracted from representatives of major classes of nitrogen-fixing microorganisms and systems except those from nodulating non-leguminous plants. A more detailed account of nitrogenase appears in the chapter on non-symbiotic nitrogen-fixing bacteria. Nitrogenase from bacteroids has been prepared and assayed for its property of acetylene

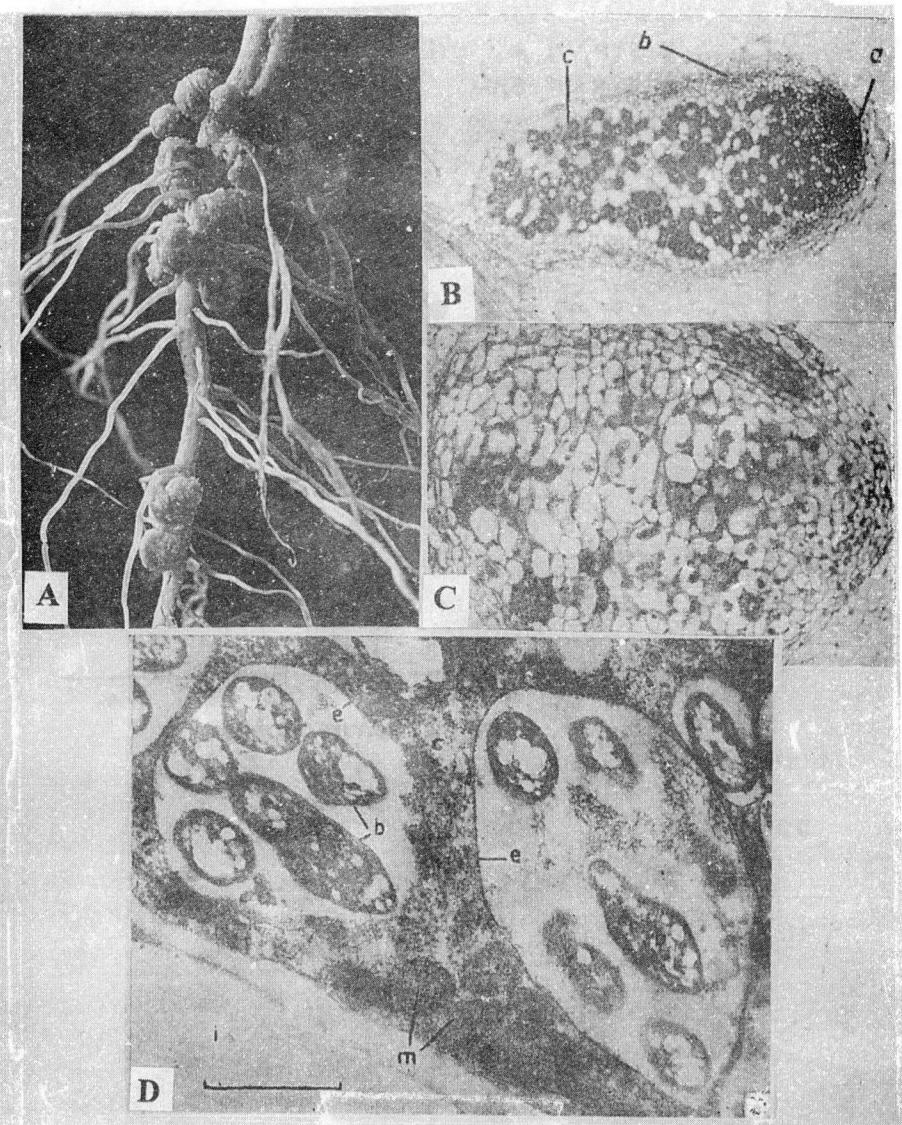

Fig. 61 Form and structure of root nodules in legumes: A—a well formed bunch of effective root nodules of soybean; B—section of an effective nodule in a 3 week old *Trifolium pratense* showing meristem (a), vascular trace (b) and bacteroid filled cells (d); C—a section of an ineffective *T. pratense* nodule showing the absence of bacteroid tissue; D—an electron micrograph of a thin section of a soybean nodule (*Glycine max*) showing bacteriods (b) in groups within membrane envelopes (e) in the host cytoplasm (c). Host mitochondria (m) are in the periphery of the host cell and (i) indicates an intercellular space. The scale shows 1 μ. (Courtesy: F.J. Bergersen,. Australia).

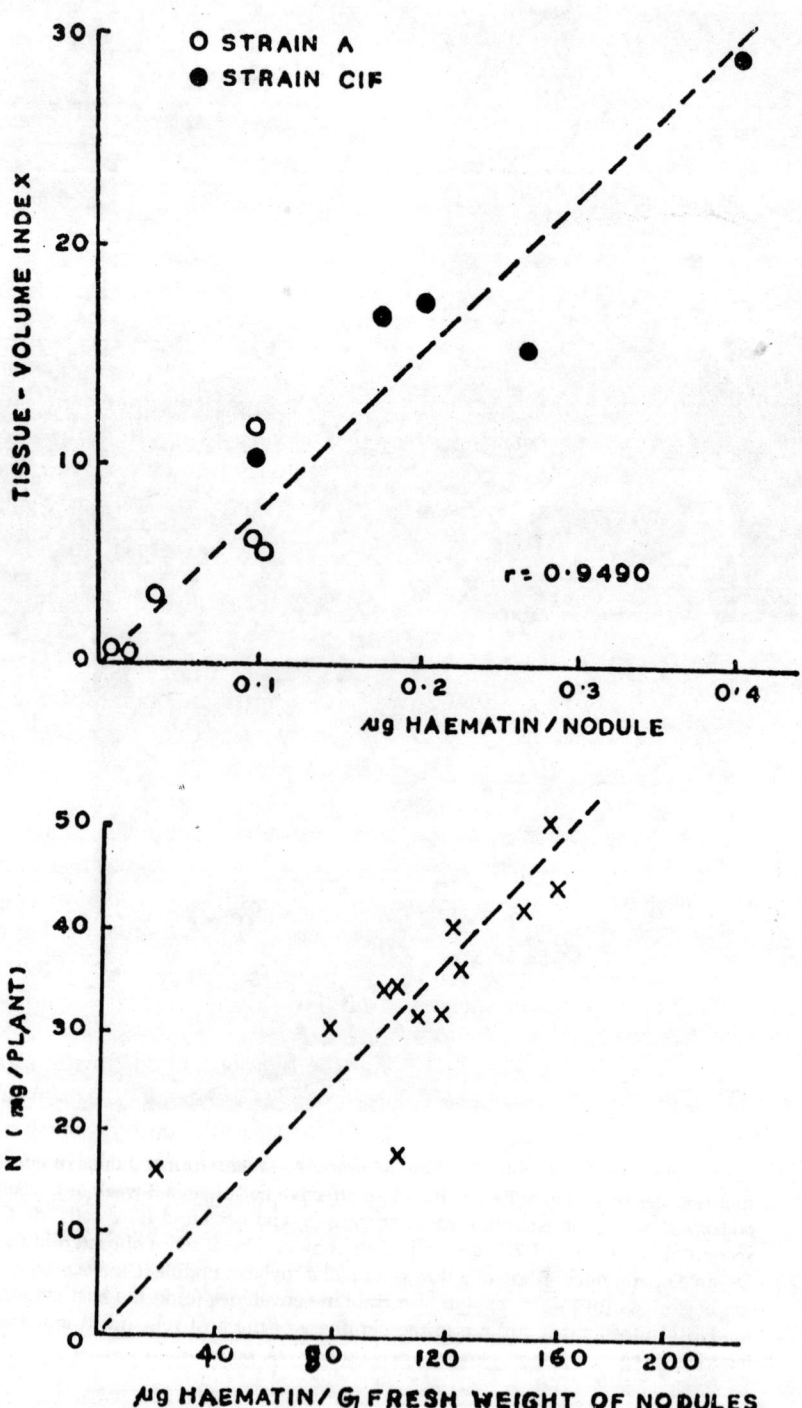

Fig. 62 *Top*: The correlation between haematin/nodule and effective central tissue volume index (bd²n) in red clover; *Bottom*: The correlation between haematin/g fresh weight nodules and nitrogen fixed by nodulated forsdag peas (from Bergersen, 1961).

Table 28 Relationship between the reduction of acetylene and N_2 by extracts and fractions of nodule bacteroids (Postgate, 1971)

Type of extract	Rate of reduction		Ratio of rate of reduction		
	C_2H_2	N_2	C_2H_2	:	N_2
	n moles per mg protein per min.				+
Crude	35.2	11.8	2.98	:	1
25–55% PPG* ppt.	76.1	27.4	2.78	:	1
Fraction 1	0.0	0.0	–		
Fraction 2	74.4	19.3	3.58	:	1
Fractions 1 and 2	607.4	192.6	3.15	:	1

*Polypropylene glycol ($P - 400$).

reduction and enzyme activity. The results of one such experiment shown in Table 28 indicate no activity in fraction 1, very low activity in fraction 2 and a striking increase in activity when fractions 1 and 2 were combined.

A red pigment akin to haemoglobin of blood is found in nodules between bacteroids and the membrane envelopes surrounding them. Leghaemoglobin, the prefix 'leg' indicating its presence in legume root nodules, is a haemoprotein having a haeme moiety attached to a peptide chain which represents the globin part of the molecule. The molecular weight of leghaemoglobin is of the order 16,000 to 17,000 only, while that of blood haemoglobin is of the order of 66,000, because leghaemoglobin has one peptide chain linked with one haeme moiety, whereas blood haemoglobin has four peptide chains, each linked with one haeme moiety. The pigment has been crystallized into two major extractable components with differing amino acid composition and molecular weights—the electrophoretically faster one containing 0.32% iron and the slower one with 0.29% iron. Minor components of leghaemoglobin have also been detected depending on the legume species. It is difficult to say whether these components represent native or degraded forms of a single fraction during the extraction procedures. The amount of leghaemoglobin in nodules has a direct relationship with the amount of nitrogen fixed by legumes (Fig. 62).

It has been variously postulated that the pigment could function: (a) as the site of nitrogen absorption and reduction, (b) as the specific electron carrier in nitrogen fixation, (c) as a regulator of oxygen supply, and (d) as carrier of oxygen. Current evidences show that leghaemoglobin plays no active role in symbiotic nitrogen fixation but functions as a biological valve in regulating the supply of oxygen to bacteroids at optimum levels conducive for proper functioning of the nitrogen-fixing system. A haemoglobin has also been characterized in cultured cells of *R. japonicum*, which however, is unrelated to leghaemoglobin of root nodules both in composition and function.

Many aspects of the biochemistry of symbiotic nitrogen fixation are not clearly understood. Nevertheless, schemes summarizing the overall

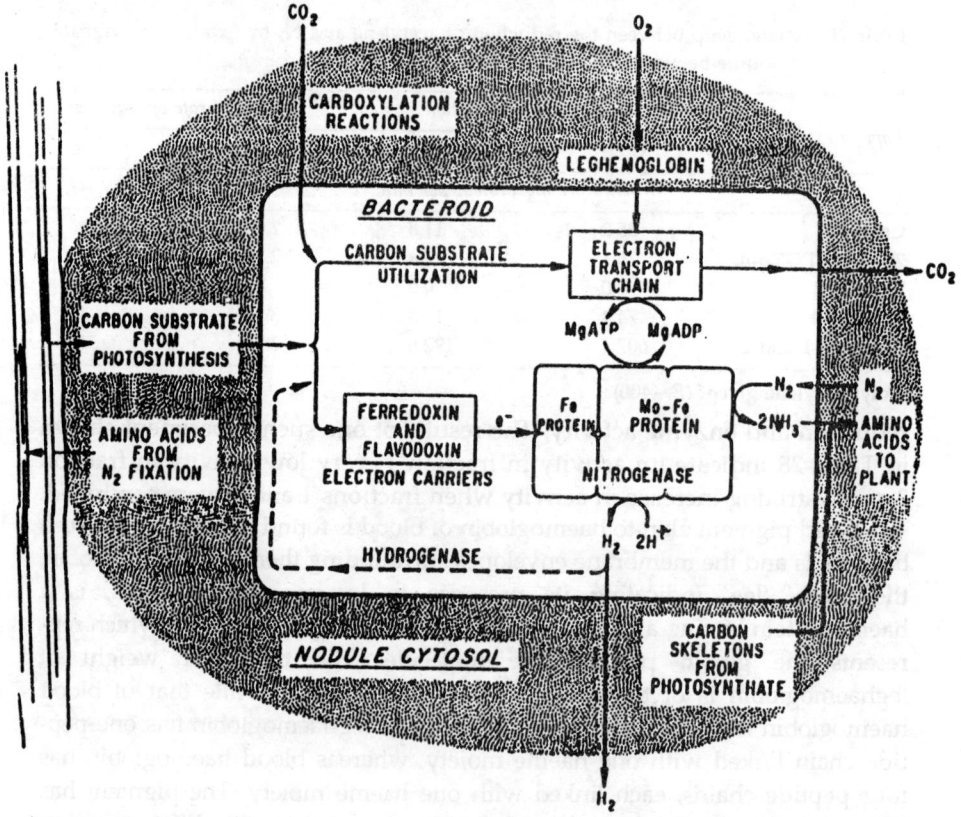

Fig. 63 A diagram illustrating the interrelationship of reactions involved in the nitrogen
fixing process in a legume nodule. (After Evans *et al.*, 1978. In *Limitations and Potentials for Biological Nitrogen Fixation in the Tropics* pp. 209–222, Eds. J. Dobereiner, *et
al.*, Plenum Press, New York).

reactions representing the results obtained with nodule homogenates and
cell-free preparations of bacteroids are shown in Figs. 63, 64. Photosynthate
from the legume provides the substrate required by nodule bacteroids for
the generation of ATP and reductant needed for nitrogen fixation. A ferredoxin similar to that of *Azotobacter* ferredoxin has been found in soybean
bacteroids which may function as an electron carrier. Since bacteroids
require oxygen for nitrogen fixation, the oxygen bound to leghaemoglobin
is transported at optimum levels without interfering with the oxygen sensitivity of the nitrogenase system.

The first stable intermediate in nitrogen fixation is ammonia which
gets incorporated into glutamic acid, glutamine, aspartic acid and alanine.
The nature of intermediates between nitrogen and ammonia remains un-
clear although compounds like hydrazine, hydroxylamine, diimide, and
carbamyl phosphate have been proposed as possible intermediates.
Hydrazine is extremely toxic to living cells; besides this fact, experiments

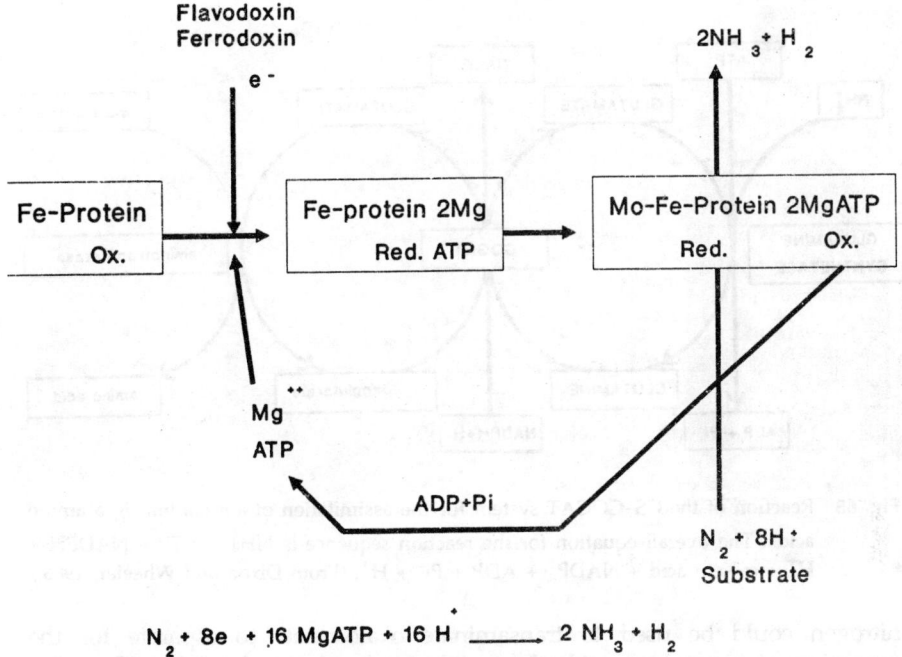

Fig. 54 A schematic diagram of general principles involved in the reduction of N2 to NH3 by the nitrogenase complex.

with ^{15}N do not indicate the likelihood of this intermediate occurring in the conversion of nitrogen to ammonia. Conclusive evidences suggesting possible sequences in the reductive pathway between nitrogen and ammonia with regard to the other intermediates are lacking.

Notwithstanding the speculations about intermediates in the reduction of nitrogen to ammonia, the binding of nitrogen by nitrogenase appears to be the crucial factor in nitrogen fixation. It has been postulated that iron (Fe) is involved in the binding of nitrogenase to nitrogen while molybdenum (Mo) is responsible for the decrease in the strength of nitrogen bonds to an optimum extent so as to facilitate reduction. A second consideration of prime importance is the mechanism whereby oxygen is excluded from the site of nitrogen fixation to protect the oxygen sensitive nitrogenase and yet maintained at optimal levels for ATP generation in bacteroids. As mentioned earlier, this function is accomplished by the leghaemoglobin in nodules which permits oxygen to enter the tissue at rates sufficient to keep the nitrogenfixation reactions at a steady level and at the same time, free oxygen levels are kept very low by the respiratory activity of the bacteroids. The ammonia fixed in cells of nodules is converted into glutamine by glutamine synthetase (GS) and then into glutamate by glutamine-oxoglutarate amidotransferase (GOGAT). A part of glutamate

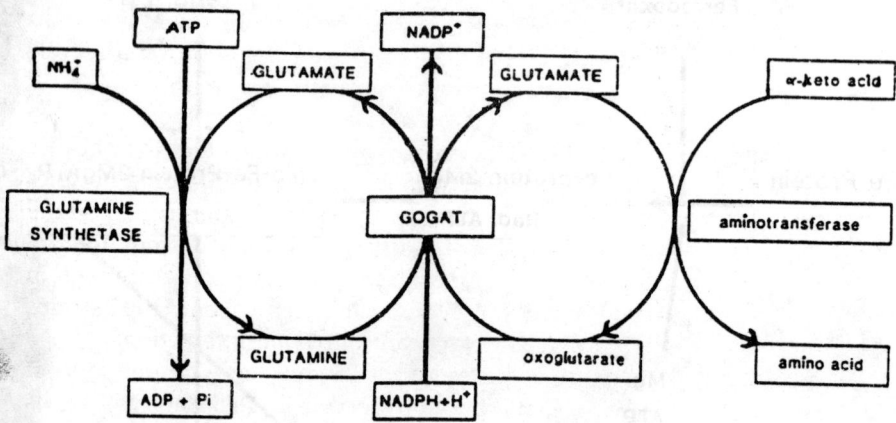

Fig. 65 Reaction of the GS–GOGAT system for the assimilation of ammonium into amino acids. The overall equation for the reaction sequence is NH_4^+ + ATP + NADPH + H^+ + α-Keto acid + $NADP^+$ + ADP + Pi + H^+. (From Dixon and Wheeler, 1986.)

nitrogen could be used to transaminate oxaloacetate to asparate for the formation of asparagine with the amido group from glutamine (Fig. 65).

Secondary reactions in the nodules lead to the formation of either amides (asparagine, glutamine) or the ureides (allantoin and allantoic acid). Based on these solutes traceable in the xylem sap, nodulating N_2 fixing plants can be broadly divided into amide exporting or ureide exporting species. These generalizations hold good with exceptions—*Vigna radiata* exports largely arginine whereas in some non-legumes (*Alnus* and *Myrica*) citrulline is the product in the xylem sap. Examples of some species which export ureide as the major solute are *Arachis hypogea*, *Cajanus cajan*, *Glycine max* and *Cyamopsis tetragonoloba*. On the other hand, *Cicer arietinum*, *Lens culinaris*, *Pisum arvense* and *Vicia sativa* can be cited as examples in which ureides are the minor solutes of the xylem sap. However, there are plants like *Lathyrus sativus*, *Trifolium repens* and *Vicia faba* where no ureides have been detected which renders the demarcation of legumes based on the nature of fixed nitrogen they export from the root to the shoot as arbitrary. High levels of ureides in xylem sap are usually associated with effective nodulation and high rates of fixation.

Nitrogenase-Hydrogenase Interrelationship

Nitrogenases from all known sources and systems catalyze hydrogen production to varying degrees during nitrogen-fixation. One of the important factors influencing the magnitude of hydrogen evolution in N_2 fixing systems is the operation of hydrogen recycling enzyme known as uptake hydrogenase. It is known that most of the nitrogen-fixing bacteria recycle the hydrogen produced during nitrogen fixation but many strains of rhizobia lack an effective uptake hydrogenase. In nature, there are both hydrogenase positive (Hup$^+$) and negative (Hup$^-$) strains and the nodules formed by these strains differ in the rate of H_2 evolution. Some nodules evolve more hydrogen whereas others evolve less. Under experimental conditions inoculants made by Hup$^+$ strains increase yield from 16–32% over Hup$^-$ strains accompanied by increased total nitrogen accretions in shoots and grains varying from 8 to 49%. Such benefits from Hup$^+$ strains have been clearly demonstrated in soybean-*Rhizobium* interaction.

The nitrogenase reaction requires about 28 mol. of ATP per mol. of N_2 utilized. It has been calculated that recycling 1.28 mol. of H_2 is equivalent to 2.56 mol. of ATP which means that about 9.1% of the 28 mol. of ATP could be recovered for utilization in reducing N_2 to NH_3 by Hup$^+$ strains which otherwise would have been wasted in H_2 evolution by Hup$^-$ strains. These assumptions are open to further investigations and precise elucidation.

Stem Nodulating Legumes

There are two genera of legumes which are known to have nodules on stem—*Aeschynomene* and *Sesbania* (Fig. 66). Of the 150–250 species of *Aeschynomene*, nodules on stem have been reported in *A. indica*, *A. aspera*, *A. elaphroxylon*, *A. villosa*, *A. evenia*, *A. paniculata* and *A. afaspera*. These reports have come from India, Mali, Ghana, Venezuela, Brazil, U.S.A., Japan, Zimbabwe, Java, South Africa, Zambia, South America, Puerto Rico and Argentina. In India, *A. indica* and *A. aspera* occur widely in waterlogged situations. The genus *Sesbania* has 170 species which are annual as well as perennial and grow in warmer regions of both the hemispheres. *S. rostrata* is an annual plant which bears nodules on stems as well as on roots. It was first described a few years ago by a group of scientists in Senegal as a plant equivalent (or higher) to soybean in nitrogen fixation to the extent of 200 kg N_2/ha in 50 days by virtue of the stem nodules. Subsequently, the plant has been tested at the International Rice Research Institute, Philippines and other places such as India and Thailand and found to establish well.

What merits the attention of rice agronomists is the green manure potential of this plant. From published reports, the increase in the yield of rice by the application *S. rostrata* green manure could be as high as 3.7 t/ha. This increase is more than that obtainable by 60 kg N/ha of chemical fertilizer application which can only result in an increase in rice

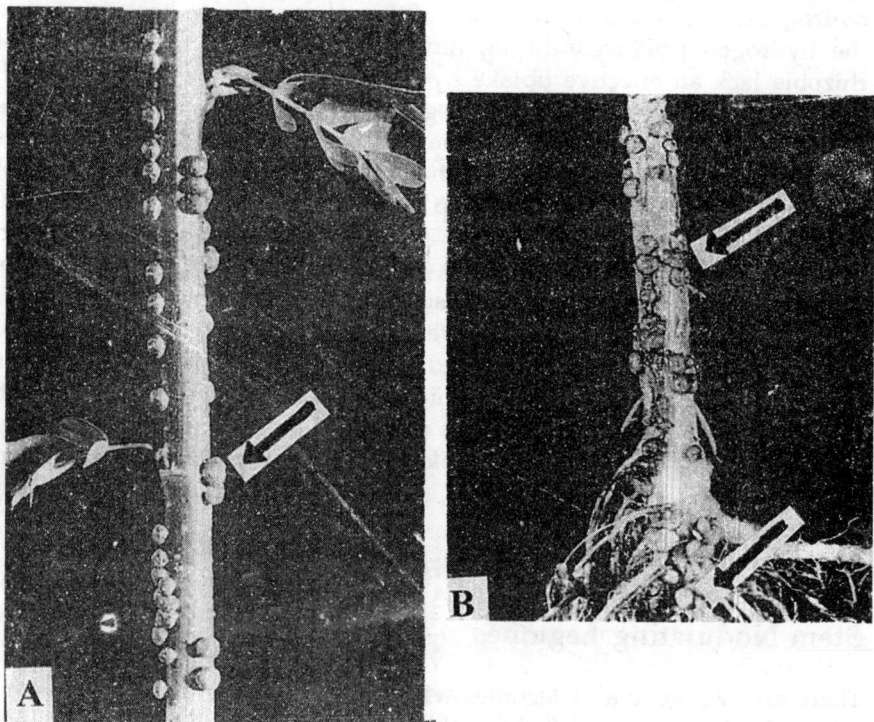

Fig. 66 Stem nodules: A—on *Sesbania rostrata*, the arrow pointing to a pair of nodules; B—on *Aeschynomene indica*, the arrow at the top pointing to a group of nodules on stem and the arrow at the bottom pointing to a group of nodules on roots.

yield to the extent of 1.7 t/ha. In Coimbatore, India, *S. rostrata* application alone yielded 2350 kg/ha rice in comparison with the application of *Azolla filiculoides* which resulted in the production of 2030 kg/ha rice.

There are other green manure and green leaf manure crops which have been used in rice cultivation by several generation of rice farmers in India and other countries. In India, species of *Crotalaria, Indigofera, Lathyrus, Pongamia, Tephrosia, Trifolium* and *Sesbania* have been used for a long time. The high N content of these crops and their ability to get easily degraded in wet rice soils of Asia are the main criteria for adopting a given plant species as a suitable green manure candidate of choice. *Astragalus sinicus*, a species widely used in China, has 108 kg N/ha whereas *S. rostrata* contains 267 kg/ha. The commonly used Indian green manure species, *Sesbania aculeata* has around 96–122 kg N/ha. In Senegal, using ^{15}N labelled fertilizers, a comparison was made between stem and root nodulated

S. rostrata and root nodulated *S. sesban* to see if the additional facility provided by stem nodulating habit conferred greater nitrogen-fixing capacity. It was calculated that *S. rostrata* fixed 83–109 kg N_2 ha^{-1} whereas *S. sesban* fixed only 7–18 kg N_2 ha^{-1} under similar conditions of growth reflecting the superiority of the stem nodulating legume.

The nitrogen content of rice grain and the organic matter content of the soil have been found to increase by green manuring of fallow soil prior to transplanting rice. Other attributes such as the availability of zinc, the hydraulic conductivity, water holding capacity and aggregate stability of soil also appear to be improved by green manuring. In addition, the practice also prevents loss of soil-nitrogen by decreasing denitrification processes. These benefits are offset by certain limiations: green manuring is labour intensive and sometimes necessitates the cultivator to set apart land for growing green manure species which otherwise could have produced a cash crop. In other words, the economical considerations of the green manure practice are not clear especially in situations where fertilizer nitrogen can be afforded.

Fundamental research on stem nodulating green manure species has been directed mainly to an understanding of the rhizobiology and genetics of the microsymbiont producing stem nodules. The species of *Rhizobium* inducing both stem and root nodules is designated as *Azorhizobium caulinodans*. The bacterium is capable of high levels of nitrogen fixation in the free living state up to 1500 nM C_2H_4/mg protein/hour at an unusually high oxygen tension of 3 per cent. Stem nodulation is extensive (50 g/plant) and acetylene reduction activity corresponds to 550–600 μm C_2H_2 reduced per hour. Besides these attributes nodulation on stem in *S. rostrata* and nitrogen fixation *in situ* are relatively insensitive to levels of combined nitrogen in the soil. The regulation of *nif* gene expression in *A. caulinodans* is complex and shares regulatory features common with those understood for *Rhizobium meliloti* and *Bradyrhizobium japonicum*.

Factors Affecting Nodulation

Temperature and Light

Critical investigations on the effect of root temperature on nodulation and infection processes carried out with temperate species of legumes such as clovers (*Trifolium* spp.) grown on agar slopes in environment controlled growth cabinets have shown that temperatures below 10°C retard root hair infection by *Rhizobium*, whereas temperatures higher than 24°C promote the same. The most congenial range of temperature for bacteroid tissue formation in nodules appears to be 20–30°C, although nitrogen fixation remains unchanged in the range of 12–32°C. These results are subject to inherent variations between strains of *Rhizobium* and host cultivars.

Among the tropical legumes, effects of day temperature on root nodulation have been studied in soybean (*Glycine max*) and Bengal gram (*Cicer arietinum*) in pot trials using selected strains of both the symbionts in environment controlled and illuminated growth cabinets. One of the bacterial strains was most effective at 33°C on soybean, while others showed no difference in effectiveness at 21°C. In Bengal gram, notwithstanding the variations noticed in bacterial strain effects with regard to temperature, it was noteworthy that none of the bacterial strains produced nodules at root temperatures beyond 32°C. Nitrogenase activity was higher in the temperature range of 24 to 33°C with little activity beyond the upper limit. These results appear significant for a crop such as Bengal gram which is grown in India mostly under rainfed conditions where day temperatures do not exceed 32°C.

Studies on the effects of combinations of photoperiods and environmental temperature on nodulation of cluster clover (*Trifolium glomeratum*) have revealed that initial nodulation was delayed as the day length was increased and that photoperiods influenced the formation, size and the number of nodules on the root system (Table 29). Apart from products of photosynthesis, which obviously influence nodulation, experiments with soybean seedlings have shown that cotyledons supply some essential factor for nodulation in seedlings. Removal of cotyledons from clover seedlings soon after germination has been shown to delay, reduce or prevent infection of root hairs and subsequent nodulation.

Table 29   The influence of different combinations of photoperiods and temperatures on symbiosis of *T. glomeratum* with *R. trifolii* in test-tubes (from Subba Rao, 1971)

Treatment	Age at which first nodule appeared	At 65 days (Av. of 10 tubes)				
		Number of leaves	Number of nodules	Height (cm)	Weight (mg)	Flowering at
A. 8 hr L 15°C	8.5	7.7	6.3	2.2	11	No flowering
B. 8 hr L 20°C	8.0	7.0	3.8	3.2	9	"
C. 16 hr L 15°C	8.5	6.8	3.7	3.4	14	52 days
D. 16 hr L 20°C	8.9	6.1	2.8	2.7	11	43 days
E. 16 hr L 15–20°C	10.6	6.0	3.3	4.3	15	55 days
F. 24 hr L 15–20°C	10.0	5.0	3.0	2.0	11	38 days and seed setting 45 days
S. Em.	0.228	0.11	0.44	0.21	0.78	
C.D. at 1% level	0.64	0.36	1.24	0.58	2.18	

Combined Nitrogen

Leguminous plants are unique among flowering plants by virtue of the fact that they can obtain their nitrogen requirements either through inorganic nitrogenous fertilizers and/or through symbiosis in root nodules of nodulating species. Both beneficial and detrimental effects of mineral nitrogen on growth and yield of nodulating legumes have been observed. In spite of seemingly divergent results obtained under field conditions, work carried out under controlled environmental conditions has shown that mineral nitrogen beyond certain levels interfere with infection of root hairs, number of nodules, structure of nodules and the amount of nitrogen fixed. In ammonium nitrogen (HN_4NO_3)-treated clover plants, the bacteroids in the nodules become enlarged and synthesis of membrane envelopes gets affected. Spraying of plants with urea prevents nodulation whereas spraying with sucrose enhances nodulation and N_2 fixation. In experiments with excised roots, nitrate provided through the base (cut end) of the root, does not inhibit nodulation, whereas inhibition occurs when nitrate is supplied in the external medium in which roots are growing. Such an inhibition is offset by addition of sucrose, mannitol or L-arabinose to the external medium.

The mechanism behind inhibition of nodulation by nitrate is entirely speculative. It is not unlikely that there exists a critical balance between the available carbon and nitrogen in the root system which is offset by extraneous source of nitrate. It has also been hypothesized that nitrate is converted to nitrite in the root environment mediated by *Rhizobium* and the nitrite so formed destroys the auxin, indole acetic acid (IAA). The frequency of nodulation of lucerne is known to increase at an optimum level of IAA (10^{-8} M) in the root medium whereas KNO_3 at the rate of 140 N (ppm), has just the opposite effect and also decreases the total number of nodules produced. This nitrate-induced inhibition, however, is reversed by additions of 10^{-8}M IAA into the root region which also partially reverses the nitrate induced inhibition of curling of root hairs and formation of infection threads in them. Such interlinked facts serve to demonstrate the interesting relationship among products of photosynthesis, mineral sources of nitrogen, soil reaction (pH) and growth promoting substances which are known to operate in the root region during different stages of symbiosis in nodulating legumes.

Hydrogen Ion Concentration

Leguminous plants grow less luxuriantly in acid media than in neutral or slightly alkaline conditions which could indirectly be due to lowered colonization of *Rhizobium* in soil and in rhizosphere leading to inadequate nodulation. The poor colonization of *Rhizobium* in rhizosphere due to acidity of the root medium could be overcome by heavy inoculation with

:

the bacterium as has been demonstrated in lucerne grown in solution cultures. Such a stimulation is purely mechanical and does not involve any multiplication of the organism in the rhizosphere or encouragement of infection of root hairs by rhizobia and subsequent nodulation.

The number and size of nodules may be affected by the reaction of the substrate on which legumes grow. Acid conditions of soil result in deficiencies of calcium, magnesium and potassium. Often, soil acidity may lead to reduced uptake of molybdenum which can be corrected by liming. Soil amendments with ammonium nitrate or calcium carbonate counteract the limiting effects of low pH and thus increase the yield of legumes.

Mineral Nutrition

Calcium stimulates nodulation when present as chloride or sulphate, although high levels of sulphate in the from of magnesium salt do not have the same effect as demonstrated from studies on subterranean clover. The beneficial effect of calcium chloride or sulphate may be attributed to the calcium ion. The effect of calcium ion may be either on the *Rhizobium* or on the legume partner of the symbiosis. In *R. trifolii*, increasing the calcium level has no effect on its growth at any pH. Similarly, various combinations of calcium and hydrogen ions produce no effect on the legume partner as well. Thus, the depressing effect on nodulation caused by poor calcium status and acidity of a substrate remains to be explained. The answer probably lies in understanding the effect of these factors on the pre- and post-infection stages of *Rhizobium* in legume roots.

It has been established that molybdenum is indispensable for symbiotic nitrogen fixation and stimulates the nitrogen-fixing activity of the nodular tissue. Nitrogen fixation by nodulated lucerne and white clover in agar cultures is enhanced by additions of molybdenum. This has also been demonstrated in field experiments with subterranean clover in the absence of combined nitrogen. The accumulation of molybdenum is higher in the nodular tissue than in other parts of the plant.

The essentiality of cobalt for symbiotic nitrogen fixation has been demonstrated in soybean, lucerne and subterranean clover under controlled conditions. The amount of total nitrogen in soybean plants inoculated with *R. japonicum* supplied with cobalt was about 12 times more than that of the control plants which did not receive the element. The concentration of vitamin B_{12} in soybean nodules and chlorophyll content of leaves could be positively correlated with the cobalt supply in the substrate on which the plants grew. In subterranean clover, addition of cobalt to sand cultures at the rate of 0.006 to 0.06 ppm resulted not only in formation of bigger nodules, but also stimulated nitrogen fixation. Under field conditions, cobalt has been found to significantly increase the yield of dry matter and nitrogen per plot of subterranean clover inoculated with *R. trifolii*, although no significant increase in nitrogen per cent was noticeable.

Several independent experiments show that the application of superphosphate to leguminous crops increases the number of nodules on roots and also improves the growth and nitrogen content of plants.

Rotation of crops with phosphatic manuring enhances the nitrogen content of soil as demonstrated at the Indian Agricultural Research Institute with berseem clover (Egyptian clover, *Trifolium alexandrinum*) where increase in soil nitrogen amounted to about 466 lb per acre over a 10-year period for 6 berseem crops. On an average, the increase in nitrogen content of soil amounts to about 78 lb per berseem crop. In a similar way, it has been calculated from experiments that the additional nitrogen fixed as a result of better growth of subterranean clover (*T. subterraneum*) due to the application of 100 lb of superphosphate could amount to 76 lb of nitrogen.

Growth Substances

Indole acetic acid (IAA) and gibberellins have been detected in root nodules. The nodules contain more IAA than the roots adjoining them. The effects of growth substances on nodulation are variable. Some substances promote nodulation while others retard it, depending on the concentration of the chemical used. These findings have come from experiments with excised roots as well as intact plants.

Lower levels of IAA are conducive to early nodulation whereas higher doses result in stunting and other morphogenetic effects on roots. Naphthalene acetic acid inhibits nodulation in subterranean clover whereas p-chlorophenoxy-isobutyric acid stimulates nodulation. Growth retarders such as B-nine (N-dimethylamino succinamic acid) and benzimidazole alone or in combination with IAA reduce the size of roots and the number of nodules in cowpea. Other growth substances such as kinetin and gibberellic acid do not induce any change in nodulation.

Among B-vitamins, thiamine, pyridoxine, biotin and riboflavin have no stimulatory effect on nodulation whereas meso-inositol increases the number of nodules and the percentage of nodulating roots on the root system. On the other hand, vitamin B_{12} depresses nodulation. Certain other unknown factors present in coconut water and extracts of alfalfa seeds also promote nodulation.

Colchicine, which induces polyploidy in plants is also known to increase (at lower concentrations of 10 mg/l) the number of nodules on the roots of legumes such as lucerne and clovers.

Amino acids such as valine, cystine, aspartic acid, alanine, tryptophane and creatin at 100 ppm depress nodulation of red clover in aseptic cultures of intact plants. Since these studies have been done at one fixed concentration of such compounds, further studies with a range of concentrations of individual metabolites are necessary before any generalizations are made with regard to their influence on nodulation.

The effects of antimetabolites on nodulation and growth of leguminous plants have also been investigated. The number of nodules in birdsfoot trefoil (*Lotus corniculatus*) is increased by indole 2-phenyl-n-butyric acid, D- and L-leucine, barbituric acid, oxythiamine and quercetin, although L-picolinic acid, a niacin antagonist, prevents nodule formation in spite of the compound possessing no detrimental influence on the growth of *Rhizobium*. Interestingly enough, another niacin antimetabolite, pyridine-3-sulfonate, does not exert an inhibitory effect in the same legume species.

Genetical Factors of the Host

Different varieties of the same legume are known to respond differently to a given *Rhizobium*, especially with regard to the number of nodules produced on them. In studies on the role of hereditary factors in root nodulation, individual non-nodulating plants or those plants resistant to infection have been used to conduct inter-specific hybridization experiments. The results of experiments in this direction show that resistance in red clover to *Rhizobium* infection can be attributed to a recessive factor acting in conjunction with a cytoplasmically transmitted component. These resistant plants do not nodulate with any other strain of *Rhizobium* under bacteriologically controlled conditions or with strains present in unsterile soil. Neither resistance nor susceptibility can be transmitted through grafting. The roots of plants resistant to infection no doubt stimulate the multiplication of *Rhizobium* in the root region followed by curling of root hairs, although the hairs contain no infection threads in them. This kind of resistance attributable to a single recessive factor is also present in soybean and has been introduced into other soybean varieties. Such resistant lines of plants can serve as controls in judging the symbiotic effectiveness under similar experimental conditions.

Grafting experiments done to obtain nodules on pea roots grafted to non-legumes such as buckwheat and horsebean roots grafted to *Nasturtium*, show that nodules can be formed only on the roots of legumes.

Early or late formation of nodule has been shown to be heritable in a variety of clovers. Early nodulating habit is advantageous in legume establishment particularly in annual species grown in nitrogen-starved soils. All hybridization and selection work with sparsely and abundantly nodulating lines of subterranean clover and other species of *Trifolium* have shown that nodule number is inherited in a complex manner.

The nuclei of nodule cells are generally tetraploid or of a higher ploidy. Polyploidy affects the number of nodules differently in different species. Autotetraploids of *T. subterraneum* form fewer nodules than diploids. Polyploids of *T. repens*, *T. ambiguum* and soybean have more nodules than diploids of the corresponding species. The influence of ploidy on nodule number may depend on the conditions under which plants grow. Strains of bacteria isolated from tetraploid red clover produce more nodules and are generally more effective on tetraploid than on diploid lines. However, substantial evidence is lacking to show that the genotype of one symbiont directly affects the other.

As stated earlier, stimulation of nodulation in lucerne and several species of *Trifolium* by pre-treatment of roots with colchicine has been reported. The mode of action of colchicine is not known. Perhaps, the mitotic poison increases the number of tetraploid cells in the root.

Divergent results have come forth on the effects of irradiation of seeds with physical mutagens on the nodulation status of plants raised from such irradiated seeds. Strong irradiation with gamma rays not only decreases nodulation and inhibits nitrogen fixation in pea plants but also results in the mutation of bacteria inside the nodules. On the contrary, seed irradiation of about 800 r over a period of seven days doubles the number of nodules and lateral roots in *Trifolium alexandrinum* and *Trigonella foenum-graecum*.

Influence of Pesticides

In modern agriculture, application of pesticides is an accepted practice towards controlling pests and diseases of plants and some of these chemicals may influence the microbiological processes in soil. Pot trials conducted at the Indian Agricultural Research Institute, New Delhi, reveal that the recommended doses of thiram, captan, PCNB, disyston, thimet and cerasan do not interfere with nodulation or grain yields of pulse crops.

On the other hand, herbicides influence nodulation and nitrogen fixation in legumes. Field trials with birdsfoot trefoil (*Lotus corniculatus*) using 2,4-DB and dalapon show that 2,4-DB, alone or in combination with dalapon reduces nodulation and tends to decrease the efficiency of nitrogen fixation. Dalapon appears to enhance the inhibitory action of 2,4-DB on nodulation. Autoradiographs indicate that the herbicide is translocated rapidly and could be detected in leaves and nodules in 12 hours after feeding with 2,4-DB-1-C^{14}. Fractionation of excised nodules reveal the presence of radioactivity in bacteroids, the highest being in the soluble portion. Dichlorodiphenyl trichloro-ethane (DDT) when applied to soil does not affect the leghaemoglobin content of green gram (*Phaseolus aureus*) up to 40 ppm. On the other hand, at 1–10 ppm levels, the insecticide appears to stimulate the pigment content.

Competition Problem

Because legumes have been grown in soil for many years repeatedly, it is logical to expect a plethora of rhizobial strains living saprophytically in soil waiting for occupying sites on the root system to produce nodules that are efficient or inefficient in N_2 fixation. In many soils the inefficient strains dominate leading to stiff competition when a superior efficient strain is introduced in bulk into the soil as an inoculant. If soils have fewer native strains, the chances are bright for the superior introduced strain to occupy space on the root system leading to improvement in N_2 fixation and grain yield in the legume under cultivation. The reverse is true if the soil has multitude of native inefficient strains which begin to compete and dislodge the introduced strain for root occupancy. This situation is commonly referred to as the competition problem, more serious with the promiscuous *Rhizobium* spp. (cowpea miscellany) inhabiting tropical soils. It is always said that the competition problem is the number one limiting factor in maximizing yields of legumes. A solution to this problem has not been easily forthcoming but there have been several leads to the issue under consideration and they are outlined below:

1. Field experiments have shown that in soybean (*Glycine max*) and red clover (*Trifolium subterraneum*) cultivation, massive inoculation with a superior strain many times over a period of several years (nearly 1000 times the soil population of native rhizobia) may help to dislodge the native strains.

2. Bulk inoculation of soil rather than seed inoculation with superior rhizobia offers better dispersion in the entire root zone thereby offering greater chances for root occupancy by the introduced strain.

3. A strategy to develop mutants of rhizobia to alter the competitive ability of a strain has been attempted but so far has yielded negative results. However, some information has been gained regarding the mutated locus providing a lead to further investigations.

4. By the use of reporter genes inserted into a *Rhizobium* sp., the introduced strain in soil can be monitored. Some examples are the insertion of luciferase gene (luc) into *R. metiloti* and *R. leguminosarum* bv *trifolii* rendering the strain bioluminescent, the construction of *lac* Z fusions in *Rhizobium* sp. facilitating detection of the *lac* Z gene products, namely the enzyme-B-galactosidase on X-gal plates [X-gal is a chromogenic substrate (5 bromo-4-chloro-3-indolyl-B-D-galactosidase)] and the construction of gus A fusions (for expression of B-glucoronidase) and estimating the tagged strain by simple colorimetry.

5. Using genetically engineered strains for better nodulation and yield in commercial *Rhizobium* inoculants. Some examples of success in this direction to improve competitiveness of a strain and produce better yield in the field are: (a) the insertion of the trifolitoxin gene from the inefficient strain T 24 of *R. leguminosarum* by *trifolii* into an effective strain so that the potent antirhizobial compound trifoliotoxin can minimize the numbers of native rhizobia, (b) by insertion of additional copies of *nif* A and *dct* ABD genes in *R. melitoli* strain RMBPC-2, (c) the construction of *R.*

leguminosarum and *R. meliloti* strains by incorporating the *Bacillus thuringiensis* sub-sp *tenebrionis* endotoxin gene (*Cry III*) so as to protect the nodule from nodule eating larvae of the insect *Sitona,* (d) the construction of strains of *Rhizobium* with the ability to oxidize the H_2 produced in the N_2 fixation reaction and funnel or recycle these wasteful electrons to the productive main reaction of converting N_2 into NH_3, as exemplified by the success of this venture in the case of soybean inoculated with H_2 positive strains of *Badyrhizobium japonicum,* and (e) by cloning the genes for antibiotic production and resistance into efficient rhizobia to impart competitive advantage for nodulation over indigenous native strains.

Rhizobiotoxin

Evidence exists that a certain type of chlorosis in soybean is caused by a phytotoxin (low molecular weight amino compound) produced in nodules of the affected plants. This toxin is produced by several strains of *Rhizobium japonicum* in pure cultures as well as in nodules. In pure culture, addition of yeast extract and casaminoacids promotes the production of the toxin. The interesting feature of this type of soybean chlorosis is that nodules of chlorotic plants continue to fix nitrogen while elaborating the toxic metabolite.

Salinity and Alkalinity

Saline and alkaline soils are widespread in many parts of the world. In India, such soils prevail in the Indo-Gangetic basin in the north and are not suitable for raising food and fodder legumes. Salinity is caused by the accumulation of soluble salts of calcium, magnesium and sodium, mostly as chlorides and sulphates. Alkalinity is due to the predominance of carbonates and bicarbonates. Application of large quantities of gypsum can ameliorate these soils and render them fit for cultivation. Alternatively, pelleting leguminous seeds with gypsum can also help in the establishment of legume crops.

Some fundamental studies carried out with lucerne plants (*Medicago sativa*) under bacteriologically controlled conditions on agar slopes have revealed that *Rhizobium* can tolerate higher levels of all the above-mentioned salts than the corresponding host plant, which reflects the importance of evolving cultivars of leguminous crop plants capable of withstanding salinity rather than attempting to introduce salt-resistant bacterial strains into soil.

In lucerne, higher salinity/alkalinity levels decrease the formation of root hairs and the deposition of mucilage around the roots. Since it is now known that formation of a mucilaginous matrix around innumerable hairs on the root system is a necessary prerequisite in the rhizosphere of a legume for successful entry of rhizobia into roots, i' is postulated that salinity/alkalinity in the root region impedes certain steps in the pre-infection stages of legume-*Rhizobium* symbiosis.

Legume Inoculation

The practice of applying artificially prepared cultures of rhizobia to leguminous seed before sowing can be referred to as legume inoculation. This practice is known since the beginning of this century. Agar based cultures were used for a long time which was later replaced by soil based ones. Finely ground and neutralized peat is now being generally used in Australia and U.S.A. as a carrier in the preparation of legume inoculants. Peat as a carrier has decided advantages over agar or soil. Besides possessing high moisture holding capacity and organic matter content, so essential for better shelf life of bacterial cultures, peat improves the survival of rhizobial cells on the seed coat, especially under dry soil conditions.

In Australia, peat is harvested, dried in the field and ground to pass a 200 mesh sieve. Peat is generally acidic in nature and hence is neutralized by adding sufficient $CaCO_3$. The neutralized peat is packed in low density, 0.05 mm gauge polythene bags and sterilized by gamma rays at a dose of 5.0×10^6 rads. Australians have found that sterilization by gamma radiation is generally superior to autoclaving for 4 hr at 121°C in promoting the growth of rhizobia.

Rhizobia are grown on yeast extract mannitol broth in suitable fermentor vessels until the numbers of rhizboia in the broth have reached the minimum standard or even more than the standard figure. The acceptable minimum Australian standard for rhizobial count in the broth culture is 500×10^6 viable rhizobia/ml, although in practice the numbers usually reach the range of $1000-4000 \times 10^6$ viable cells/ml. If the broth is added to unsterilized peat, mechanical mixers are used to mix peat with the broth so as to provide a moisture content of 45–50% on a wet weight basis. The peat is sieved through a coarse sieve to remove lumps and then matured for 4 days at 26°C in trays covered with polythene and packaged according to convenience in polythene bags. The polythene bags are made from polythene sheeting of 0.0015 inch or 0.0375 mm. If the broth is added to sterilized peat, the peat is initially packaged in polythene bags and the required quantity of broth is added through a syringe to provide a moisture content of 60%. The puncture is sealed by adhesive tape, the contents mixed by rolling in hand, incubated at 26°C for two weeks and then stored at 4°C. The minimum Australian standard for rhizobia in peat are $10^8 - 10^9$ viable cells/g of peat at manufacture and $10^7 - 10^8$ cells/g of peat during the full shelf life of the culture.

The Australian method of using cultures grown on agar is to make a suspension of the bacterial cells in water. This suspension may be directly applied to seed or improved by the use of 10% sugar or 40% neutral gum arabic in the suspending fluid. If peat based cultures are used, 25 g of culture is added to 100 ml of water or a solution of sugar and gum arabic. The resultant slurry is used for application on seed.

Pelleting of inoculated seed with lime (finely divided $CaCO_3$) or rock phosphate improves survival of rhizobia on seed and hence secures better root nodulation under adverse soil conditions. For pelleting, a sticker is generally used. The recommended stickers are pharmaceutical fine grade gum arabic at approximately 40% level, 5% methyl ethyl cellulose (cellofas A) and methyl hydroxypropyl cellulose (methofas) or carboxy methyl cellulose. It is necessary that finely ground calcium carbonate must pass through 300 mesh. Small quantities of seeds may be handled in dishes for pelleting. For larger quantity of seeds, a concrete mixer may be used. Initially, the seeds are mixed thoroughly in the peat-adhesive slurry. Finely ground $CaCO_3$ is then added to the inoculated seed while it is still wet and rolled evenly so as to obtain uniform pelleting of lime over the seed. The pelleted seed (Fig. 67) may be sown immediately and if absolutely necessary may be stored up to 2–3 weeks at temperatures not exceeding 18°C.

In U.S.A., large quantities of legume inoculants are produced for internal use as well as for exporting to other countries. The usual method followed in manufacture of inoculants is as follows: Inocula are transferred from agar slants into starter flasks containing yeast extract mannitol broth. After 4 days growth, the culture from starter flasks is transferred into a small seed-tank fermentor. At this stage, yeast extract mannitol broth is formulated in a battery of large production fermentors, pH adjusted to 7.0, sterilized, cooled and kept ready for use. The contents of seed tank fermentors are transferred to production fermentors. The temperature range for fermentation is 30–35°C, depending on the species of rhizobia. Aeration is done by forcing sterile air through porous carborundum or stainless steel spargers at the bottom of the fermentors. A cell population of 5×10^9 can be attained in 96 hours which is again dependent on the amount of initial inoculum added.

A minimum of 10^9 cells per ml is needed in the broth for being used in the preparation of peat cultures. In U.S.A., the broth is sprayed to powdered, neutralized and flash-dried peat (partial sterilization) while the mixture is being agitated in a ribbon or paddle-type batch mixer. After mixing, the inoculant is spread in thin layers on floor for 48 to 72 hours at 22 to 24°C. The product is then milled to break up aggregates, finely pulverized and packed in polythene bags. The general standards advocated are to provide massive inoculum and sound recommendations for the use of the inoculant since various factors affect the number of viable cells in peat.

In India, peat-like material is available in the Nilgiri valley to the extent of 5.5 million tonnes as estimated by the soil survey department. An unknown quantity is also known to occur in the Kashmir Valley. Lignite is

Fig. 67 Soybean seed—(1) uninoculated; (2) inoculated with peat-based culture of *Bradyrhizobium japonicum*; (3) pelleted with lime after inoculation; (4)—chickpea plants *(Cicer arietinum)* uninoculated, and (5)—inoculated with *Rhizobium*.

another carrier which is widely used and it is estimated that about 3 million tonnes of mined lignite is available annually from Neyveli lignite mines in Tamil Nadu. Powdered peat or lignite is neutralized with $CaCO_3$

(passing through 200 mesh sieve) and is autoclaved at 15 lb pressure for 4 hours before use. Upon cooling, the broth from shake cultures (or from a fermentor) is poured into powdered peat and mixed by hand or a mixer in such a way that the finished product retains 40% moisture. After curing for few hours at room temperature the product is packed in polythene packets (Fig. 68). Inoculant slurry is prepared in 5% aqueous solution of jaggery or sugar before being applied to seed. In practice, the following bacterial counts have been achieved while making peat based inoculants at the Indian Agricultural Research Institute: (1) the minimum count for broth is 10×10^7/ml and maximum count being 80×10^9, (2) the minimum peat count at manufacture is 100×10^6 and the maximum being 90×10^8, (3) the minimum peat count at peak shelf life (4 weeks) is 10×10^8 and the maximum being 10×10^{10} at 12 weeks, and (4) the minimum count on seed at sowing is 10^3 per seed and the maximum being 10^5 per seed.

Yield Response to *Rhizobium* Inoculation in India

The results of experiments done on all-India level have highlighted the following important facts: (1) Soybean responds spectacularly to *Rhizobium* application and grain yields are often increased up to 50% over uninoculated controls since our soils are deficient in specific bacteria capable of nodulating soybean (Table 30); (2) Depending on the agro-climatic conditions and the variety planted, significant increases in yield over control (Table 31) could be expected with arhar (*Cajanus cajan*), chickpea or Bengal gram (*Cicer arietinum*) and masur (*Lens culinaris*); (3) At certain sites, even in neutral soils, where conventional simple *Rhizobium* inoculations have failed, pelleting inoculated seeds with lime or charcoal could significantly increase yields of red gram (*Cajanus cajan*); and (4) Though responses to *Rhizobium* inoculation with regard to other crops are rather variable, it may still prove beneficial in certain regions. Inoculation is so inexpensive that some farmers have taken to inoculating all pulses and fodder crops such as berseem and lucerne with *Rhizobium* as a routine practice to insure against crop failure.

Some Common Methods Used in the Study of Legume Root Nodulation

Isolation of *Rhizobium*

After washing the root system of the test leguminous plant in running water, a well-formed, healthy pinkish nodule on the tap root is carefully cut out with a portion of the root attached to the nodule. The nodule is

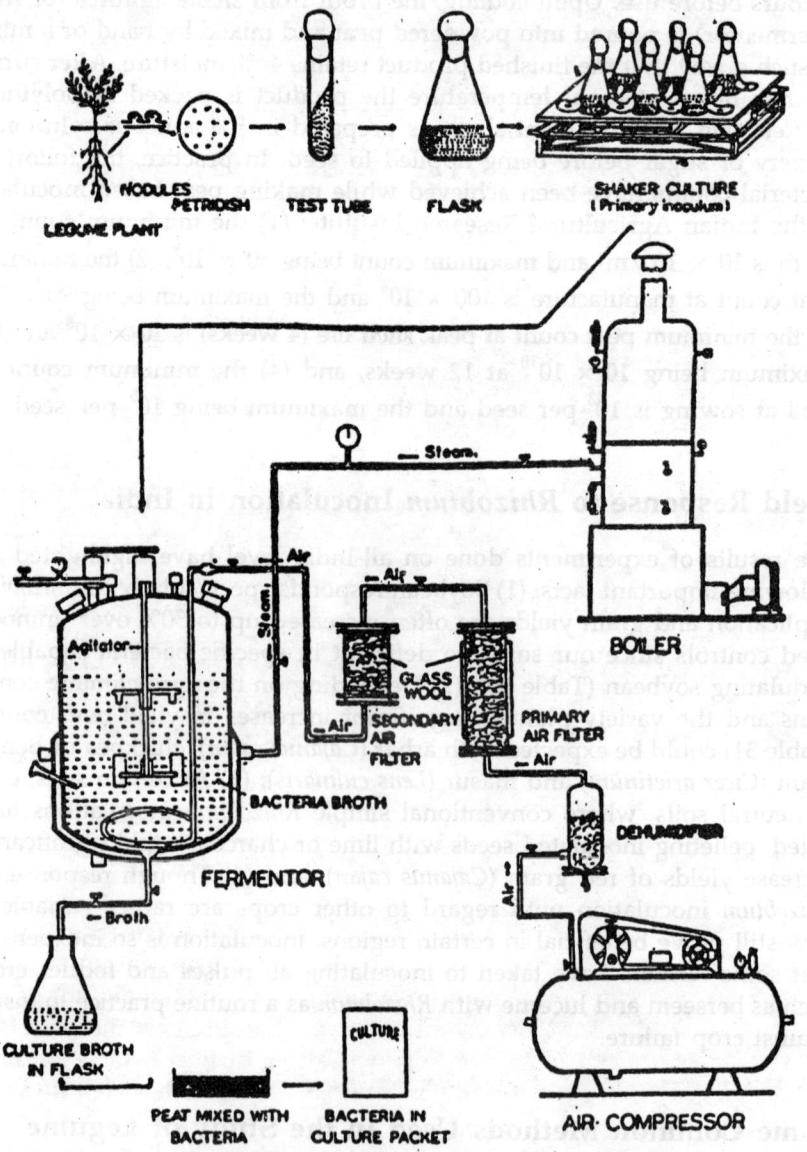

Fig. 68 A schematic presentation of processes involved in the mass production of rhizobia.

surface sterilized for 5 minutes in 0.1% mercuric chloride in water and repeatedly washed with sterile water to get rid of the chemical. The nodule is then washed in 70% ethyl alcohol for 3 minutes followed by more washings

Table 30 Effect of seed inoculation with *Rhizobium japonicum* (IARI strains, SB6 + SB16) on the yield of soybean (*Glycine max*); mean performance of 1972, 1973 and 1974 field trials (from V.R. Balasundaram, unpublished data)

Locations (variety)	Yield (kg/ha)		% increase in yield due to inoculation
	Uninoculated	Inoculated	
Delhi (Bragg)	1498	1883	25.7
Pantnagar (Bragg)	1308	1988	52.0
Bangalore (Davis)	1722	2442	41.8

Table 31 Effect of seed inoculation with rhizobial culture on the yield of various grain legumes in Tarai soil (pH 7.3)

Treatment	Cajanus cajan (arhar)		Cicer arietinum (Bengal gram)		Lens culinaris (lentil)	
	Yield (q/ha)*	Per cent increase over uninoculated control	Yield (q/ha)*	Per cent increase over uninoculated control	Yield (q/ha)*	Per cent increase over uninoculated control
Uninoculated (control)	11.3		10.5		8.7	
Inoculated with IARI culture	13.5	19.47	12.7	20.94	11.5	32.20
40 kg N/ha	13.2	16.82	11.8	12.38	12.1	39.10

−1q = 100 kg

with sterile water. The nodule is now crushed with a sterile glass rod in a small aliquot of sterile water. Serial dilutions of the suspension are then made so as to obtain sparse and distinct colonies when an aliquot of the appropriate dilution is plated on yeast extract mannitol agar medium (see Appendix). At the end of one week's incubation, distinct colonies of bacteria are picked up and transferred to agar slants for further identification (Fig. 69).

It may so happen that several strains of *Rhizobium* of varying effectiveness, often associated with a related genus *Agrobacterium*, are isolated from the same plant. Obviously, further screening of such isolates becomes an essential step.

Tests for Distinguishing from *Agrobacterium*

(a) Congo red medium: It has been found that on Congo red incorporated agar medium (2.5 ml of a 1% solution per litre of yeast extract mannitol

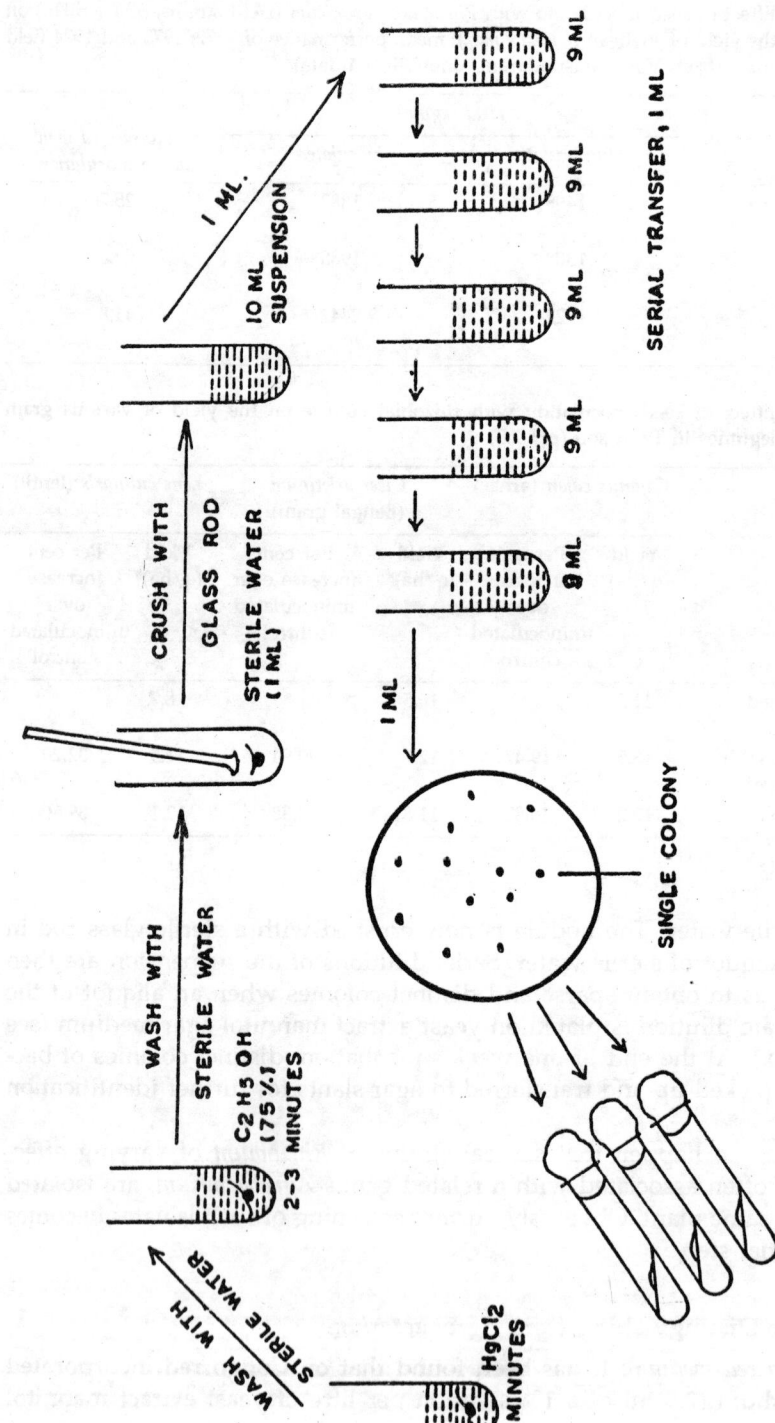

Fig. 69   Stepwise procedure for isolation of nodule bacteria from root nodules.

agar), rhizobia stand out as white, translucent, glistening, elevated and comparatively smaller colonies with entire margins in contrast to stained colonies of agrobacteria.

(b) *Hofer's alkaline broth:* Agrobacteria grow at higher pH levels, while rhizobia are unable to do so. Therefore, growing bacterial isolates on an alkaline medium (K_2HPO_4, 0.5 g; $MgSO_4$, 0.2 g; NaCl, 0.1 g; $CaCO_3$, 0.05 g; yeast extract, 1.0 g; mannitol, 10 g; and water, 1000 ml; pH adjusted to 11.0 by adding approximately 28 ml of N NaOH and 1 ml of 1.6% thymol blue) would serve as a useful criterion for distinguishing the two allied genera.

(c) *Lactose agar:* Agrobacteria utilize lactose to form a reduced product, ketolactose, through the enzyme ketolactase. Benedict's reagent is prepared as follows: 173 g sodium citrate and 100 g anhydrous sodium carbonate are dissolved in 66 ml distilled water; 17.3 g crystalline copper sulphate is dissolved in 100 ml distilled water. The latter solution is added to the former with constant stirring, the mixture filtered if not clear and made up to 1000 ml with distilled water. The reagent (may be stored indefinitely) is poured over agar medium containing lactose (10 g/litre) on which nodule bacteria are growing. The formation of yellow coloration due to Cu_2O indicates the presence of agrobacteria.

Tests for the Ability of *Rhizobium* Isolates to Nodulate (Fig. 70)

Apart from the existence of *Rhizobium* isolates with a range of effectiveness, occurrence of non-nodulating and also pathogenic strains among the isolates obtained by plating extracts of nodules cannot be ruled out. Therefore, testing strains for their ability to nodulate and for their effectiveness under sterile conditions is of prime importance.

(a) *Test-tube method for small seeded legumes:* Agar-agar, vermiculite or sand can be used as the substrate for growing plants. Any nitrogen-free plant nutrient (e.g., Jensen's nutrient for seedlings which has the following composition: $CaHPO_4$, 1.0 g; K_2HPO_4, 0.2 g; $MgSO_4 \cdot 7H_2O$, 0.2 g; NaCl, 0.2 g; $FeCl_3$, 0.1 g; agar-agar, 8.0 g; and distilled water, 1000 ml; pH—6.8) is used to make agar slants in 2.5 × 15 cm or 4 × 20 cm test tubes, depending on the size of the seed. Seeds are surface sterilized with suitable agents and sown on the substrate in tubes. On germination, they are inoculated with a suspension of the bacterial isolate under test prepared in a small aliquot of the nutrient solution. At regular intervals, the moisture content of the tube is checked and the nutrient solution replenished when necessary. During the experiment which may run for several weeks, the plants are examined and nodulation data obtained. A general criterion for distinguishing the effectiveness is to note the relative dry weight of nodulated plants, provided the replication in the experiment is large enough to accommodate variations.

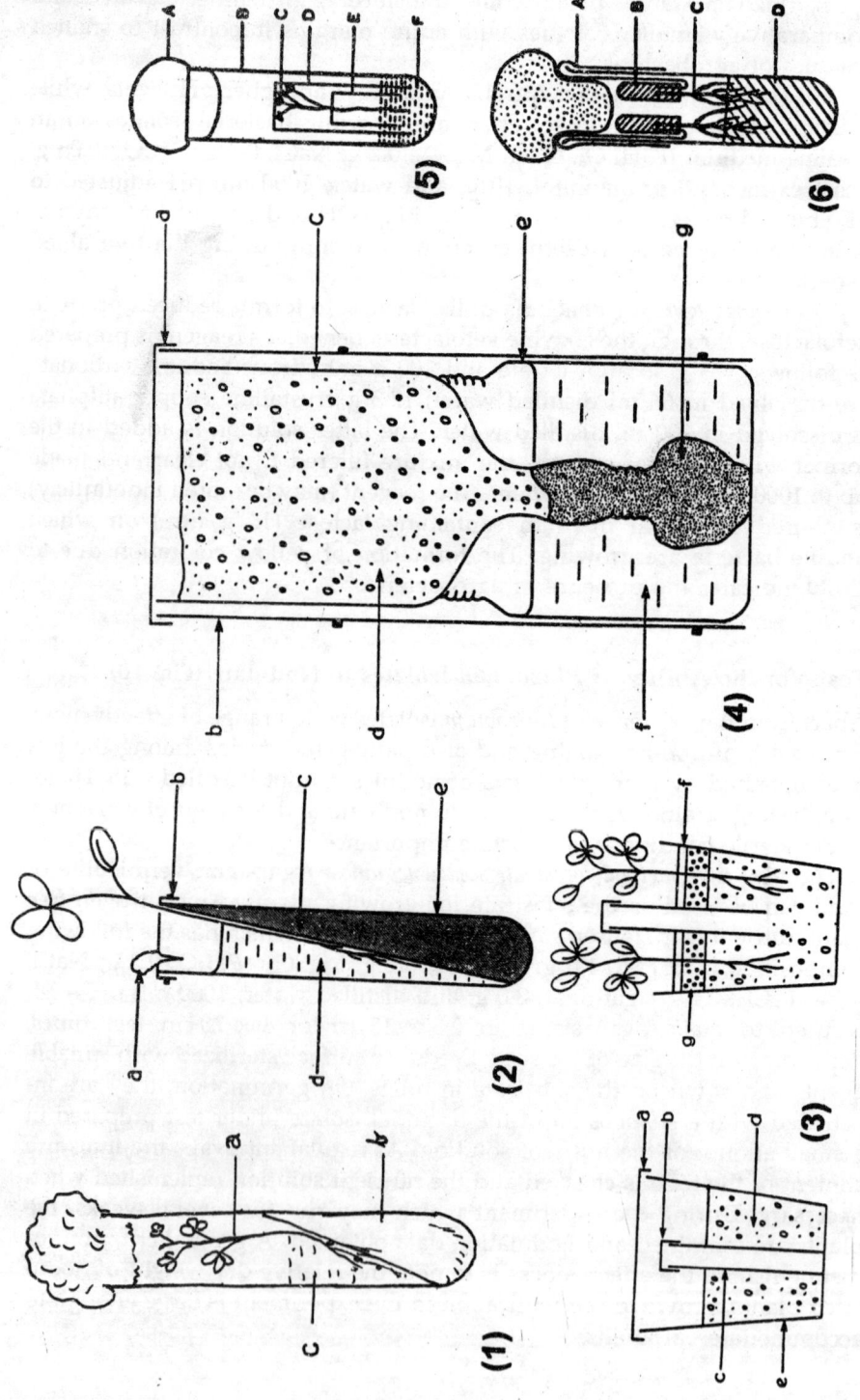

The above method [Fig. 70 (1)] restricts the growth of seedlings in the closed test tube. Hence, the tube method was modified to allow the growth of seedlings in the air by covering the rimless test tube prepared as mentioned earlier with an aluminium foil cap with a hole covered with a cotton plug. When the seedling reaches the aluminium cap, the leaves are gently placed above the tube rim by adjusting the cap to allow the plant to grow freely in air as shown in the Fig. 70 (2). Watering can be done by removing the cotton plug periodically under aseptic conditions.

(b) *Plastic cup method:* By using a plastic cup and an open ended glass tube for watering, seedlings can be grown on sand and watered aseptically. The plastic cup is initially covered with a Petriplate and later when the seedlings become big, sand mixed with paraffin wax can be poured to keep the assembly in aseptic conditions as shown in Fig. 70 (3).

(c) *Testing for large seeded legumes:* Bacterial isolates are tested for efficiency by growing the inoculated legume in sterilized sand with nitrogen-free plant nutrient solution. At the end of the desired period, data on the appearance of plants in terms of colour and vigour, the number of nodules formed on the root system, dry weight and nitrogen content of the plants are collected. During the experiment, it is necessary that temperature and light conditions under which the plants grow should be kept at optimum levels conducive for maximum performance of legume–*Rhizobium* symbiosis.

A standard assembly known as Leonard Jar, consisting of a round bottomless screw-capped bottle and a glass jar designed to permit automatic and continuous water flow to the growing root system [Fig. 70(4)] can be used for the routine testing of isolates of *Rhizobium* for efficiency in symbiosis. Alternatively, sterilized sand or soil in earthenware or porcelain pots can also be used to grow large seeded legumes.

Fig. 70 Testing nodulating ability of *Rhizobium* isolates: (1) Jensen's closed tube showing a—seedling, b—agar base, c—1/4 strength nitrogen-free nutrient liquid; (2) Gibson's tube showing a—cotton plug, b—aluminium foil cap, c—1/4 strength nutrient solution solidified with agar, d—1/4 strength nutrient solution, e—rimless tube; (3) The plastic cup method showing a—glass Petri dish lid, b—metal cap, c—open ended glass tubing for watering, d—sand, e—plastic cup (Nutman, unpublished); (4) A Leonard jar: a—glass Petri dish lid; b—brown paper wrapping the assembly and held on with rubber bands; c—inverted bottomless wine or beer bottle; d—sand; e—jar (500 ml capacity); f—nutrient solution; g—foam wick or cotton wick (from Dye, 1979); (5) a simple technique to follow infection of legume roots by *Rhizobium* which can be continuously used for observation at different days of incubation—(A) cotton plug, (B) boiling tube, (C) small seed plant (clover), (D) microscope slide, (E) cover slip to hold the plant intact; the four corners of the cover slip are attached to the slide by means of araldite, and (F) nitrogen-free nutrient medium (Nutman personal communication). (6) Technique for culture of excised roots—(A) holder of stainless steel wire, (B) organic nutrients in agar, (C) excised root, and (D) coarse sand with inorganic nutrients and *Rhizobium* inoculum (from Bunting and Horrocks, 1964).

(d) *Infection test:* Seedlings may have to be examined for the density of infection with a view to improve the strains for virulence and early nodulation. This is done by growing plants on agar slants containing nitrogen-free medium and inoculating them with desired strains. At regular intervals, seedlings are fixed in 4% formalin or acrolein and examined later under a microscope for infection threads in root hairs and emergence of nodule primordia on roots. Depending on the extent of infection, isolates may be rated for virulence. A simple device for continuous observation of the progress of infection has been designed which has proved useful in many laboratories [Fig. 70 (5)].

(e) *Nodulation in excised roots:* In excised root cultures, the growing root of a 2–4 day old legume seedling (grown aseptically) is cut with a sharp razor blade and the blunt end implanted in a small vial containing organic constituents (glycine, B-vitamins and sucrose) in solidified agar. The tip of the excised root is suspended in nitrate-free mineral salt medium in a Petri dish or in a test-tube into which a suspension of *Rhizobium* is added. In this way, the organic constituents are provided to the growing root through the cut end and the growing end of the root receives mineral constituents along with *Rhizobium* culture. A diagrammatic sketch of the technique is shown in Fig. 70 (6). Nodules develop after 2–3 weeks when lateral roots are fairly well developed.

(f) *Tissue and cell cultures:* The use of tissue cultures in plant morphogenetic studies is now well known. Explants of roots, stem and apical meristem can be grown into an undifferentiated callus mass on a suitable basal medium. New plantlets could be raised from the callus mass in test-tubes. When such plantlets are transplanted to soil, they grow into mature plants bearing flowers and fruits. Tissue culture techniques have had extensive use in plant disease control, especially in the control of seed-borne viruses.

In recent years, root calli have been used to study the interactions of plant tissues with *Rhizobium* or *Azotobacter*. In this method, a mass of undifferentiated callus or a suspension of isolated cells or protoplasts is inoculated with nitrogen-fixing bacteria such as *Rhizobium* or *Azotobacter* and incubated under controlled environmental conditions. Samples are analysed periodically for nitrogenase activity by the acetylene reduction technique. Samples of calli or cells could also be cut into thin sections and observed under light or electron microscope.

Such studies have revealed the following facts and possibilities: (1) Infection thread-like structures and bacteria or bacteroid containing cells could be seen in callus tissues or isolated cells inoculated with *Rhizobium;* (2) Definite nitrogenase activity could be detected in inoculated calli or cells indicating the establishment of symbiosis in such plant cell-bacteria associations; (3) when cells of *Azotobacter* and carrot were grown on nitrate-free media, not only the bacterial cells entered isolated cells of carrot but such cells showed nitrogenase activity; and (4) If such studies are extended further, the day may not be far off when 'nitrogen-fixing' plants are raised as clones through plantlets regenerated from bacteria incorporated cells or tissues.

Determination of Total Nitrogen by Kjeldahl Method

This method involves the conversion of nitrogen in biological materials into $(NH_4)_2SO_4$ by digestion with H_2SO_4 followed by distillation of NH_3 in an alkaline medium. The ammonia is collected in sulphuric acid of known strength (0.05 N) which is back titrated with standard sodium hydroxide solution. While the method is adequate for soils, plant materials and active nitrogen fixers, it is not sensitive enough to measure less than 1 mg of nitrogen.

The biological material (about 1 g) is placed in a 250 ml Kjeldahl flask and digestion mixture is added to it. The digestion mixture consists of the following: 1 g of a well-ground mixture of 20 g $CuSO_4$ $5H_2O$ and 1 g Se. To one part of this mixture 20 parts of anhydrous Na_2SO_4 or K_2SO_4 are added. Five g of this catalyst mixture is added to the digestion flask along with 5 ml of mercuric sulphate solution (12 ml concentrated H_2SO_4 in 100 ml water in which 10 g red mercuric oxide is dissolved). The Na_2SO_4 raises the boiling temperature of H_2SO_4, $CuSO_4$ and Se, accelerates the rate of digestion while the mercuric salt helps in the digestion of methylamines and prevents loss of N_2 which may occur if only Se is present. Fifteen ml concentrated H_2SO_4 and a few glass beads are then added. The digestion flask is heated at first gently until all the water is removed and charring is completed. The heat is then gradually increased so that the solution is brought to constant boiling with slight bubbling. After complete clearing, boiling is continued gently for another 15–20 mins. By this time whole of the nitrogen might have been converted into $(NH_4)_2SO_4$. The flask is allowed to cool, about 25 ml distilled water is added, the contents transferred to a 100 ml volumetric flask and the volume made up with distilled water.

Hoskins steam distillation apparatus is commonly used for distillation. Measured quantity of the digested material (10–25 ml, depending on the N_2 content of the material) is taken in the distillation flask and 10–15 ml of 40% NaOH added to the sample. The flask containing sulphuric acid and indicator solution is kept under the condenser of the distillation apparatus. Toshiro's indicator (0.25 g of methylene blue, 0.375 g of methyl red and 300 ml of 95% ethanol) or a mixed indicator (0.099 g bromocresol green and 0.066 g of methyl red in 100 ml of ethanol) could be used advantageously. Heating should be carefully regulated to prevent sucking back of the sulphuric acid.

After sufficient distillate has been collected, the sulphuric acid is back titrated with standard 0.05 N NaOH (1 ml of 0.05 N H_2SO_2 = 0.7 mg ammonium N). The colour changes from green to pink or purple depending on the indicator.

Acetylene Reduction Technique

The enzyme nitrogenase, in addition to reducing N_2 to NH_3, can reduce certain other compounds like acetylene. This assay, formulated independently by Dilworth in 1966, Koch and Evans in 1966, Sloger and Silver in 1967, Stewart *et al.* in 1967 and Hardy and associates in 1968, is based on the nitrogenase-catalyzed reduction of C_2H_2 to C_2H_4. The advantage of the technique includes facility for large number of *in situ* assays, economy, simplicity, effective gas exchange prior to incubation, easy removal of the gaseous end products and sensitive analysis with a flame ionization detection system after separation by gas chromatography. The technique is 10^3 times more sensitive than ^{15}N tracer technique.

Samples to be tested are incubated in serum vials, syringes, plastic bags or suitable containers with purified acetylene (C_2H_2) under standard temperature, incubation period and illumination. At the end of incubation, the gas from the container is removed for ethylene (C_2H_4) analysis by gas chromatographic fractionation followed by detection by flame ionization or colorimetry. After correction for the small amount of C_2H_2 present as an impurity in C_2H_4 the amount of nitrogen fixed is determined by the extent of reduction of C_2H_2 to C_2H_4. Moles of N_2 fixed is obtained by division of moles of C_2H_2 reduced to C_2H_4 by a factor of 3. This factor is based on the ratio of the requirement of 6 electrons for reduction of N_2 and 2 for C_2H_2. The method is extremely useful for routine field evaluations.

The Use of ^{15}N

While acetylene reduction test offers a rapid though indirect way of measuring nitrogenase activity and thus the ability of a biological system to fix atmospheric nitorgen, any critical evaluation of results must pass the ^{15}N test. The technique formulated by Burris and Wilson in 1957, involves the use of ^{15}N as a tracer. The amount of nitrogen fixed by a system is determined as the percentage increase of the tracer in samples incubated with the labelled gas. An enrichment of 0.015 or greater atom per cent ^{15}N in the sample (expressed as 0.015 atom per cent excess) is usually taken as an indication of nitrogen fixation. This method has been used extensively in the analyses of nitrogen fixation in whole cells and in cell-free extracts. The technique is time consuming and requires the use of expensive mass spectrometer and ^{15}N tracer but is 10^3 times more sensitive than the conventional Kjeldahl method.

Serology

This test makes use of the antigen-antibody reaction. An antigen is any substance which induces the production of antibodies upon introduction into the blood stream of an animal. The antigenic substances are proteins or polysaccharides and bacteria can also serve as an antigenic substance. Antibodies are proteins produced by plasma cells in response to antigens with the unique capability of binding specifically tc the antigen which induced their formation. The bacterial cell may contain many antigenic components. Those antigens which are present on the surface or in the cell are known as somatic or 'O' antigens while those present in the flagella are known as flagellar or 'H' antigens. Somatic antigens may be either heat-labile or heat-stable, the latter mostly found on the surface of cell and are considered to be strain specific. When identical antigens are present in bacterial strains, they are known as serologically identical bacteria. If bacterial strains contain some antigens common among them, they are designated as serologically related. On the other hand, serologically unrelated bacterial strains possess unidentical antigens. A series of laboratory tests are available to distinguish bacterial strains for the above-mentioned relationships and they come under the broad category of agglutination, precipitation and complement fixation.

Among these tests, the precipitation or precipitin reaction is commonly used to distinguish between strains of rhizobia. In this test, a reaction takes place between a soluble antigen and a solution of its homologous antibody which is manifested by the formation of a visible precipitate at the interface of the reactants. Greater accuracy and efficient separation of components in mixtures of antigen and antibodies can be obtained by allowing the reactants to diffuse together on an agar gel. For this purpose, a double-diffusion method in two dimensions devised by Ouchterlony has been commonly used. In this method, wells are cut on agar in Petri dishes and the reactants diffuse from wells. When the reactants meet midway, they react and form precipitin bands. Based on the characteristics of bands, the homologous or heterologous nature of the bacterial strains could be determined. The outline of the procedure, as followed in the Division of Microbiology, Indian Agricultural Research Institute, New Delhi, for distinguishing rhizobial strains is detailed below:

1. The mother culture is grown in yeast extract mannitol agar or liquid medium at 28°C ± 1°C for 5 to 8 days, depending on whether the bacterium is slow or fast growing.
2. The bacterial cells are harvested in sterile physiological saline (0.85% NaCl) and the cell suspension centrifuged at 15,000 r.p.m. to obtain a pellet.
3. The pellet is transferred to a screw capped tube in minimum amount of physiological saline and stored below 0°C. This will serve as the antigen.

4. The antigen is diluted so as to carry 10–15 µg protein/ml or 10^6 cells/ml and one ml mixed thoroughly in 1/2 ml of Difco Bacto adjuvant or sterilized paraffin oil.

5. The antigen is injected intramuscularly into a healthy rabbit (minimum weight 1.5 kg), at the upper part of the hind leg. Similar injections are repeated thrice at weekly intervals.

6. Four weeks after the first injection, 1 ml of the antigen (10^6 cell/ml) is again injected intravenously without the adjuvant, through the marginal vein of the ear to boost the titre value of the antibody.

7. After 1 week, the rabbit is made to bleed through the marginal vein of the other ear and 5 ml of blood collected in a small beaker. The blood is incubated at 35°C for an hour to coagulate and stored in a refrigerator above 4°C overnight when the serum stands as a supernatant.

8. The supernatant (straw coloured) is separated and centrifuged at 3000 r.p.m. for 5 minutes. This is the antiserum which should be tested initially against the reference antigens (that is, the antigen described in columns 1 to 3 above). If bands due to precipitin reaction are formed (see procedures in column 10–14), the rabbit may be immediately bled and sizeable amounts of blood collected, coagulated, antiserum separated and stored in vials (0.2 to 1 ml aliquots) in a deep freezer, until further use.

9. At this stage, agar plates are prepared for immunodiffusion tests, as follows: One hundred ml of agar suspension (1.5%) is prepared in physiological saline by steaming in an autoclave and an equal volume of 0.05% sodium azide solution is also prepared in physiological saline without heating. Both the solutions are mixed gently (to avoid air bubbles) in a measuring cylinder and the volume made up to 200 ml with distilled water. This will mean that the mixture will contain: agar: sodium azide: sodium chloride in the ratio of 0.75%: 0.025%: 0.85%. The mixture is poured in Petriplates and allowed to solidify on a uniformly levelled table so as to obtain a gel of 4 mm thickness.

10. Equidistant holes are punched in a hexagonal array on agar in Petriplates with the help of a cork borer of 4 mm. diameter and the agar plugs removed under mild suction by a Pasteur pipette using a water pump. The bottoms of the holes are sealed by pouring droplets of molten agar through a Pasteur pipette.

11. The centre well of the Petri plate is filled with the reference antiserum and the reference antigen filled into two wells across the centre well. In each of the remaining wells, antigens of other isolates whose serological relationships have to be determined are filled. To differentiate thermostable surface antigens from thermolabile flagellar or intracellular antigens, the cell suspension is steamed for 30 minutes in an autoclave and used.

12. The Petri plates are incubated in a moist air-tight plastic box at constant temperature (below 25°C) for 4 to 6 days. The formation of precipitin bands around the centre well is observed. If the bands formed due to the reference strain and a strain under test are identical in nature, position, and happen to coalesce without the formation of spur, it may be concluded that the strain under test is serologically identical with the reference strain, failing which the strain, under test may be either partially related or totally unrelated to the reference strain (Fig. 71).

Immunofluorescence (IF)

The general principles in the use of fluorescence microscopy and fluorescent antibody techniques have already been outlined in the chapter on Rhizosphere. The same principles may be applied to study the fate of individual strains of rhizobia in soil, especially with the use of monoclonal antibodies.

Antibiotic Resistant Markers

It is not easy to trace the fate of introduced inoculum in field soil amidst a large population of resident indigenous rhizobia. However, in an experimental set-up, rhizobia may be distinguished by patterns of intrinsic or induced antibiotic resistance. Two of the antibiotic markers commonly used are resistance to rifamycin and streptomycin. Rifamycin binds to one of the subunits of RNA polymerase and thus prevents the transcription of genes to messenger RNA. Streptomycin binds to a ribosomal subunit and prevents protein synthesis. In both the instances, antibiotics alter the molecules. The altered molecules function as specific 'markers' and help in distinguishing rhizobial strains by developing resistance to any one of these antibiotics. In a mixed population of rhizobia, only those introduced strains which are marked to either rifamycin or streptomycin will grow when plated on antibiotic containing agar medium. However, it should be borne in mind that the antibiotic resistant markers must remain stable and unchanged in symbiotic performance and characteristics that are not common to indigenous soil bacteria.

Enzyme-Linked Immunoabsorbant Assay (ELISA)

Currently ELISA is being used for 'nodule typing' to determine the 'nodule occupancy' of a selected strain introduced to seed surface or soil. For this purpose, nodules can be oven-dried and stored. The technique is particularly useful for field testing of mixed infections in nodules without the use of microscopes, facilitated by easily transportable ELISA kits. There

a) Reaction of Identity

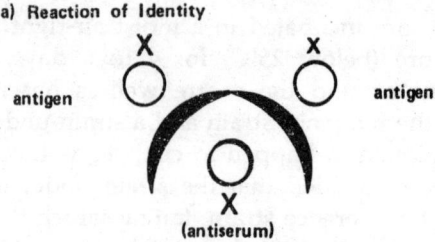

antigen   X   X   antigen

X
(antiserum)

b) Reaction of Partial Identity

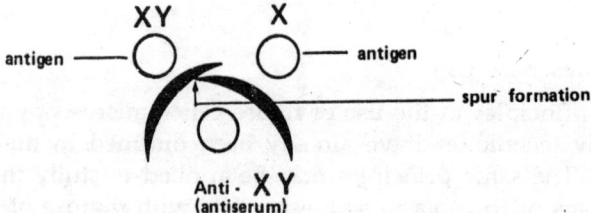

XY   X

antigen ——   —— antigen

—— spur formation

Anti · XY
(antiserum)

Two antigenic components are present here. Antigen
possesses specificity not possessed by the other antigen

c) Reaction of Non-identity

X   Y

antigen ——   —— antigen

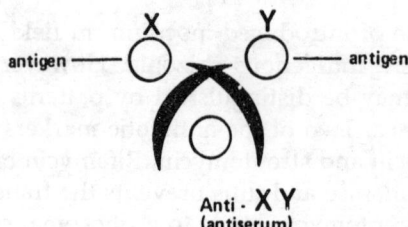

Anti · XY
(antiserum)

Antigen X and antigen Y do not possess common antigens
and the antibody possesses specificity for both.

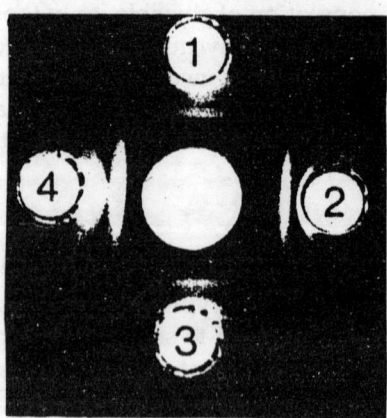

Fig. 71 Immunodiffusion reaction in gel showing precipitation bands. Antiserum appears
in centre well and antigens in surrounding wells. From Somasegaran and Hoben,
1985 of NifTAL, USA).

have been improvements to the technique by the use of fluorescent substrate and monoclonal antibodies.

The method uses the earlier stated procedure for developing the required antisera in rabbits followed by bleeding and purifying the antisera. The antisera are then conjugated with alkaline phosphatase enzyme. The root system of the desired nodulated plant is gently pulled out of soil, washed, the nodules numbered and individually squashed in the wells of a microtitre plate. The plate is then incubated after the addition of the antisera enzyme conjugate in a substrate and colour developed with o-nitrophenol phosphate.

The traditional methods so far described which relate to antigen-antibody reactions and resistance to known antibiotics are expensive, labour intensive and lack sufficient specificity to identify strains of rhizobia in soil without doubt. There are however, some newer methods based on molecular biology that have advantages over earlier methods. These methods include the introduction of marker genes and identification of DNA sequences that are characteristic of particular strains.

The Use of Marker Genes

The methods using marker genes are simple, inexpensive and involve observation of colours by the unaided eye. The marker must be highly sensitive, affordable and usable in the field with no background activity either in the bacterium or in the phase in which they are studied such as soil or plant roots. The marker gene selected must have no or very little interference with the physiology or genetics of the host. The strains constructed with marker genes are indeed genetically engineered organisms (GEOs) and must not possess characteristics that are prone to spread in any environment. In other words, the construct must pass the legislative regulatory measures of GEOs of a country in which they are used.

A number of marker genes are available for their potential use in rhizobial competition studies. They include the *Gus A* gene coding for B-glucuronidase (GUS) hydrolysing a variety of glucuronide substrates to give coloured fluorescent products; the *LacZ (YA)* gene coding for B-galactosidase hydrolysing a variety of galactoside substrates to give coloured fluorescent compounds; the *phoA* gene coding for alkaline phosphatase hydrolysing a variety of phosphate substrates to give coloured fluorescent compounds; the *XylE* gene coding for Catechol 2,3-dioxygenase capable of converting colourless catechol to 2-hydroxymuconic semialdehyde which is bright yellow in colour; *Vio*-operon coding for violacein biosynthetic enzymes capable of producing violacein, a purple pigment from endogenous tryptophan; *tfd A* gene coding for 2,4-dichlorophenoxyacetate monoxygenase capable of converting phenoxyacetate to phenol which can be measured by gas chromatography or by the development of a red dye on reaction with 4-aminoantipyrene; *luxAB, luc* genes coding for luciferase enzyme (firefly enzyme) whose activity leads to light production on an aldehyde substrate containing oxygen and reducing equivalents.

Among these marker genes, the *gusA* gene merits consideration because the method involving this gene satisfies the several criteria listed above to qualify as an easy and acceptable tool for rhizobial ecological studies. Gus marker gene can be introduced into *Rhizobium* through a plasmid or as a piece of DNA (transposon) the latter becoming integrated into the chromosome of the bacterium. The procedure for assay can be outlined as follows: A donor plasmid or transposon element having Gus A gene is mated with a pure culture of desired rhizobial strain → plate on selective media → the marked rhizobial strain grows on selected plates → use the marked rhizobial strain as an inoculant → grow plants → harvest nodulated roots → place washed nodulated roots in a solution of 5-bromo-4-chloro-3-indolyl-D-glucuronide (X-glc A). The GUS enzyme cleaves X-glcA to release an indoxyl derivative which on dimerization gives rise to a stable indigo precipitate that is blue in colour → count blue nodules. The method is useful for quantitative enumeration of selected rhizobial strains in a mixture of strains in the laboratory or in the field and to track inoculant strains on seeds or in carriers.

PCR Fingerprinting of DNA

The technique of amplification of specific DNA sequences by polymerase chain reaction (PCR) and its use in detection of microorganisms has already been described (see Chapter 3). The technique enables the amplification of characteristic fingerprints of DNA bands from bacterial genomes. The precision and reproducibility of these fingerprints allows perfect discrimination between different strains of the same species of *Rhizobium*. These patterns are known to be stable and conserved even in clonal descendants of the same strain that had been propagated independently for some years. Therefore, patterns of DNA (fingerprints) are excellent tools for verification and identification of strains.

A present, DNA fingerprinting has been done on strains of rhizobia grown in pure culture and also on strains that were recovered from nodules induced by single strain inoculum in sterile media. The method has not been used to detect rhizobial strains in field conditions where multiple strains may be involved in producing different nodules on roots. The envisaged method involves collecting field grown plants or plants grown in sterilised medium inoculated with a known strain of *Rhizobium* and separating the nodules from roots. The nodules are surface sterilized followed by extraction of crude DNA. The polymerase chain reaction

(PCR) is carried out to obtain amplified DNA fragments which are separated by gel electrophoresis on an acrylamide gel. The gels are stained and photographed to reveal patterns of DNA.

Nodulated Leguminous Trees

The bulk of tree species come from the sub-families caesalpinioideae and Mimosoideae of the family leguminoseae. The other sub-family papilionoideae has about 1000 tree species. Uprooting trees for studying nodulation is a difficult task and in adult trees nodules are often scarce and get dislodged from the main root. These nodules are often black in colour, resembling callus growth or tumours. They have various shapes because they are often indeterminate in growth. The mode of entry of *Rhizobium* into roots is generally through root epidermis at the point of origin of lateral roots.

Not all tree legumes are nodulated and if nodulated many do not possess extensive root nodules. Non-nodulated leguminous trees have extensive root system that are often deep rooted to tap nutrients. Non-nodulating *Acacia albida* and *Cercidium* sp. grow excellently in dry areas of Africa. Trees such as *Tamarindus indica* and *Ceratonia siliqua* grow and survive for over 100 years inspite of the fact they remain non-nodulalted, apparently deriving nitrogen from soil.

Measurement of actual amounts of N_2 fixed by nodulating tree legumes is a tricky proposition. Eventhough acetylene reduction measurements are often unreliable, they indicate low fixation *in situ* for many trees except for *Leucaena leucocephala* where fixation of 110 kg N/ha^{-1} in a low rainfall area in Tanzania has been reported. Only few reports on natural abundance of ^{15}N method which is more reliable than acetylene reduction method, are available. In Tanzania for instance *Prosopis* sp. was found to fix 25–30 kg $N/ha^{-1}/year^{-1}$, which had nodules on roots at a depth 4–5 m.

Rhizobium isolates from tree species are generally slow growing and have been assigned to the genus *Bradyrhizobium*. Wide variations in *Rhizobium* specificity have been observed in isolates from tree species when examined by conventional cross-inoculation grouping or by numerical taxomony.

There have been reports of nodulation of non-legumes by rhizobia. These include the well substantiated root nodulation of *Parasponia* spp. and the not so well authenticated reports of nodulation of *Zygophyllum* and *Tribulus* spp.

In agroforestry, co-cultivation of leguminous trees with other plants have been found to be beneficial. In Nigeria and other African countries, it has been observed that maize contained significantly more nitrogen in plots where *Rhizobium* inoculated *Leuceana* plants were also grown when compared to plots where maize was grown along with *Leuceana* plants which were not inoculated with *Rhizobium*. Superior growth of non-leguminous trees such as *Quercus rubra*, *Liquidambar styraciflua*, *Liriodendron tulipifera* (yellow poplar) and *Juglans regia* (black walnut) was observed in areas previously planted with nodulated leguminous tree *Robinia pseudo-acacia* (black locust) for 23 years, thereby indicating the ability of this leguminous tree to leave behind nitrogenous rich manure behind in soil. Many more examples of this kind are available in literature mostly from temperate forests. In developing tropical countries, nodulated leguminous trees such as *Prosopis* spp. *Leucaena leucocephala* (Su-babul), *Acacia nilotica*, *Dalbergia sissoo*, *Pongamia glabra*, *Pterocarpus indicus*, *Acacia auriculiformis*, *Acacia mangium*, *Calliandra calothyrsus*, *Sesbania bispinosa*, *Albizzia* spp. and others are planted to reclaim marginal denuded land. In mixed cropping or as ground cover, herbaceous legumes such as *Stybosanthes*, *Calapagonium* and *Pueraria* have been planted in tree plantations to replenish soil nitrogen.

Very few reports are available on mycorrhizal habit among trees. Many species of trees from ceasalpinioideae have been shown to be ectomycorrhizal. There appears to be widespread occurrence of AM fungi among tree species. Studies have been made on the interaction of AM fungi (*Glomus*, *Acaulospora*, *Gigaspora*) with *Rhizobium* isolates from *Leucaena leucocephala* and benefits due to the associative effects to these two symbionts on growth and phosphorus nutrition of the plant recorded. Similar results have been obtained with *Stylosanthes* sp. Since many studies have been carried out mainly from temperate regions, it is necessary to obtain factual data on tree nodulation and benefits accured by tree plantations in single or mixed cropping under tropical conditions to facilitate generalizations on the benefits of nodulated trees to the ecosystem.

Selected References

Acharya, C.N., Jain, S.P. and Jha, J. 1953. Studies on the building up of soil fertility by the phosphatic fertilization of legumes. Influence of growing berseem on the nitrogen content of the soil. *J. Indian Soc. Soil Sci.*, 1, 55–64.

Ahmed, S. and Evans, H.J. 1961. The essentiality of cobalt for soybean plants grown under aseptic conditions. *Proc. Natn. Acad. Sci., U.S.A.*, 47, 24–36.

Alexander, M. 1961. *Introduction to Soil Microbiology*. John Wiley & Sons, Inc., New York and London.

Allen, O.N. 1959. *Experiments in Soil Bacteriology*. Burgess Publishing Co., Minneapolis, Minn., U.S.A.

Berger, J.A., May, S.N., Berger, L.R. and Bohlool, B.B. 1979. Colorimetric enzyme-linked immunosorbent assay for the identification of strains of *Rhizobium* in culture and in the nodules of lentils. *Appl. and Environ. Microbiol.*, 37, 642–646.

Bergersen, F.J. 1960. Incorporation of ^{15}N into various fractions of soybean root nodules. *J. Gen. Microbiol.*, 22, 671–677.

Bosworth, A.H., Breil, B.T. and Triplett, E.W. 1993. Production of the anti-rhizobial peptide, trifolitoxin, in sterile soils by *Rhizobium leguminosarum* bv. *trifolii* T24. *Soil Biol. Biochem.* 25, 829–832.

Bosworth, A.H., Williams, M.K., Albrecht, K.A., Hankinson, F.R., Kwiatkuwski, R., Beynon, J., Ronson, C.W., Cannon, F., Wacek, T.J., and Triplett, E.W. 1994. Alfalfa yield response to inoculation with recombinant strains of *Rhizobium meliloti* carrying an extra copy of *det* and/or modified *nif* A expression. *Appl. Environ. Microbiol.*, 60, 3815–3882.

Bothe, H., deBruijn, F.D. and Newton, K.E. Eds. 1988. *Nitrogen Fixation: Hundred Years After*, Gustav. Fischer, Stuttgart.

Brockwell, J. and Bottomley, P.J. 1995. Recent advances in inoculant technology and prospects for the future. *Soil Biol. Biochem.*, 27, 683–697.

Bunting, A.H. and Horrocks, J. 1964. An improvement in the Raggio technique for obtaining nodules on excised roots of *Phaseolus vulgaris* L. in culture. *Ann. Bot.*, 28, 229–237.

Burns, R.C. and Hardy, R.W.F. Eds. 1975. *Nitrogen Fixation in Bacteria and Higher Plants*, Vols. I and II. Springer Verlag, Berlin.

Burris, R.H. 1969. Progress in the biochemistry of nitrogen fixation. *Proc. R. Soc.*, 172 B, 339–354.

Burris, R.H., Eppling, J.J., Wahlin, H.B. and Wilson, P.W. 1943. Detection of nitrogen fixation with isotopic nitrogen. *J. Biol. Chem.*, 148, 349–356.

Chopra, C.L. and Subba Rao, N.S. 1967. Mutual relationships among bacteroids, leghaemoglobgin and nitrogen content of Egyptian clover and gram. *Arch. Mickrobiol.*, 58, 71–76.

Dadarwal, K.R., Singh, C.S. and Subba Rao, N.S. 1974. Nodulation and serological studies of rhizobia from six species of *Arachis*, *Pl. Soil* 40, 535–544.

Date, R.A. 1970. Microbiological problems in the inoculation and nodulation of legumes. *Pl. Soil.*, 32, 703–725.

Dazzo, F.B., Yanke, W.E. and Brill, W.J. 1978. Trifollin—A *Rhizobium* recognition protein from white clover. *Biochem. Biophys. Acta.*, 539, 276–286.

De Ley, J. 1968. DNA base composition and classification of some more free-living nitrogen-fixing bacteria. *Antonie van Leeuwenhoek*, 34, 66–70.

Dreyfuś, B.L. and Dommergues, Y.R. 1981. Nitrogen fixing nodules induced by *Rhizobium* on stems of tropical legume *S. rostrata. FEMS Microbiol. Lett.* 10, 313–317.

Dobereiner, J., Burris, R.H. and Hollaender, A. Eds. 1978. *Limitations and Potentials for Biological Nitrogen Fixation*. Plenum Press, New York.

Dudman, W.F. 1971. Antigenic analysis of *Rhizobium japonicum* by immunodiffusion. *Appl. Microbiol.* 21, 973–985.

Dye, M., Skot, L., Mytton, L.R., Harrison, S.P., Dooley, J.J. and Cresswell, A. 1995. A study of *Rhizobium leguminosarum* biovar *trifolii* populations from soil extracts using randomly amplified polymorphic DNA profiles. *Can. J. Microbiol.* 41, 336–344.

Ellfolk, N. 1960a. Crystalline leghaemoglobin. I. Purification procedure. *Acta. Chem. Scand.*, 14, 609–616.

Ellfolk, N. 1960b. Crystalline leghaemoglobin. II. The molecular weights and shapes of the two main components. *Acta. Chem. Scand.*, 14, 1819–1827.

Ellfolk, N. 1961. Crystalline leghaemoglobin. III. Amino acid composition of the two main components. *Acta. Chem. Scand.,* 15, 545–554.

Elkan, G.H. 1984. Taxonomy and metabolism in *Rhizobium* and its genetic relationship, pp. 1–38, In *Biological Nitrogen Fixation.* Ed. M. Alexander, Plenum Press, New York.

Fred, E.B., Baldwin, I.L. and McCoy, E. 1932. *Root Nodule Bacteria and Leguminous Plants.* Univ. Wisconsin, Madison, Wis., U.S.A.

Gibson, A.H. 1967. Physical environment and symbiotic nitrogen fixation. IV. Factors affecting early stages of nodulation. *Austr. J. Sci.,* 20, 1087–1104.

Gibson, A.H. and Newton, W. 1981. *Current Perspectives in Nitrogen Fixation. Aust. Acad. Sci.*

Graham, P.H. 1963a. Vitamin requirements of root nodule bacteria. *J. Gen. Microbiol.,* 30, 245–248.

Graham, P.H. 1963b. Antigenic affinities of root nodule bacteria of legumes. *Antonie van Leeuwenhoek,* 29, 281–291.

Graham, P.H. 1964. An application of computer techniques to the taxonomy of the root nodule bacteria of legumes. *J. Gen. Microbiol.,* 35, 511–517.

Hallsworth, E.G. 1958. Ed. *Nutrition of the Legumes.* Butterworth's Scientific Publications, London.

Hardy, R.W.F. and Gibson, A.H. Eds. 1977. *A Treatise on Dinitrogen Fixation.* Section IV, Agronomy and Ecology, Wiley, New York.

Hollaender, A. Ed. 1977. *Genetic Engineering for Nitrogen Fixation.* Plenum Press, New York.

Lakshmi Kumari, M. and Subba Rao, N.S. 1973. Root hair infection and nodule to lateral root relationship in lucerne seedlings as influenced by dalapon and amitrole. *Arch. Mikrobiol.,* 93, 175–178.

Lakshmi Kumari, M., Singh, C.S. and Subba Rao, N.S. 1974. Effect of salinity and alkalinity on the early phases of infection in lucerne (*Medicago sativa* L). *Pl. Soil.,* 40, 261–268.

Lakshmi Kumari, M., Biswas, A., Vijayalakshmi, K., Narayana, H.S. and Subba Rao, N.S. 1974. Effect of certain water soluble herbicides on legume *Rhizobium* symbiosis. *Proc. Indian Natn. Sci. Acad.,* 40B, 528–534.

Maier, R.J. and Triplett, E.W. 1996. Toward more productive, efficient and competitive nitrogen-fixing symbiotic bacteria. *Critical Rev. in Plant Sci.,* 15, 191–234.

't Mannetje L. 1967. A re-examination of the taxonomy of the genus *Rhizobium* and related genera using numerical analsysis. *Antonie van Leevwenhoek,* 33, 477–491.

Mes, M.G. 1959. Influence of temperature on the symbiotic nitrogen fixation of legumes. *Nature,* London, 184, 2032–2033.

Ndoye, I. and Dreyfus, B. 1988. N_2 fixation by *Sesbania rostrata* and *Sesbania sesban* estimated using ^{15}N and total N difference methods. *Soil. Boil. Biochemistry,* 20, 209–213.

Newton, W.E. and Orme-Johnson, W.H. Eds. 1980. *Nitrogen Fixation.* University Park Press, Baltimore, U.S.A.

Nutman, P.S. 1976. Ed. *Symbiotic Nitrogen Fixation in Plants.* International Biological Programme, 7, Cambridge Univ. Press, London.

Philips, D.A. 1992. Flavonoids: Plant signals to soil microbes. *Recent Advances in Phytochemistry,* 26, 201–231.

Quispel, A. 1974. Ed. *The Biology of Nitrogen Fixation.* North Holland Publishing Co., Amsterdam.

Raggio, M. and Raggio, N. 1962. Root nodules. *Ann. Rev. Pl. Physiol.*, 13, 109–128.

Raggio, M., Raggio, N. and Torrey, J.G. 1957. The nodulation of isolated leguminous roots. *Amer. J. Bot.*, 44, 325–334.

Ranga Rao, V., Subba Rao, N.S. and Mukherji, K.G. 1973. *In vitro* effects of some growth regulators on *Rhizobium. J. Gen. Microbiol.*, suppl., 19, 55–58.

Ranga Rao, V., Sopory, S. and Subba Rao, N.S. 1974. Establishment of symbiosis *in vitro*, between *Rhizobium* and Pea (*Pisum sativum*) root callus. *Curr. Sci.*, 43, 503–505.

Rinaudo, G., Dreyfus, B. and Dommergues, Y.R. 1982. Influence of *Sesbania rostrata* green manure on the nitrogen content of the rice crop and soil. *Soil Biol. Biochem*, 15, 111–113.

Rolfe, B.G. and Gershoff, P.M. 1988. Genetic analysis of legume *nodule initiation. Ann. Rev. Plant Physiol. and Plant Mol. Biol.*, 39, 297–319.

Sahlman, K. and Fahraeus G. 1963. An electron microscope study of root hair infection by *Rhizobium. J. Gen Microbiol.*, 33, 425–427.

Schmidt, E.L., Bakole, R.O. and Bohlool, B.B. 1968. Fluorescent antibody approach to the study of rhizobia in soil. *Jour. Bacteriol.* 95, 1987–1992.

Sethi, R.P. and Subba Rao, N.S. 1968. Inhibitory or stimulatory effect of soil fungi on rhizobia. *J. Gen. Appl. Microbiol.*, 14, 325–327.

Sharma, P.K., Anand, R.C. and Lakshminarayana, K. 1991. Construction of Tn 5 tagged mutants of *Rhizobium* spp. (*Cicer*) for ecological studies. *Soil Biol. & Biochem.* 23, 881–885.

Subba Rao, N.S. 1971. The role of photoperiod and temperature in determining the efficiency of legume-*Rhizobium* symbiosis in cluster clover (*Trifolium glomeratum*). *Proc. Indian Acad. Sci.*, 74B, 106–111.

Streeter, J.G. and Smith, R.T. 1998. Introduction of rhizobia into soils—problems, achievements and prospects for the future. pp. 45–64. In *Microbial Interactions in Agriculture and Forestry*, Eds. N.S. Subbarao and Y.R. Dommergues, Oxford and IBH Publishing Co. Ltd., New Delhi.

Somasegaran, P. and Hoben, H.J. 1985. *Methods in Legume-Rhizobium Technology*, NiFTAL, USA.

Subba Rao, N.S. 1972. Rhizobia and nodulation. *Curr. Sci.* 41, 1–9.

Subba Rao, N.S. 1976. Field response of legumes in India to inoculation and fertilizer application. In *Symbiotic Nitrogen Fixation in Plants*, pp. 255–268. Ed. P.S. Nutman, International Biological Programme, Vol. 7, Cambridge Univ. Press, London.

Subba Rao, N.S., Lakshmi Kumari, M., Biswas, A. and Singh, C.S. 1974. Salinity and alkalinity in relation to legume-*Rhizobium* symbiosis. *Proc. Indian Natn. Sci. Acad.*, 40B, 544–547.

Subba Rao, N.S. 1979. Ed. *Recent Advances in Biological Nitrogen Fixation*. Oxford & IBH Publishing Co., New Delhi.

Subba Rao, N.S. 1988. Ed. *Biological Nitrogen Fixation—Recent Developments*. Oxford and IBH Publishing Co. New Delhi.

Subba Rao, N.S. Ed. 1982. *Advance in Agricultural Microbiology*. Oxford & IBH Publishing Co., New Delhi.

Subba Rao, N.S. and Dommergues, Y.R. Eds. 1998. Microbial Interactions in Agriculture and Forestry (Vol. I). Oxford and IBH Publishing Co. Ltd., New Delhi.

Subba Rao, N.S. and C. Rodriguez-Barrueco. 1993. Ed. *Symbioses in Nitrogen Fixing Trees.* Oxford and IBH Publishing Co., New Delhi.

Subba Rao, N.S., Tilak K.V.B.R. and Singh, C.S. 1980. Root nodulation studies in *Aeschynomene aspera. Pl. Soil,* 56, 491–494.

Subba Rao, N.S. and Yatazawa, M. 1984. Stem nodules. pp. 101–110. In *Current Developments in Biological Nitrogen Fixation.* Ed. N.S. Subba Rao, Oxford and IBH Publishing Co., New Delhi.

Triplett, E.W. and Sadowsky, M.J. 1992. Genetics of competition for nodulation of legumes. *Ann. Rev. Microbiol.* 46, 399–428.

Vachhani, M.V. and Murty, K.S. 1964. *Green Manuring for Rice.* Bull. No. 4, Central Rice Research Institute, Cuttack, India.

Van Schreven, D.A. 1959. Effect of added sugars and nitrogen on nodulation of legumes. *Pl. Soil.,* 11, 93–112.

Vincent, J.M. 1970. *A Manual for the Practical Study of the Root Nodule Bacteria.* IBP Hand Book No. 15, Blackwell Scientific Publications, Oxford.

Wilson, J.K. 1964. Over five hundred reasons for abandoning the cross-inoculation groups of the legumes. *Soil Sci.,* 58, 61–69.

Wilson, K.J. 1995. Molecular techniques for the study rhizobial ecology in the field. *Soil Biol. Biochem.* 27, 501–504.

Yao, P.Y. and Vincent, J.M. 1969. Host specificity in the root hair 'Curling factor' of *Rhizobium* spp. *Austr. J. Biol. Sci.,* 22, 413–423.

9. Actinorhizal Plants (*Frankia*-Induced Nodulation)

Root Nodules Caused by *Frankia* (Actinorhizal Symbiosis)

Apart from legumes nodulated by bacteria of the genus *Rhizobium*, roots of some plants belonging to diverse families are also nodulated by members of actinomycelates (tentatively classified in the genus *Frankia*) and fix considerable amounts of nitrogen. Alder trees (*Alnus* spp.) are known to recuperate soil nitrogen in temperate forest ecosystems. This is a well-recognised fact all over the world in forest management. Under field conditions, nodules of *Alnus* and *Casuarina* occur in clusters attaining a diameter of 5 to 6 cm somewhat resembling a tennis ball (Fig. 72), often weighing up to 444 kg dry weight of nodules/ha. Field estimates have shown that alder trees increase the nitrogen content of soil by about 61.5 to 157 kg N/ha/year and *Casuarina* trees release about 60 kg N/ha/year which indicate the importance of root nodule bearing non-leguminous plants in the overall nitrogen economy of soil.

The term actinorhiza and actinorhizal plants is derived from 'actino' for the actinomycete *Frankia* named after its discoverer Frank in the 1880s and "rhiza' for the plant root bearing the nodules formed by symbiosis.

In developing countries, deforestation for fuel has rendered the land barren and continuous deforestation of the same land in overpopulated regions of such countries has resulted in soils which remain deficient in nitrogen, the most important element for the normal growth of plants. One of the least expensive and non-polluting ways to replenish the lost soil nitrogen is reafforestation by planting self-supporting nitrogen-fixing trees. In Pennsylvania, U.S.A., reafforestation of mine spoils have been done by planting nitrogen fixing red alder (*Alnus rubra*) inoculated with nodule forming *Frankia*. In New Quebec, Canada, alder plants (*Alnus* spp.) have been planted on a large scale to fill dam dykes. In Senegal, Egypt, and the coastal region of India and China nitrogen-fixing *Casuarina* spp. have been planted on a large scale to contain and stabilise sandy tracts which have made inroads to agricultural land.

There are 24 genera from 8 Angiosperm families which have been described to possess actinorhizal root nodules. These genera with families mentioned in parentheses, from most primitive to most advanced orders,

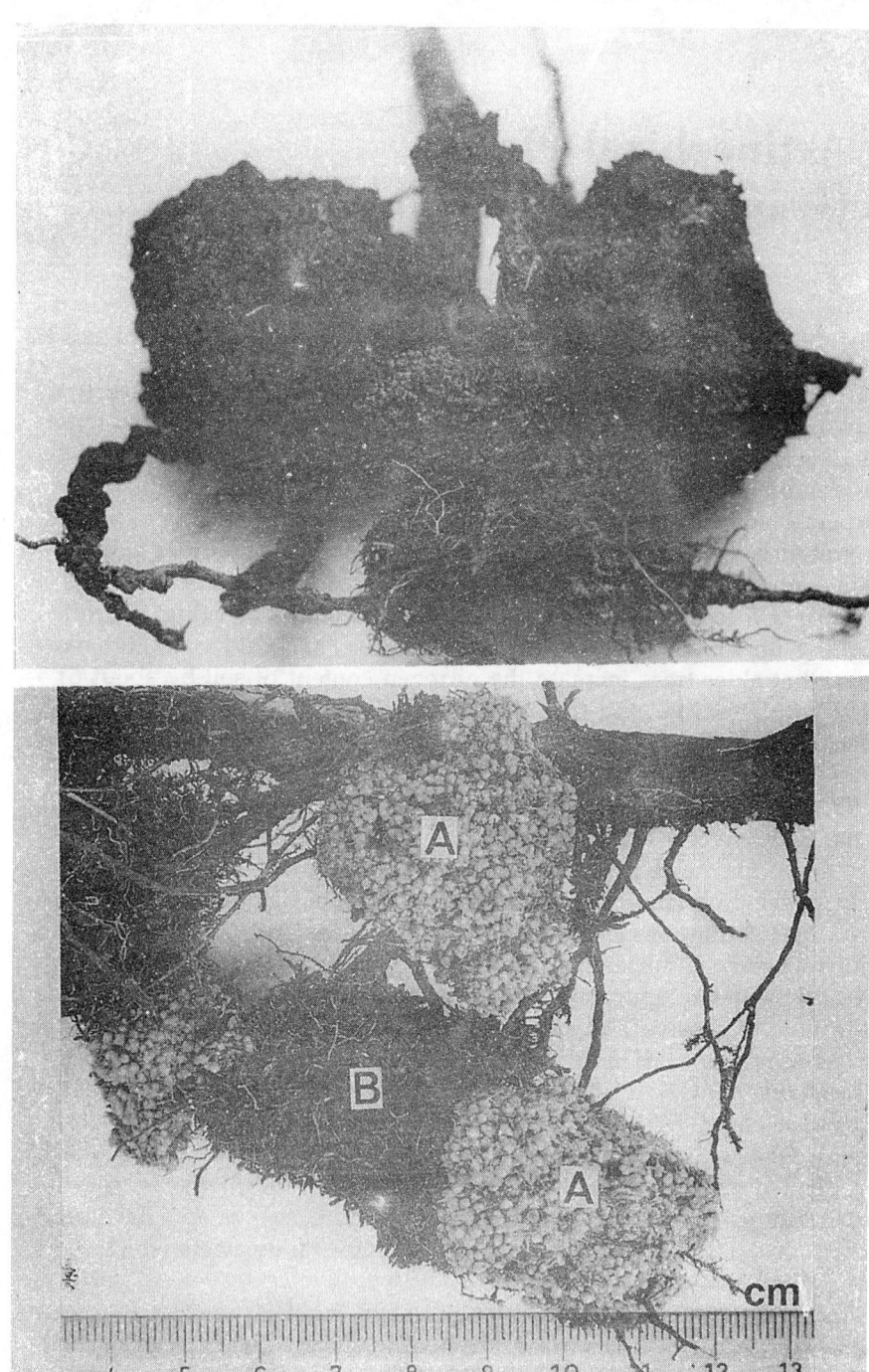

are *Coriaria* (Coriariaceae), *Cerocarpus, Chamaebatia, Cowania, Dryas, Pur-shia, Rubus* Rosaceae, *Datisca* (Dastiscaceae), *Comptonia, Myrica* (Myricaceae), *Alnus* (Betulaceae), *Elaeagnus, Hippophae, Shepherdia* (Elaeagnaceae), *Coenothus, Colletia, Discaria, Kentrothamnus, Retamnilla, Talguenea, Trevoa* (Rhamnaceae), *Allocasuarina, Casuarina* and *Gymnostoma* (Casuarinaceae). The genera *Casuarina* for tropical and sub-tropical regions and *Alnus* (*A. rubra., A. glutinosa, A. crispa, A. jorullensis, A. acuminata*) for temperate regions stand out as excellent examples for the benefits they provide to the ecosystems by way of nitrogen inputs. They can adapt themselves to grow under most diverse environmental conditions and geographical zones. *Casuarina* species (*C. equisetifolia. C. cunninghamiana, C. littoralis, C. stricta, C. junghuniana, C. glauca* and *C. torulosa*) provide substantial fuel and building materials in tropical countries while alders provide the most utilised hard wood as well as bark for paper industries in temperate regions. The microsymbiont in actinorhizal root nodules is an actinomycete and has been collectively designated as *Frankia*.

Cross-inoculation Experiments

It may be recalled that cross-inoculation groups in nodulated legumes was based on the ability of rhizobia to nodulate a group of legume species. The same yardstick has been used in attempts to classify actinorhizal plants with much less success than in leguminous plants. Earlier studies were carried out by inoculating plants with nodule suspensions and with the advent of pure cultures of *Frankia*, similar studies have been carried out with laboratory grown cultures. The objective of cross-inoculation experiments has been to understand whether host-*Frankia* strain specificity exists in nature.

In general, plants from a given species were nodulated effectively by inocula prepared from crushed nodules of the same species. Exceptions were however noticed; for instance, *Casuarina stricta* inoculum was unable to nodulate *Casuarina cristata, C. cunninghamiana* and *C. torulosa*. Similarly, inoculum prepared from *Casuarina glauca* and *Coriaria myrtifolia* nodule suspensions did not nodulate *Coriaria japonica*. Likewise, *Myrica pilulifera* nodule homogenates did not nodulate *M. gale*. Other examples cited in literature have been that *Alnus* nodule suspensions could not nodulate *Hippophae* and *Coriaria* species but showed low level of infectivity towards *Elaeagnus angustifolia* and *Myrica faya*. Similarly, *Coriaria myrtifolia* in-

Fig. 72 Actinorhizal root nodules: Top—an excellent specimen of *Alnus glutinosa* nodules from the laboratory of C.R. Barruecco of Salamanca, Spain; Bottom—Nodules formed on the roots of a 10 year old *Casuarina equisetifolia* tree. A—young nodule lobes; B—degenerating nodule lobes. (courtesy, Diem and Dommergues, France).

oculum was not infective on *A. glutinosa,* and *E. angustifolia* inoculum showed low infectivity on *A. glutinosa.* Alnus nodule extracts did not produce nodules on *Casuarina equisetifolia* and similarly *Casuarina* nodule extracts were non-infective on *Myrica* and *Coriaria* plants.

From extensive inoculation studies with pure cultures obtained from *Alnus, Casuarina, Allocasuarina, Gymnostoma, Datisca, Elaeagnus, Hippophae, Shepherdia, Myrica, Comptonia, Ceanothus, Colletia, Cerocarpus, Cowania* and *Purshia* on host plants such as *Alnus glutinosa, A. rubra, Casuarina equisetifolia, Elaeagnus angustifolia, Hippophae rhamnoides* and *Myrica cerifera,* the following four host specificity groups were defined: 1. strains nodulating members of *Alnus* and *Myrica* 2. strains nodulating members of *Casuarina* and *Myrica* 3. strains nodulating members of *Elaeagnaceae* and *Myrica,* and 4. strains nodulating only members of *Elaeagnaceae.*

The above generatizations and many other similar reports tend to indicate that the so called cross-inoculation groups do not correlate with plant taxonomic groups and therefore may not be a sound procedure to classify *Frankia* isolates.

Isolation and Characteristics of *Frankia*

After many unsuccessful attempts, Callaham and associates in Torrey's laboratory in the USA first reported the successful isolation of *Frankia* from *Comptonia peregrina* in 1978. Since that time, many more strains have been isolated from several species of non-leguminous plants: *Alnus glutinosa, A. rhomvifolia, A. rubra, A. incana, A. nitida, A. viridis, Elaeagnus umbellata, E. commutata, E. angustifolia, Hippophae rhamnoides, Colletia cruciata, Purshia tridentata, Cowania mexicana, Ceanothus americanus, Cerocarpus pennsylvanica, C. ledifolius, Casuarina* spp., *Allocasuarina lehmanniana* and *Gymnostoma papuanum.*

Older actinorhizal nodules are generally big and are subjected to contaminants which interfere with the isolation of pure cultures of *Frankia.* This was the reason why earlier workers had missed real *Frankia* and ended up by isolating contaminants. Young nodules that are not suberized are preferred to older ones. The contaminants are excluded by surface sterilization with osmium tetroxide under mild vaccum in a fume hood to avoid injury by the toxic nature of the surface sterilant. The nodules are then washed repeatedly and disrupted or dissected to expose nodular cells to release the sparse number of *Frankia* cells. The process of maceration or disruption of nodules may often release toxic phenolic compounds that may suppress *Frankia* growth. Hence some workers have passed the homogenates through activated charcoal to get rid of the polyphenolic compounds. To concentrate *Frankia* cells, the homogenates are passed through nylon membranes or subjected to sucrose-density centrifugation.

The use of enzymes such as cellulases and pectinases were adopted in the maiden attempt at isolation of *Frankia* but future workers have not found the need for using them.

For *Casuarina* nodules the following procedure has been successfully adopted: After removing the outer layers, the nodules are surface sterilized in 3.0 per cent osmium tetroxide, cut into lobes or small pieces, washed in sterile water repeatedly and pieces transferred to a vial containing the recommended Q mod liquid medium (see appendix) containing 0.3 per cent agar. The vials are incubated at 25–28°C upto 2 months discarding those vials that develop growth of fast growing contaminants. When *Frankia* growth becomes clearly noticeable in some nodule lobes, such colonies are subcultured repeatedly and tested for nodulation on aseptically grown seedlings of *Casuarina*.

Diem and Dommergues from France provide the following protocol for the isolation of *Frankia* endophytes from root nodules: Clear the nodule of extraneous organic matter, soil and dirt under running water by frequent examination under a dissecting microscope. Fragment the nodule into individual lobes, sterilize the lobes by immersion into a 3.0 per cent aqueous solution of osmium tetroxide for 1–4 min. according to the nodule mass and age, wash in sterile distilled water several times and cut the nodule lobes into 0.1–0.5 mm^3 pieces with the help of a sterile scalpel. Transfer these nodule pieces into bottom layer of 1.5 per cent of nutrient medium in a petri dish (yeast extract dextrose medium or Q mod medium or casamino acids and sodium pyruvate medium or Q mod medium with activated charcoat/tween 80—see Appendix for the composition of media). Addition of cycloheximide at a concentration of 50 µg/ml may be useful for preventing fungal contamination. Pour 3 ml of the same medium over the layer containing nodule pieces, thereby providing microaerophilic conditions and facilitating *Frankia* growth which can be periodically checked under a dissecting microscope. Seal the Petri dishes with paraffin and incubate at 28–30 degree centigrade. After 4 weeks, colonies of *Frankia* generally appear at the edge of nodule pieces.

Frankia colonies appear to come up on plates easily when activated charcoal is added to the medium, apparently due to the elimination of toxic phenolic compounds present in the nodular tissue. Solid agar medium is preferable to liquid medium, especially when concentrated nodule homogenates are incorporated into the upper layer of double layered pour plate. In this way, fast-growing contaminants can be eliminated. Lipids from roots encourage the growth of *Frankia*. Although Callaham's original medium was complex, rather more simpler media are now available for the isolation of *Frankia*. In liquid culture, *Frankia* growth results in ellipsoidal or spherical colonies of 0.5–1.00 mm in diameter after clustering together and sticking at the bottom of the container. It is not clear which portion of the nodular material whether hyphae, spores, sporangia, vesicles or other propagules gives rise to *Frankia* colonies. What is described as polymorphism appears to exist for cultures isolated from nodule segments. These colonies may be diffuse on agar plate with a loose network of hyphae around the centre, or compact with a dense network of hyhae bearing many sporangia. *Frankia* colonies arising from a single sporangiophore may be regarded as genetically pure and the polymorphism exhibited may be a multi-strain effect.

Cultural Characteristics of *Frankia*

Frankiu has a long lag phase up to 14 days, slow exponential phase and autolysis of most vegetative structures without any stationary phase which makes it difficult to grow the organism on a mass scale. Besides being microaerophilic, it has a temperature requirement of 28–30 degree centigrade and pH requirement of 6.0–7.0. Succinate is probaby a good carbon source for *Frankia*. Spores are often produced in culture as well as in nodules. Two types of nitrogen-fixing root nodules have been identified in actinorhizal plants based on the extent of spore formation by *Frankia* within nodules. They are called spore (–) for nodules where spores are absent or few and spore (+) for those containing many spores.

Frankia grows slowly and often takes 2 months to show up in culture. They are known to exhibit polymorphism of colonies ranging from starfish, diffuse or compact shapes. The formation of round, cylindrical or highly irregular compartmentalized sporangia intrahyphally or terminally, filled with spores in submerged culture is unique to the genus *Frankia*. The hyphae are poorly branched, may be colourless or pigmented depending upon the nature of the medium. Round, cylindrical, stipitate vesicles are formed in nitrogen-free media. These swollen tips of hyphae (vesicles) assume various shapes ranging from pear, club or filamentous types and are regarded as the sites of nitrogen fixation (like heterocysts of blue-green algae) both in culture media and within nodules. The exception to this rule is *Casuarina* nodules which lack vesicles and where other mechanisms operate for nitrogen fixation. Vesicles in general possess an intrinsic oxygen protection mechanism to sustain continued nitrogenase activity (Fig. 73).

Taxonomy of *Frankia*

Speciation in *Frankia* has been a difficult proposition and none of the criteria (physiological, genetical or otherwise) have been satisfactory. As mentioned earlier, infectivity criterion has also been misleading and informed workers in the field have agreed that species naming in *Frankia* should be deferred until a clear picture emerges. Presently, designation of

Fig.

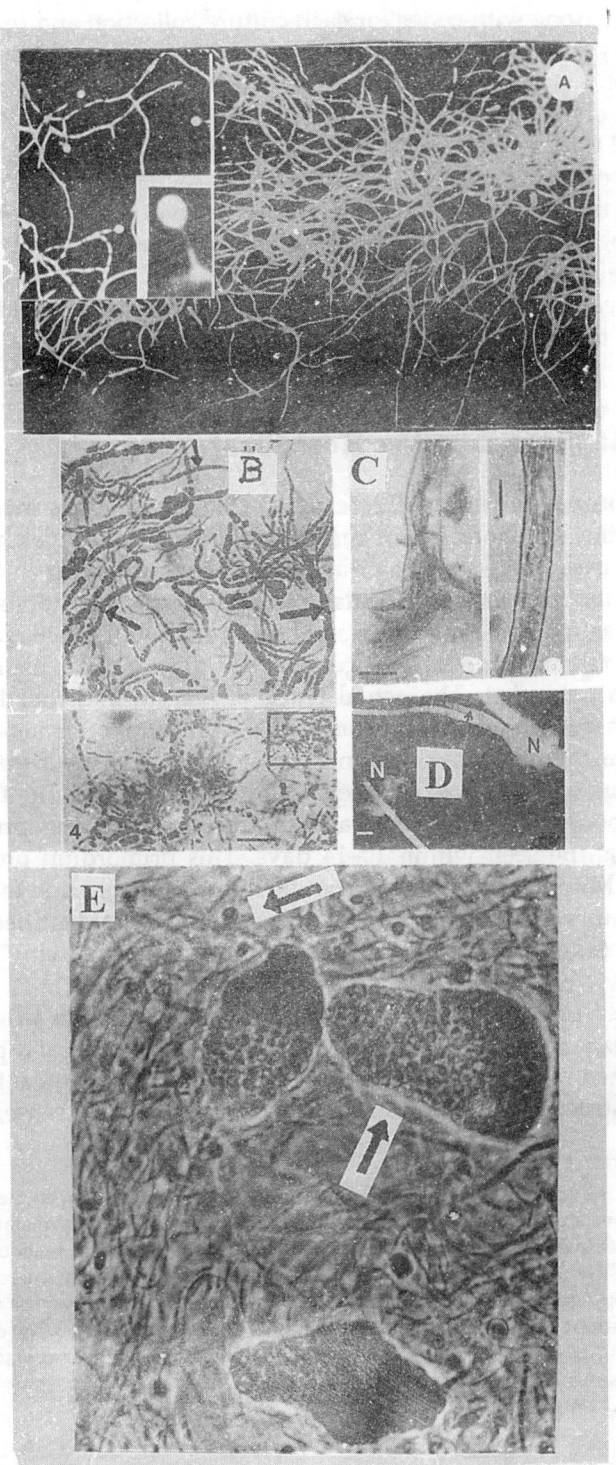

a strain is done with codes for each culture collection and upto 10 numerical digits for each research group, the first two numbers reflecting the source of a strain. For example, the University of Leiden would be ULN, the University of Laval ULQ and so on followed by 00 for soil, 01 for *Alnus*, 02 for *Casuarina* etc; HFP 020203 indicates Harvard Forest Petersham and 02 refers to *Casuarina*, the remaining digits referring to the number given to the isolate by the culture collection centre. Because there are only few recognised centres for *Frankia* research, the system appears to be a workable method of finding out the source of a particular culture and infact, a catalogue of recognised *Frankia* cultures in the world is constantly updated to help research workers.

Entry of *Frankia* to the Host Plant

Frankia propagules normally occur in soils of all types including sandy soils on the sea shore where casuarinas grow profusely as a sand binder and fuel tree. The events that have been shown to occur in the root zone of *Casuarina* after *Frankia* inoculation under axenic conditions can be summarized as follows: *Frankia* cells get embedded in a mucilage layer in the root region or the spores may get attached to root hairs, the root hairs get deformed or curled. The actual entry of *Frankia* into root hairs has not been seen but hyphae are seen as simple or multiple threads often branching inside the deformed hair in a host derived cell wall material that is continuous with the root hair cell wall (encapsulation). The threads could be seen penetrating the cortex and in some root sections, pre-nodule formation can be seen within 10–14 days. This primordiun protrudes from the root. Mitotic activity is rapid at this time probably due to growth substances, auxins, cytokinins, gibberellins and many unidentified substances. The infected cells and the nucleus appear enlarged with a prominent nucleolus.

Sooner or later, lateral roots in the vicinity of the primary nodule primordium appear, their meristems undergo branching and progressively get infected with *Frankia* resulting in the formation of a typical adult nodular structure referred to as a 'rhizothamnion'. In a sense, actinorhizal

---

Fig. 73 Morphological features of *Frankia* in culture medium: A—open-mesh microcolony from an exponential phase; the inset shows the same strain in stationary cultures with vesicles and also an enlarged vesicles with stalk and bulbous nature of the vesicle (courtesy J. Schwenke, France); B-septate hyphae with intercalary sporangia (top figure) and the same strain in the rhizophere of *C. equisetifolia* (bottom figure); C—root hairs of *C. equisetifolia* with infection threads; D—3 week old nodules (N) of *C. equisetifolia* with nodular roots; E—enlarged version of *Frantia* showing bulbous vesicles and large sporangia indicated by black arrows. (courtesy Diem, Gauthier and Dommergues of France).

root nodule is essentially a modified lateral root. The nodule consists of a central vascular bundle surrounded by the endodermis and a cortical parenchyma with pockets of infected cells scattered all over. Ineffective nodules have few infected cells with only hyphae and lack vesicles whereas effective nodules have both sporangia and vesicles with active cytoplasm. Haemoglobin-like substances whose role in actinorhizal N_2 fixation is not clear have been encountered in *Alnus, Hippophae, Myrica* and *Casuarina* nodules. However, the red colour of nodules in some other actinorhizal plants is due to pigments of the anthocyanin type. Under hot humid coastal areas in Senegal casuarinas are known to produce aerial nodules on stem which fix nitrogen.

Structure of Actinorhizal Nodule

There are two types of structural organisation in actinorhizal root nodules—the *Alnus* type and the Casuarina type (Fig. 74). *Alnus* type of nodules have many lenticels on nodules providing ventilation while *Casuarina* type do not have ventilatory organs. Instead, those lateral roots that escape infection by *Frankia* develop into nodule roots serving as antennae for oxygen diffusion. Both the types of nodules differ from legume root nodules. Legume root nodules are characterized by a central uniformly infected bacteroid and leghaemoglobin containing zone surrounded by a tight inner cortex that limits gas diffusion with the vascular bundles lying outside the inner cortex. In actinorhizal nodules, however, there is a central vascular bundle surrounded by a cortex in which several pockets of *Frankia* inhabiting zones can be seen containing vesicles (Fig. 75) barring those belonging to *Casuarina* types which have suberized cells containing the hyphal endophyte with swollen tips. The suberized cells in *Casuarina* type of nodules are impervious to air and hence provide protection to nitrogenase and to the swollen hyphal tips which are believed to be the sites of nitrogen fixation. These differences are diagrammatically depicted in Fig. 76.

Nitrogen Fixation and Assimilation

Unequivocal evidence has been presented using ^{15}N enriched gas to demonstrate that *Casuarina* nodules in intact plants are capable of fixing nitrogen. The nitrogenase enzyme in *Alnus glutinosa* has properties similar to legume nodule nitrogenase, except the fact that sources of ATP and reductant have not been identified. The nodules have an efficient uptake hyrogenase system in *Alnus* and *Casuarina*. Nitrogenase activity is host as well as *Frankia* strain dependent, especially on the morphological state of *Frankia* whether in the form of spores or hyphae. Nitrogenase has been

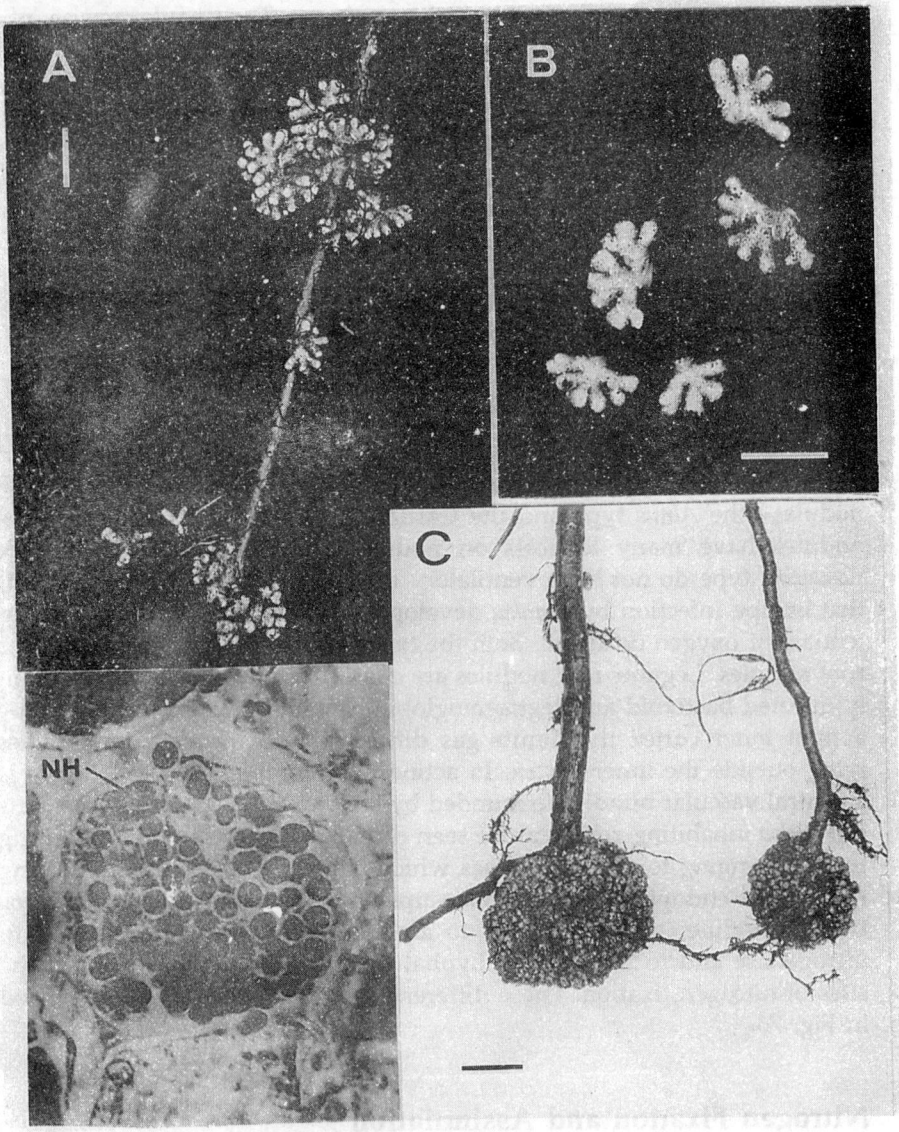

Fig. 74 *Alnus*-type root nodule.
A coralloid root nodule with arrested growth of the apical meristem.
(A) Relatively young root nodules of *Alnus glutinosa in situ*; Water-culture plant.
(B) Detached and divided *Alnus glutinosa* root nodules showing the dichotomous branching of the nodular lobes.
(C) Field collected root nodules of *Alnus glutinosa* showing their globular structure and the size of aged root nodules.
Bar scale = 1.0 cm.
(D) Cells of parenchyma of *A. glutinosa* nodule showing vesicles of the endophyte
(Courtesy, J.H. Becking Wageningen).

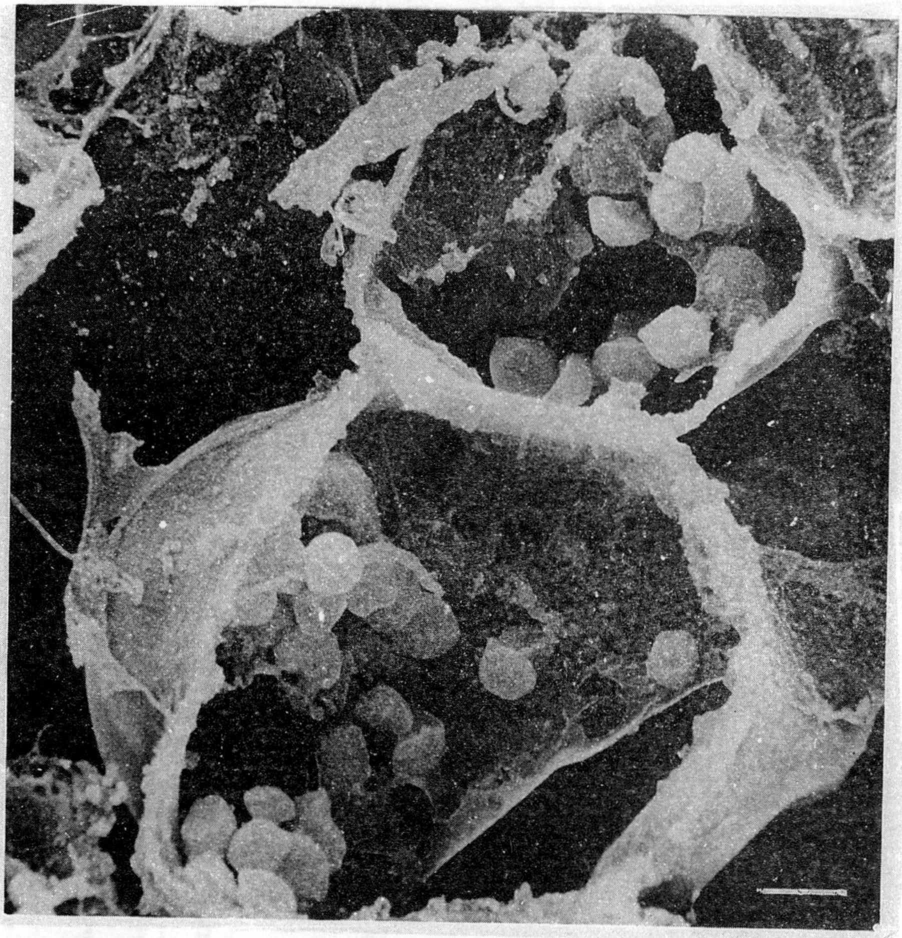

Fig. 75 Scanning electron micrograph of two host cells in an *Alnus glutinosa* nodule showing vesicular structures of the endophyte. (Courtesy, J.H. Becking Wageningen.) Bar scale = 5.0 µm.

detected in vesicles as well as hyphae but abundance of vesicles coincides with high nitrogenase activity. In 9 out of 21 genera of actinorhizal plants nodules have been encountered with many sporangia having plenty of spores designated as spore$^+$ nodules. In other plants sporangia with less abundant spores have been seen, referred to as spore$^-$ nodules. In general, Sp$^-$ nodules are far superior to Sp$^+$ nodules in nitrogen fixation and high plant biomass production.

Two hypotheses have been suggested to indicate the possible mechanism of ammonia assimilation on lines similar to the ones operating in cyanobacterial heterocysts involving GS and GOGAT enzymes. One of

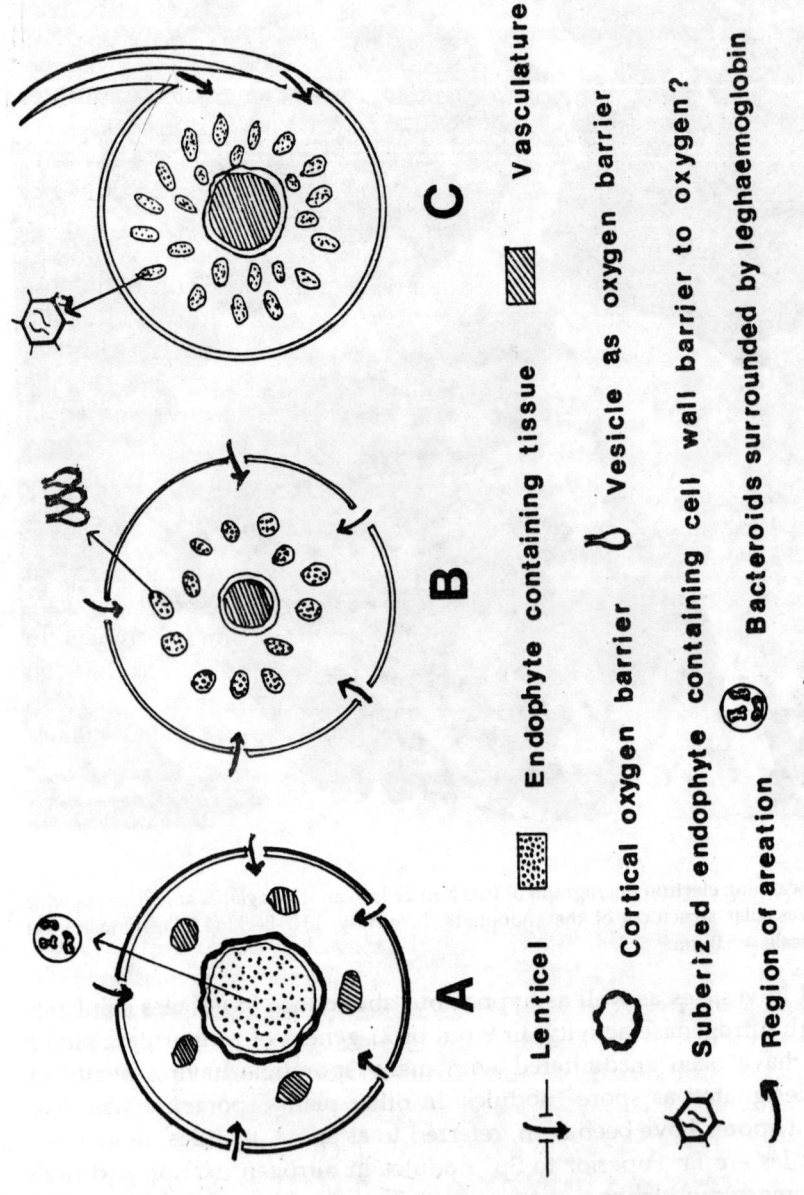

Fig. 76   Diagrammatic representation of how internal structure of root nodule of a typical legume and a typical actinorhizal plant differ: A—legume; B—Alnus; C—Casuarina showing upwardly oriented nodule lateral roots: (after Tjpkema, 1998 and Silvester et al, 1990.)—Not to scale.

Lenticel   Endophyte containing tissue   vasculature

Cortical oxygen barrier   vesicle as oxygen barrier

Suberized endophyte containing cell wall barrier to oxygen?

Region of areation   Bacteroids surrounded by leghaemoglobin
regulating oxygen supply

these assumes that vesicles have the potential to produce glutamine, which could be transferred to vegetative hyphae through the constricted stem cell of the vesicles. In the hyphae, glutamine would be converted by GOGAT to glutamate with one of the resulting glutamates going back to the vesicles to act as an ammonia acceptor for repeating the reaction. In the second hypothesis, it is assumed that the GS is not active in ammonia assimilation in vesicles which leads to accumulation of the fixed product in the hyphae and surroundings where it would be assimilated by the GS-GOGAT system, presumably aided by the high affinity ammonia permease present in nitrogen starved hyphae which helps in mopping up all free ammonia.

Evidences for the fixation of nitrogen in non-leguminous root nodules have come from two angles: (1) Long-term experiments with nodulated and nodule-free plants grown on nitrogen-free rooting media: Nodulating *Alnus glutinosa* can accumulate as much as 300 mg nitrogen per plant in contrast to nodule-free plants which exhausted the nitrogen in the cotyledons and made little further growth. When plants were raised in nitrogen-free mineral medium, total nitrogen content of nodulated (N) and non-nodulated (NN) plants at the end of 17 to 23 weeks were as follows: (1) *Myrica cerifera*—84.5mg (N)/1.0 mg (NN) per plant; *Myrica gale*—158.0mg (N)/1.0 mg (NN) per plant; *Alnus glutinosa*—492.0mg (N)/0.5mg (NN) per plant. (2) By the use of ^{15}N: In experiments with nine species of *Myrica*, nodules and apparently nodule-free roots of the same plants were exposed to ^{15}N labelled gas. Samples were analysed for enrichment in nodules and roots. Convincing evidence was obtained for nitorgen fixation in nodules which invariably showed higher accumulation of ^{15}N in nodulating species of *Alnus*, *Hippophae*, *Casuarina*, *Shepherdia* and *Ceanothus*.

Field Experiments on the Benefit of *Frankia* Inoculation

Earlier field experiments were carried out with fresh nodule homogenates but later experiments have been done using pure cultures of *Frankia*. To overcome the limitation in obtaining bulk inoculum because of the slow growing characteristic of *Frankia*, continuous culturing of *Frankia* for field experiments with *Casuarina* have been adopted. *Frankia* cells have also been entrapped in alginate beads or montmorillonite or kaolite clays and used for inoculation in the field in Senegal on sandy soil deficient in nitrogen with the object of improving the biomass and nitrogen content of *Casuarina*.

Mycorrhizae and Actinorhizal Plants

One or more species of the following genera of actinorhizal plant species are known to possess either ectomycorrhiza (EM) or arbuscular mycorrhiza (AM) or both: *Alnus, Casuarina, Comptonia, Myrica, Elaeagnus, Shepherdia, Purshia, Cerocarpus, Dryas, Coriaria* (both EM and AM) and *Discaria, Hippophae, Ceanothus, Rubus, Datisca, Colletia* (AM). In general, the AM symbionts belong to the genera *Glomus, Gigaspora* and *Acaulospora*.

AM inoculation of *Casuarina equisetifolia, Hippophae rhamnoides* and *Ceanothus velutinus* proved beneficial in enhancing the uptake of phosphorus, nodulation and nitrogen fixation.

Proteoid Roots

The formation of dense cluster rootlets on elongating lateral roots is characteristic of *Casuarina equisetifolia* (Fig. 77). Even though *Azospirillum* and AM fungi could be isolated from such roots, no clear cut evidence has comeforth to pinpoint that microorganisms are responsible for the formation of proteoid roots. The function of these roots may be to absorb phosphorus from the vicinity of root system more efficiently than the lateral roots.

Genetics of *Frankia*

The genome size of *Frankia* is twice that of *E. coli* and like *Streptomyces* measures 10,000 kilo bases. The G + C per cent of *Frankia* strains so far quantified lie between 68 and 72 which is rather high. In some strains of *Frankia* sp., plasmids ranging from 8kb to 190kb have been detected. It has been shown that in *Frankia* sp., isolated from *Alnus glutinosa*, at least some of the *nif* genes are located on a large indigenous plasmid of 190kb. Comparisons with *nif* H sequences of other nitrogen-fixing microorganisms such as *Anabaena, K. Preumoniae* and *Azotobacter* suggest that *Frankia nif* genes are similar to these microorganisms.

Of late, DNA probes have been used to study the competitiveness for nodule formation of *Frankia* strains in *Alnus glutinosa* and *A. incana*. Applying the analytical data of 165 r RNA sequence from several *Frankia* strains, two major *Frankia* divisions have been distinguished—those infective on *Elaeagnus* and those infective either on *Alnus* or *Casuarina*. One particular strain from *Purshia tridenta* was however grouped outside these two divisions. This information has also helped in confirming the identity of several atypical non-infective isolates of *Frankia* and also in grouping of strains infective on *Casuarina equisetifolia*.

Rhizobium-induced Root nodulation of *Parasponia*

This unique association between a *Rhizobium* and a non-leguminous plant has been demonstrated in the genus *Parasponia* (also described as *Trema*),

Fig. 77 Cluster or Proteoid roots in *Casuarina equisetifolia*—A—under natural conditions in soil and B—under laboratory conditions in a perlite substrate in pot in a growth chamber.

belonging to the family Ulmaceae in 3 out of 5 species of the genus—*P. rugosa*, *P. parviflora* and *P. andersonii*. The structure of the root nodule resembles that of legume root nodules. However, unlike the infection thread in leguminous root nodules, in *Parasponia* root nodules rhizobia are very rarely released from infection threads which grow and fill the host cell. The isolation of *Rhizobium* from nodules is easy and nodulated plants exhibit demonstrable nitrogenase activity.

Leaf Nodules

The occurrence of leaf nodules is confined to the families of Rubiaceae and Myrsinaceae. Among the several genera of the family Rubiaceae reported to form leaf nodules (*Pavetta*, *Chomelia*, *Psychotria*), the genus *Psychotria* has received considerable attention from many workers. Whether or not the bacterial endophyte in the leaf nodule of this genus fixes nitrogen in association with the plant is a debatable point. Earlier reports claiming nitrogen fixation have later been disproved by experiments with nodulated leaves from plants grown in media free of combined nitrogen. Detached *Psychotria* leaves bearing leaf nodules were tested for nitrogenase activity by the acetylene reduction technique and also by exposing to an atmosphere containing ^{15}N labelled nitrogen gas to determine the extent of nitrogen fixation. The results showed no evidence of nitrogen-fixation. Experiments with nodulated leaves of several species of *Pavetta*, either by the use of acetylene reduction technique or by the use of ^{15}N have also not shown any clear-cut indications of nitrogen-fixing ability.

The identity of the endophyte of *Psychotria* nodules remains still in doubt in spite of repeated investigations. Isolates of bacteia from leaf nodules have been variously named *Mycobacterium rubiacearum*, *Mycoplana rubra*, *Flavobacterium* sp., *Bacterium rubiacearum*, *Phyllobacterium rubiacearum* and *Klebsiella rubiacearum*. If fixation of nitrogen is not attributable to leaf nodules of *Psychotria*, the function of symbiosis may lie in the ability of the endophyte to produce phytohormones, a suggestion which has been put forth by some investigators. Some improvements in the growth of endophyte-free plants have been observed following the application of indole acetic acid and gibberellic acid. Evidence exists which suggests that cytokinins may be produced in leaf nodules. Retention of chlorophyll by leaves is a characteristic feature commonly used in the bio-assay for cytokinins. In keeping with this attribute for cytokinin activity, it is noteworthy that naturally occurring senescent and yellow leaves of *Psychotria* retain chlorophyll around leaf nodules indicating the presence of cytokinins. Moreover, under experimental conditions when leaf-nodule discs from *Psychotria* were placed on oat leaf segments, the discs caused the retention of chlorophyll in oat leaves below them even after 6–8 days of incubation indicating cytokinin activity.

The symbiotic association between bacteria and species of *Ardisia* of the family Myrsinaceae in leaf nodules has been investigated since the beginning of this century by several investigators. Using definitive techniques (acetylene reduction and ^{15}N methods), it has been clearly shown that *Ardisia* symbiosis has no capacity to fix molecular nitrogen. Alternate explanations for the significance of symbiosis based on growth factor (cytokinin-like substances) production have been put forward by recent workers in the field.

Dioscorea macroura, a member of the family Dioscoreaceae possesses leaf glands at the acuminate apices of leaves which are inhabited by bacteria whose identity is not known. The nitrogen content of leaf tissue containing glands is higher than the remainder of the leaf which indicates the ability of the symbiotic association to fix molecular nitrogen although tests with ^{15}N have not been carried out.

Mycorrhizal associations in members of Ericaceae and Orchidaceae, more particularly in *Calluna vulgaris* and *Neottia nidusavis* have been implicated in nitrogen fixation but not substantiated by ^{15}N tests.

Gymnosperm Nodules

Several species of *Podocarpus* possess numerous small nodules on the root system. The nature of the endophyte in the nodule has been variously described in the literature but it is now clear that the most common endophyte is a non-septate fungus resembling the fungal component of endotrophic mycorrhizae. The ability of the nodulated root system to fix nitrogen is a debatable point but recent studies have provided evidences to demonstrate that nodule-bearing roots reduce acetylene at a slow rate and accumulate ^{15}N rather feebly when exposed to labelled nitrogen gas. Such results point out that root nodules of *Podocarpus* are not as vigorous in nitrogen fixation as their counterparts in angiosperms. Mycorrhizal roots of many conifers, especially those of the genus *Pinus* were claimed earlier as nitrogen fixers but recent tests using acetylene reduction method and ^{15}N enrichment technique have yielded results contrary to earlier findings. However, more critical studies are necessary to prove or disprove this point.

Selected References

Becking, J.H. 1970. Plant endophyte symbiosis in non-leguminous plants. *Pl. Soil,* 32, 611–654.

Becking, J.H. 1971. The physiological significance of the leaf nodule of *Psychotria. Pl. Soil,* sp. vol., 361–374.

Becking, J.H. 1979. Root nodule symbioses between *Rhizobium* and *Parasponia* (Ulmaceae), *Pl. Soil,* 51, 289–296.

Bond, G. 1971. Root nodule formation in non-leguminous angiosperms. *Pl. Soil,* sp. vol., 317–324.

Bond, G. 1974. Root nodule symbiosis with actinomycete-like organisms. In *The Biology of Nitrogen Fixation,* pp. 342–378, Ed. A Quispel, North Holland Publishing Co., Amsterdam.

Callaham, D., Del Tredici, P. and Torrey, J.G. 1978 Isolation and cultivation in *vitro* of the actinomycete causing root nodulation in *Comptonia. Science,* 199, 899–902.

Diem, H.G. and Dommergues, Y.R. 1988. Isolation, characterization and cultivation of *Frankia.* In *Biological Nitrogen Fixation—Recent Developments.* pp. 227–254. Ed. N.S. Subba Rao, Oxford & IBH Publishing Co., New Delhi.

Gardner, I.C. 1965.. Observation on the fine structure of the endophyte of root nodules of *Alnus glutinosa* (L.) Gatertn. *Arch. Mikrobiol.,* 51, 365–383.

Newcomb, W., Peterson, R.L., Callaham, D. and Torrey, J.G. 1978. Structure and host-actinomycete interaction in developing root nodules of *Comptonia peregrina. Canad. J. Botany,* 56, 502–531.

Normand, P. and Lalonde, M. 1986. The genetics of actinorhizal *Frankia:* A review. *Pl. Soil,* 90, 429–453.

Normand, P., Simonet, P. and Bardin, R. 1988. Conservation of *nif* sequences in *Frankia. Mol. Gen. Genet.* 213, 238–246.

Quispel, A. The endophytes of root nodules in non-leguminous plants. In *the Biology of Nitrogen Fixation,* pp. 499–520. Ed. A Quispel, North Holland Publishing Co., Amsterdam.

Quispel, A. and Burggraaf, A.J.P. 1988. Infection, initiation and structure of actinorhizal root nodules. In *Biological Nitrogen Fixation—Recent Developments.* pp. 225–281. Ed. N.S. Subba Rao, Oxford and IBH Publishing Co., New Delhi.

Rodriguez-Barrueco, C. and Subramanian, P. 1988. Host specificity in *Frankia* symbiosis. In *Biological Nitrogen Fixation—Recent Developments.* pp. 283–310. Ed. N.S. Subba Rao, Oxford and IBH Publishing Co., New Delhi.

Schwintzer, C.R. and Tjepkema, J.D. Eds. 1990. *The Biology of Frankia and Actinorhizal Plants.* Academic Press. Inc. San Diego, U.S.A.

Subba Rao, N.S. Ed. 1982. *Advances in Agricultural Microbiology.* Oxford and IBH Publishing Co., New Delhi.

Subba Rao, N.S. and Dommergues, Y.R. Eds. 1998. Microbial Interactions in Agriculture and Forestry Vol I. Oxford and IBH Publishing Co. New Delhi.

Subba Rao, N.S. and Rodriguez-Barueco. C. Ed. 1993. *Symbioses in Nitrogen-Fixing Trees.* Oxford and IBH Publishing Co., New Delhi.

Subba Rao, N.S. and Rodriquez-Barrueco, C. 1995. *Casuarinas.* Oxford and IBH Publishing Co. New Delhi.

Tarrant, R.F. and Trappe, J.M. 1971. The role of *Alnus* in improving the forest environment. *Pl. Soil*, sp. vol., 335–348.

Trinick, M.J. 1973. Symbiosis between *Rhizobium* and the non-legume *Trema aspera*. *Nature*, London, 244, 459–460.

Trinick, M.J. 1979. Structure and nitrogen-fixing nodules formed by *Rhizobium* on roots of *Parasponia andersonii*. *Canad. J. Microbiol.*, 25, 265–578.

Uemura, S. 1971. Non-leguminous root nodules in Japan. *Pl. Soil*, sp. vol., 349–360.

10. Organic Matter Decomposition

Degradation of Plant Residues

Soil can be defined as a natural medium for plant growth composed of minerals, organic materials and living organisms. While physical weathering of rocks caused by changes in temperature and the consequent chemical decomposition contributes largely to the formation of the soil, biological activities such as root growth and microbial metabolism in the soil contribute to its texture and fertility. Undoubtedly, the amount of organic matter present in any soil determines its natural suitability for plant cultivation.

Soil organic matter comprises residues of plant and animals at all stages of decomposition mediated by soil microorganisms. Various organic compounds which reach the soil by way of animal and plant residues are made up of complex carbohydrates, simple sugars, starch, cellulose, hemicelluloses, pectins, gums, mucilage, proteins, fats, oils, waxes, resins, alcohols, aldehydes, ketones, organic acids, lignin, phenols, tannins, hydrocarbons, alkaloids, pigments and other products. The size of particles in the organic matter, the nature and abundance of microorganisms involved, the extent of availability of C, N, P and K, the moisture content of soil, its temperature, pH and aeration, presence of inhibitory substances (such as tannins) etc. are some of the major factors which influence the rate of organic matter decomposition.

Plant residues contain 15–60 percent cellulose, 10–30 per cent hemicellulose, 5–30 per cent lignin, 2–15 per cent protein and 10 per cent sugars, amino acids and organic acids. Cellulose occurs in a semicrystalline form with a molecular weight of 10^6 and has glucose units with B(1–4) linkages. The individual chains of glucose are held together by hydrogen bonds. Cellulase enzyme complex decomposes cellulose into disaccharide cellobiose which is hydrolyzed by the enzyme cellobiase to glucose (Fig. 78). Hemicelluloses are various polymers of hexoses, pentoses and sometimes uronic acids with commonly occurring monomers such as xylose and mannose. Pectin is an example of hemicelluloses and is an important constituent of the middle lamella of cell walls. Pectin is degraded by the enzyme pectinase which is a complex of several enzymes. Lignin is much more complex than celluloses and is formed by chemical reaction involving phenols and free radicals without any specific order. Lignin gets encrusted on the cellulose and hemicellulose matrix. Compounds like caffeic acid and ferulic acid have structures similar to lignin and they have been used in studies on degradation of lignin.

The lignin molecule has only three elements carbon, hydrogen and oxygen. The molecule is a polymer of aromatic nuclei with either a single repeating unit or several similar units as building blocks. The repeating units range from about 200 to 1000 depending upon the origin of lignin and the methods used to determine the molecular weight.

Degradation of lignin is brought about by fungi mainly belonging to Basidiomycetes. The genera of fungi which degrade lignin as well as cellulose are *Clitocybe, Collybia, Mycena, Marasmius, Polystictus, Armillaria, Polyporus, Stereum, Ganoderma, Pleurotus, Trametes, Fomes* and *Ustulina*.

Bacteria constitute the most abundant group of microorganisms. In normal fertile soils, 10–100 million bacteria are present per g of soil. This figure may increase depending on the organic matter content of any particular soil. The bulk of soil bacteria are heterotrophic and utilize readily available source of organic energy from sugars, starch, cellulose and protein. On the other hand, autotrophic bacteria which occupy a smaller portion of the biomass in soil and use inorganic sources such as iron (*Ferrobacillus*) and sulphur (*Thiobacillus*) are not directly involved in organic matter decomposition. The number of actinomycetes may be as high as 200 million per g of soil and may increase in manured soils. Thermophilic (tolerating 50 to 65°C) forms are not uncommon in compost piles. Actinomycetes grow on complex substances such as keratin, chitin and other complex polysaccharides and thus play an active role in humus formation. Soil fungi are mostly heterotrophs and use organic residues easily but their numbers vary in soil depending on whether a species has a dominant vegetative or reproductive phase in the soil environment. Sporulating fungi such as *Mucor, Penicillium* and *Aspergillus* appear on agar plates rather profusely than non-sporulating ones. Soil algae in cultivated soils vary greatly in numbers and may contribute a small amount of organic matter through their biomass but they do not have any active role in organic matter decomposition. The microorganisms involved in the decomposition of organic matter are listed in Table 32. The end products of decomposition are CO_2, H_2O, NO_3, SO_4, CH_4, NH_4, and H_2S depending on the availability of air.

In spite of the paucity of data on biomass in relation to litter decomposition, especially on the microbial part of the complex, studies on the estimations of biomass in a deciduous woodland soil at meathop, Lancashire (brown earth with a mull humus) indicate that annual litter production on dry weight basis was of the order of 7640.0 kg/ha and the total biomass was 497.5 kg/ha (including animals and microbes). The split up figures for individual groups in the biomass were as follows in kg/ha: bacteria, 7.3; actinomycetes, 0.2; fungi, 454.0 (contrary to the general

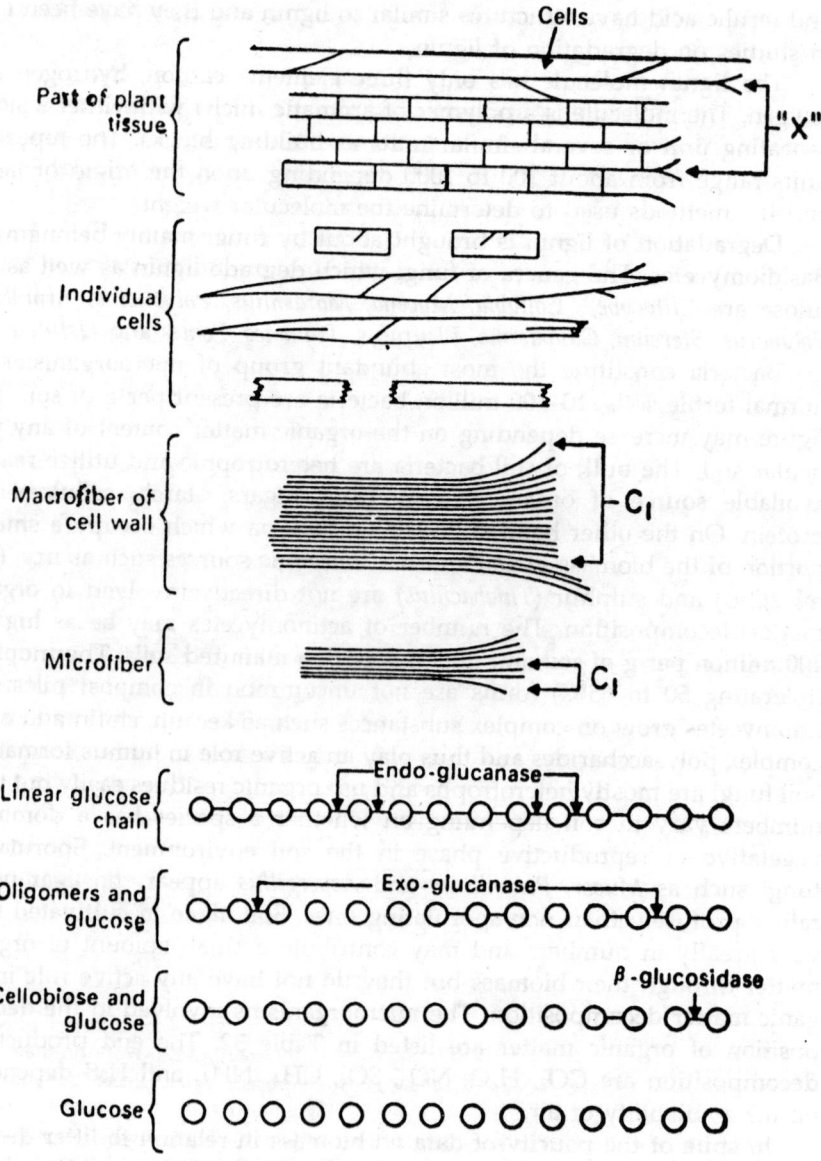

Fig. 78 Diagrammatic representation of breakdown of cellulose in lignocellulosic materials (based on recent findings). (From Chahal and Overend, 1982). Part of plant tissue showing different types of cells and the action of an unnamed enzyme "X" on middle lamella. Individual cells released by the action of "X". Some of the cells show broken

(*Contd.*)

estimates based on dilution-plating), protozoa, 1.0; nematodes, 2.0; earthworms, 12.0; Enchytraeidae, 4.0-; molluscs, 5.0; acari, 1.0; Collembola, 2.0; Diptera, 3.0; other arthropods, 6.0; total microflora, 461.5 and total microfauna, 36.0.

The relationship between organic matter and plant growth may be direct or indirect. Organic matter is a natural substrate for saprophytic microorganisms and provides nutrition to plants indirectly through the activity of soil microorganisms. It is essential for the formation of soil aggregates and hence soil structure which ultimately determines the extent of soil aeration and rooting habit of plants. Organic matter helps in the conservation of soil nutrients by preventing erosion and surface run-off of nutrients.

Humus

Apart from the transient products in the decomposition of organic matter, a dark coloured and fairly stable soil organic matter called humus with known and unknown physical and chemical properties is also an intergral part of the organic matter complex in soil. Humus can be defined as a ligno-protein complex or an amino acid-lignin complex containing approximately 45% lignin compounds, 35% amino acids, 11% carbohydrates, 4% cellulose, 7% hemicellulose, 3% fats, waxes and resins and 6% other miscellaneous substances including plant growth substances and inhibitors. However, the age and composition of the humus are dependent on its origin and environment. By radio carbon dating techniques, the age of humus in podzols has been estimated to be in the range of 1580 to 2860 years while that of chernozemic soils as 1000 years old. Among the fractions of humus, humic acids have been regarded as the oldest and most

---

Fig. 78 (*Contd.*)

ends while others show some cracks in their cell walls. C_1 acts on cellulose of cell walls and releases macrofibres (macrofibrils) and it continues to work on them to release microfibres (microfibrils). During this process single linear anhydrous glucose chains are also released. C_1 releases single linear anhydrous glucose chains from microfibre (microfibril). Endo-glucanase acts on the linear anhydrous glucose chain at random to release oligomers. During this reaction some single glucose units may also be released. Exo-glucanase acts on non-reducing ends of long linear anhydrous glucose chains and also on oligomers to release cellobiose and glucose units. The exo-glucanase which releases cellobiose from non-reducing ends is called cellobiohydrolase. β-glucosidase acts on cellobiose to yield glucose. All the reactions given above occur simultaneously and also synergistically on cellulose to release glucose units. A cellulase-system containing hemicellulases will also hydroylze the hemicelluloses to release various sugars—xylose, mannose, galactose, arabinose, etc.

Table 32 Genera of microorganisms capable of utilising different components of organic matter as reported by several workers: F—fungi; B—bacteria; A—actinomycetes

Nature of substrate in organic matter		Genera of microorganisms
Cellulose	F.	Alternaria, Aspergillus, Chaetomium, Coprinus, Fomes, Fusarium, Myrothecium, Penicillium, Polyporus, Rhizoctonia, Rhizopus, Trametes, Trichoderma, Trichothecium, Verticillium, Zygorynchus
	B.	Achromobacter, Angiococcus, Bacillus, Cellfalcicula, Cellulomonas, Cellvibrio, Clostridium, Cytophaga, Polyangium, Pseudomonas, Sorangium, Sporocytophaga, Vibrio
	A.	Micromonopora, Nocardia, Streptomyces, Streptosporangium
Hemicellulose	F.	Alternaria, Fusarium, Trichothecium, Aspergillus, Rhizopus, Zygorynchus, Chaetomium, Helminthosporium, Penicillium, Coriolus, Fomes, Polyporus
	B.	Bacillus, Achromobacter, Pseudomonas, Cytophaga, Sporocytophaga, Lactobacillus, Vibrio
	A.	Streptomyces
Lignin	F.	Clavaria, Clitocybe, Collybia, Flammula, Hypholoma, Lepiota, Mycena, Pholiota, Arthrobotrys, Cephalosporium, Humicola
	B.	Pseudomonas, Flavobacterium
Starch	F.	Aspergillus, Fomes, Fusarium, Polyporus, Rhizopus
	B.	Achromobacter, Bacillus, Chromobacterium, Clostridium, Cytophaga
	A.	Micromonospora, Nocardia, Streptomyces
Pectin	F.	Fusarium, Verticillium
	B.	Bacillus, Clostridium, Pseudomonas
Inulin	F.	Penicillium, Aspergillus, Fusarium
	B.	Pseudomonas, Flavobacterium, Beneckea, Micrococcus, Cytophaga, Clostridium
Chitin	F.	Fusarium, Mucor, Mortierella, Trichoderma, Aspergillus, Gliocladium, Penicillium, Thamnidium, Absidia
	B.	Cytophaga, Achromobacter, Bacillus, Beneckea, Chromobacterium, Flavobacterium, Micrococcus, Pseudomonas
	A.	Streptomyces, Nocardia, Micromonospora
Proteins and nucleic acids	B.	Bacillus, Pseudomonas, Clostridium, Serratia, Micrococcus
Cutin	F.	Penicillium, Rhodotorula, Mortierella
	B.	Bacillus
	A.	Streptomyces
Tannin	F.	Aspergillus, Penicillium
Humic acid	F	Penicillium, Polystictus
Fulvic acid	F.	Poria

persistent. Bacterial and algal protoplasm with their attendant biological constituents contribute in large measure to the nutritive value of humus. If decomposition by microorganisms is arrested by factors such as low temperature, anaerobiosis, low mineral content and the presence of microbial growth inhibitors such as phenolic compounds, the humus is

17

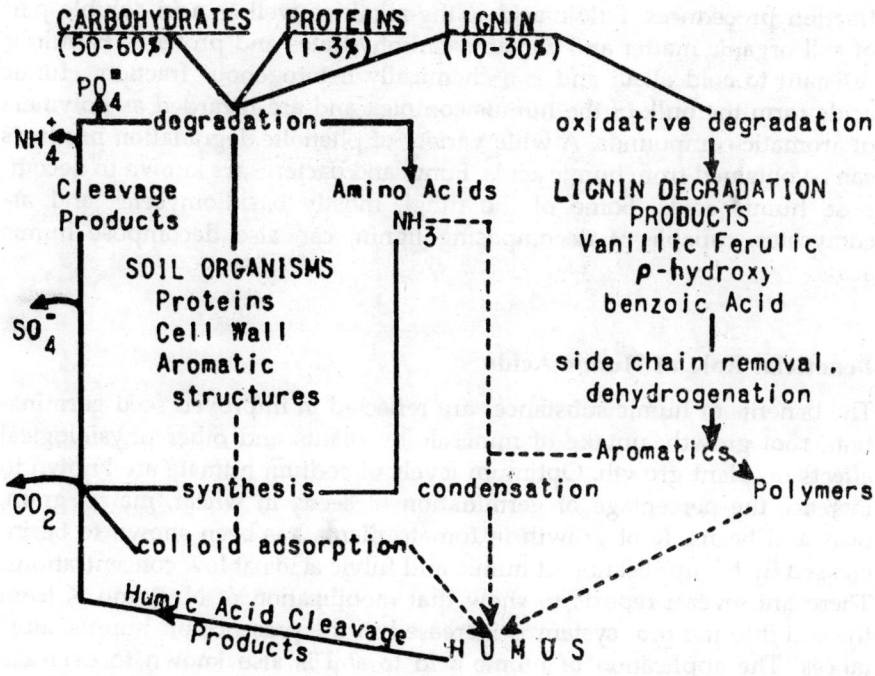

Fig. 79 Degradation of plant residues and formation of soil organic matter. (From Paul and Clarke, 1989).

known as 'raw humus'. On the other hand, 'nutrient humus' in fertile soils contains large amounts of sugars, starch and soluble energy material serving as substrates for microorganisms (Fig. 79).

While soil microorganisms in general take part in humus formation, some fungi such as *Penicillium*, *Aspergillus* and also actinomycetes produce dark humus-like substances (amino acids, peptides and polyphenols) which serve as structural units for the synthesis of humic substances. Extracts of spores of *Aspergillus niger* possess properties similar to those of humic acids. Thus in recent years, considerable evidence has come forth to show that humus is not only a biochemically derived material but also a synthetic microbial product.

Humic Acids

An understanding of the biochemistry of humus degradation has posed problems mainly due to the procedural difficulties in solvent extraction of organic matter complexes. Further, bio-degradation products of lignin are closely linked with those of humus substances. In spite of these inherent difficulties, fulvic acid, humin and humic acid have been recognised as the three major fractions by subjecting humus complexes to solvent extraction procedures. Fulvic acid is the alkali as well as acid soluble part of soil organic matter and contains carbohydrates and proteins. Humin is resistant to cold alkali and is a chemically heterogenous fraction. Humic acids form the bulk of the humus complex and are regarded as polymers of aromatic compounds. A wide variety of phenolic degradation products can be obtained from humic acids. Fungi and bacteria are known to decompose humic acids. Some of the fungi, mostly basidiomycetes and ascomycetes, capable of decomposing lignin, can also decompose humic acids.

Beneficial Role of Humic Acids

The benefits of humic substances are reflected in improved seed germination, root growth, uptake of minerals by plants and other physiological effects on plant growth. Optimum levels of sodium humate are known to increase the percentage of germination of seeds in wheat, maize, gram, peas and beans. Root growth in tomato plants has been shown to be increased by the application of humic and fulvic acids at low concentrations. There are several reports to show that mobilization of N, P and K from the soil into the root system is increased in the presence of humus substances. The application of humic acid to soil is also known to decrease phosphorus fixation in soil, particularly in calcareous soils. The uptake of trace elements by plants is increased by the application of humus substances since the latter is known to effectively chelate with trace metals, especially iron. A combination of fulvic acid and iron is known to be more effective in increasing lateral root formation in plants than iron alone. The chelating ability of humates suggests that they play a role similar to EDTA (ethylene-diamino tetraacetic acid), the well-known synthetic chelator.

Enzyme actions involved in plant metabolism have been linked with humus complexes since humic acids function as hydrogen acceptors. The growth stimulatory activity of humus complexes on wheat roots has been attributed to the noticeably increased cytochrome oxidase activity in the root system. Similarly, humic acids are known to increase the activity of glutamic acid transaminase and phosphorylase enzymes and also the synthesis of deoxyribose and ribose nucleic acids. Several reports indicate a shift in carbohydrate metabolism of plants mediated by changes in aldolase, saccharase, phosphatase and amylase by the application of humic acids. Apart from the chelating effect of humic acids on trace elements, the beneficial influence of humic acids on iron uptake by roots may also be attributed to the permeability changes in the plasma membrane as is evident from several experiments with plants grown on mineral solutions supplemented with humus substances. Experiments with foliar sprays of sodium humate have shown that the vigour and yield of certain plants can be enhanced by such treatments (Table 33).

Table 33 Effect of spraying sodium humate and hydroquinone on grain yield and nitrogen uptake by *Glycine max* var. Bragg and on yield of *Solanum lycopersicum* var. Heinz, 1370 (Average of four replications; from Varshney and Gaur, 1974)

| Treatment | Soybean | | | Tomato |
| | Grain | | | |
	Yield/plot (g)	Nitrogen (%)	Nitrogen plot uptake (g)	Yield/plot (g)
Control	200.7	6.23	12.5	661.7
Hydroquinone 10 ppm	218.5	6.15	13.4	1309.7
Hydroquinone 50 ppm	209.2	6.50	13.6	798.9
Humate 10 ppm	248.4	6.50	16.4	1383.4
Humate 50 ppm	229.0	6.60	14.6	1347.4
C.D. at 5%	6.17			157.9

Humic acids are known to influence the growth and proliferation of microorganisms. The growth of *Aspergillus niger*, *Penicillium glaucum*, *Bacillus mycoides* and *Scenedesmus* spp. is enhanced by additions of humus substances. The number of *Azotobacter* cells and the amount of nitrogen fixed by them are enhanced by the application of humic acid. Similar beneficial effects have also been noticed on the growth of *Rhizobium* and nodulation of legume roots inoculated with *R. trifolii*, *R. meliloti*, *R. leguminosarum* and *R. japonicum*. In fact, legume inoculants containing viable cells of rhizobia intended for artificial inoculation of leguminous seeds are often prepared on humus-type materials.

Mineralization and Immobilization Processes

When microorganisms grow and multiply on organic debris, carbon is utilised for building the cellular material of microbial cells with the release of carbon dioxide, methane and other volatile substances (Fig. 80). In this process, microorganisms also assimilate nitrogen, phosphorus, potassium and sulphur which get bound in the cell protoplasm. Therefore, the C/N, C/P, C/K or C/S ratios in soil are governed by the extent of organic matter utilised by soil microorganisms depending on the oxygen content and the microbial biomass at a particular stage in decomposition. Thus, three parallel processes go on during decomposition: (1) degradation of plant and animal remains by cellulases and other microbial enzymes, (2) the increase in the biomass of microorganisms which comprises polysaccharides and proteins and (3) the accumulation or liberation of end products. The term 'mineralization' is used to designate the conversion of

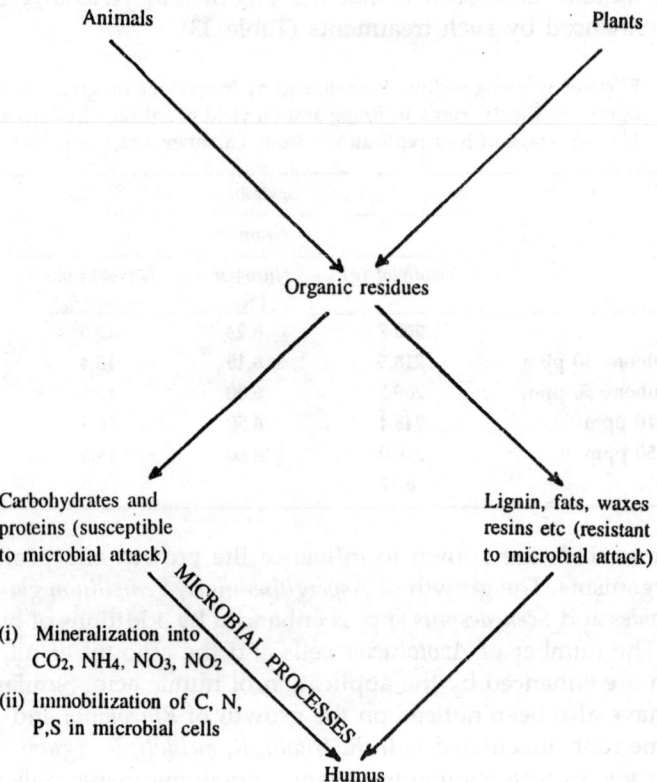

Fig. 80 Pathways in organic matter decomposition.

organic complexes of an element to its inorganic state which embodies the first of the three processes mentioned above. The second process which involves the microbial uptake of nutrients such as nitrogen, phosphorus and sulphur is opposite in magnitude to mineralization and is known as 'immobilization'. From the agronomic point of view, immobilization reduces the availability of nutrients to plant growth, the intensity of which is related to the total microbial biomass at a given time. The last of these processes provides an index of microbial activity in soil and is interlinked with nitrification and denitrification processes which are also mediated by microorganisms.

When fresh organic matter is added to soil, owing to the resistance offered by fresh residues, the number of microorganisms developing on these substrates is small and therefore, immobilization is not so rapid whereas mineralization gradually increases resulting in the accumulation of ammonia and nitrates. The rate of immobilization of nitrogen depends on the nature of the soil microflora, soil temperature, fertiliser nitrogen status and the C/N ratio of the organic matter added. Organic matter from diverse plant tissues vary widely in their C/N ratio. Optimum levels of C/N ratio in the range of 20–25 (1.4–1.7% N) seems to be ideal for maximum decomposition since there will be no immediate release of mineral nitrogen from residues over and above the amount required for microbial synthesis. In other words, a favourable soil environment is created to bring about an equilibrium between mineralization and immobilization processes. This critical balance may be upset if the C/N ratio is less than 25 when mineralization is likely to exceed immobilization leading to accumulation of ammonium and nitrate forms of nitrogen. Therefore, one reliable way of measuring mineralization in soil is to find out the total available ammonium, nitrate and nitrite forms of nitrogen at a given time since the processes of ammonification, nitrification and denitrification take place almost concurrently in arable soils.

Plant debris containing higher amounts of lignin are resistant to microbial attack and hence mineralization proceeds slowly in organic materials like sawdust. The lignin content of plant residues may serve as an index of the vulnerability of organic residues to microbial attack.

Studies with ^{15}N-tracer techniques have shown that immobilized nitrogen in microbial cells is resistant to mineralization even after prolonged periods. These results have emerged from studies with pure cultures of microorganisms grown on synthetic substrate and hence cannot be equated to field conditions where chemically fixed nitrogen in clays and organic matter cannot easily be distinguished from microbiologically immobilized nitrogen.

Effects of Residues of Crops on Plant Growth

The decomposition products of plant residues in soil may become toxic to growth of plants under certain conditions. The absence of satisfactory extraction procedures and bio-assay methods have come in the way of identifying the nature and extent of phytotoxic principles produced by plant remains which undergo decomposition. However, detrimental effects of plant residues have been detected through seed germination tests, growth of radicles and seedling injury under laboratory conditions which have been supported by field observations like stunted overall growth of plants, chlorosis, slow maturation, premature leaf abscising and failure of flowering and seed setting.

The oxygen status of the soil (aerobic or anaerobic) is the most important factor in determining the qualitative and quantitative aspects of microbially mediated bio-degradation of plant remains. Some of the phytotoxic compounds detected so far include methane, acetic, lactic, butyric, formic and other organic acids; phenolic compounds including syringaldehyde, vanillin, p-hydroxybenzaldehyde, ferulic, syringic, vanillic, p-hydroxybenzoic, p-methoxybenzoic and benzoic acids, various amino acids and many other unidentified products. These products seem to accumulate under waterlogged anaerobic surroundings whereas under normal arable soils the presence of toxic compounds is either rare or negligible.

Soil Sickness

Apples, peaches, grapes, cherries and plums are prone to suffer from soil sickness if replanted in the same soil successively. They however, recover from sickness if replanted in new soils. The symptoms differ from plant to plant and the disease syndrome has been frequently reported from Germany, Canada and U.S.A. Such plant disorders may be attributed to nutritional deficiency, pathogenic microorganisms whose identity is yet to be established and phytotoxins secreted by roots of plants or by microbial decomposition of plant residues. A cumulative effect of one or more factors may also be responsible for the disease syndrome in fruit orchards. For example, the failure of peach cultivation in California and Ontario has been attributed to a phytotoxin produced by residues of a previous crop. The barks of roots of peach contain cyanogenic glycoside, amygdalin in relatively low amounts and as such is not toxic to peach plants. The glycoside is converted to toxic components like benzaldehyde and hydrogen cyanide through the mediation of microorganisms in soil supporting the growth of peach plants. The hydrolysis of amygdalin has also been attributed to a soil nematode (*Pratylenchus penetrans*) found in peach soils which secretes an enzyme. In apple trees, one of the constituents of the bark is phloridzin which is broken down in soil to phloretin, phloroglucinol, p-hydroxyhydrocinnamic acid and p-hydroxybenzoic acid compounds which have been proved to be toxic to apple plants. *Penicillium expansum*, a normal fungal inhabitant of soil from apple plantations, produce patulin and an unidentified phenolic compound in media amended with apple residues. These compounds are also inhibitory to the growth of apple trees.

Composting

Farmyard manure is the oldest manure known to mankind and is made up of solid excreta or dung of animals, urine and plant remains which are allowed to decay with the help of soil microorganisms capable of decomposing complex organic debris into substances that are easily assimilated by plants. The manurial value of farmyard manure depends on the nature of raw materials used and the extent of decomposition by soil microorganisms.

Compositing farm residues and night soil has been practiced for long in China and India (Fig. 81 A, B). In China, the compost pits dug in soil have usually dimensions of 3.5 m × 2.5m × 1.5m (L × B × H). The pits are filled layer by layer and each layer is about 15cm thick. The bottom layer (layer No. 1) consists of green plants and aquatic weeds available on the farm followed by silt-straw mixture (layer No. 2) and animal excreta (layer No. 3). The layering is repeated until the pit is filled. Finally, a layer of mud is made on top of the pit in such a way that water of about 4 cm depth is maintained on the surface to create anaerobic conditions which helps to reduce losses of nitrogen. In a time span of about 10 weeks, the mud plaster is dismantled and the contents of the pit are turned over or mixed with superphosphate and water (if necessary). At the end of 3 months, the compost is ready for use on the farm. The compost may have a C/N ratio of 15–20 and organic matter content of 8–10 per cent. A compost is considered superior if the C/N ratio is 20 or less and the organic matter content is around 30–60 per cent.

In India, the pit method is also practiced without any water logging in an elevated place often protected by a shed (Fig. 81 C). The layering at the bottom is usually the urine-soaked bed in the cattle shed. The bed is made of farm materials such as vegetable wastes, fodder remnants, green matter etc. The bed layer is sprinkled with a slurry of cowdung and mixed with well decomposed manure from the previous batch. This sort of layering and sprinkling with cowdung slurry is repeated until the pit is filled. The compost pit sits for a period of 2–3 months within which time the contents are turned over or stirred three times.

Composting can also be done by the heap method in which the base material on a hard ground consists of hardwood materials coming from cotton and pigeonpea stalks followed by layering with farm residues such as leaves, hay and garbage. The heap can be rectangular in shape. After wetting with water, the heap is mud plastered and allowed to sit. Within a period of 2–3 months, the heap is broken, materials turned or stirred and again mud plastered. The final product becomes a heap of well decomposed organic matter.

High temperature composting is done in China by heaping alternate layers of night soil, urine, sewage, animal dung and chopped plant residues. The base material consists of hard stalks of crops and this is followed by layering with other materials. Water is added at optimum levels. The entire heap is shaped finally with mud plaster taking care to insert bamboo or maize stalks into the mud covered heap all the way to the bottom of the heap. After 24 hours, the bamboo poles or maize stalks are withdrawn to leave behind holes for ventilation. Within 4–5 days the temperature in the heap reaches 60–70°C when the holes are closed and

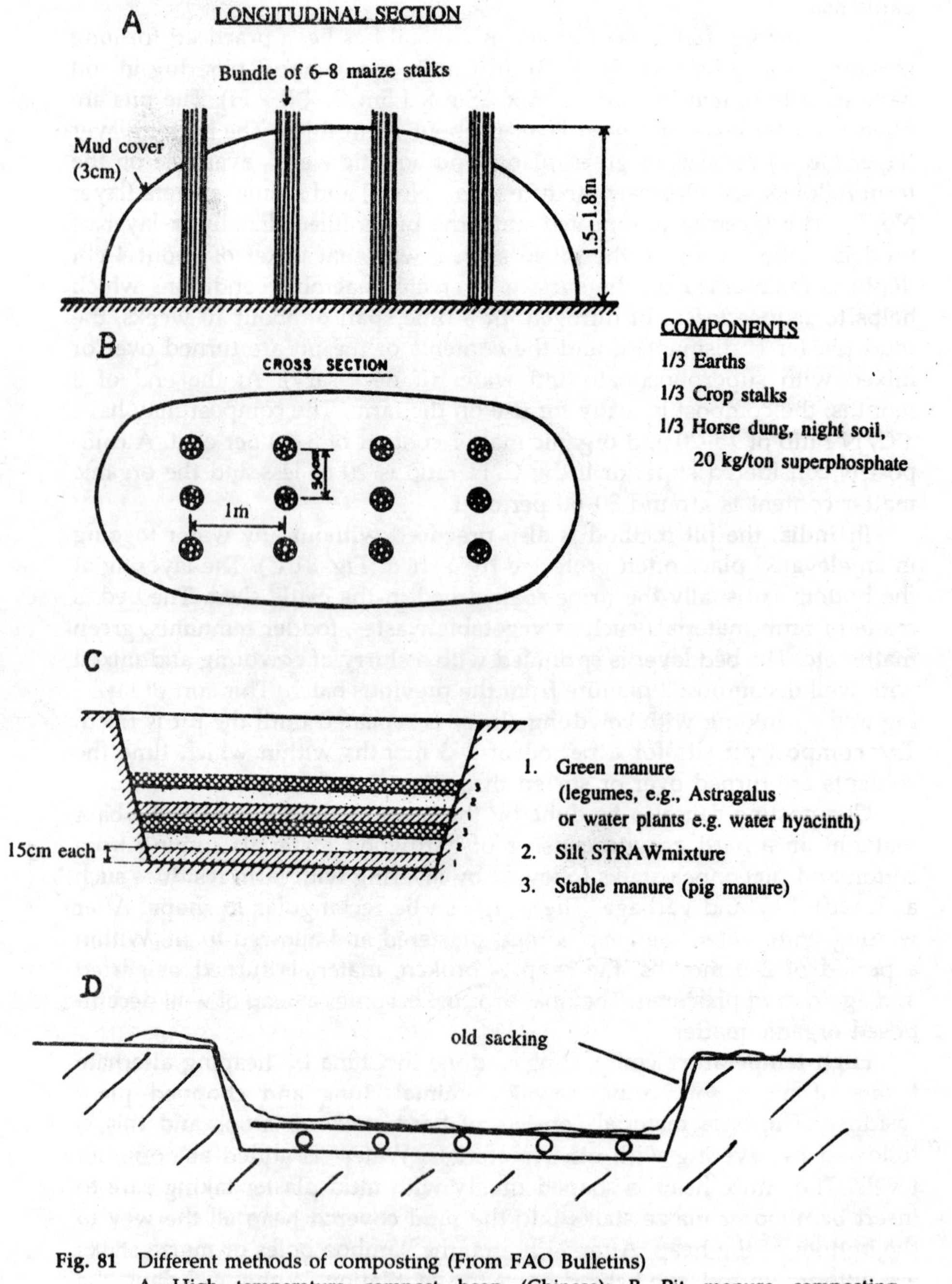

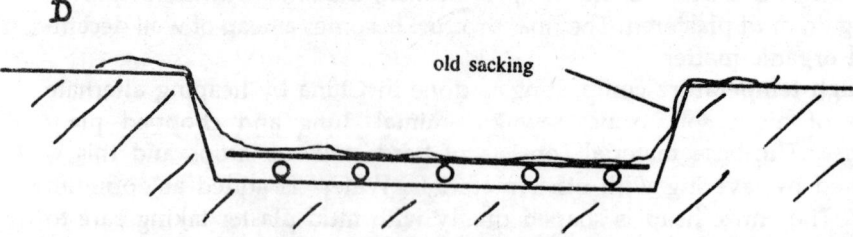

A <u>LONGITUDINAL SECTION</u>

Bundle of 6–8 maize stalks

Mud cover
(3cm)

1.5–1.8m

B

CROSS SECTION

30

1m

COMPONENTS

1/3 Earths
1/3 Crop stalks
1/3 Horse dung, night soil,
    20 kg/ton superphosphate

C

15cm each

1. Green manure
   (legumes, e.g., Astragalus
   or water plants e.g. water hyacinth)
2. Silt STRAWmixture
3. Stable manure (pig manure)

D

old sacking

Fig. 81 Different methods of composting (From FAO Bulletins)
A—High temperature compost heap (Chinese); B—Pit manure composting
(Chinese); C—Heap method of composting (Indian); D—Vermicomposting pit with
bamboo poles on a wooden basal support for drainage of water.

sealed with mud plaster. After a period of 2 weeks the mud plaster is broken and the contents mixed followed by resealing with mud plaster. At the end of 2 months, the decomposed compost free from pathogens is ready for use on the farm.

The finished compost can be enriched with finely powdered rock phosphate and inoculated with non-symbiotic nitrogen fixers such as *Azotobacter, Azospirillum* and phosphate dissolving bacterial or fungal species such as *Pseudomonas, Micrococcus, Bacillus, Flavobacterium, Pencillium, Fusarium, Asppergillus* etc.

Vermicomposting

The use of earthworms in composting process is known as vermicomposting. One kg of earthworms can consume one kg of organic materials in a day and excrete as castings that are rich in nitrate, available phosphorus, potassium, calcium and magnesium. The castings encourage growth of bacteria and actinomycetes due to aeration in soil pores.

There are about 3000 species of earthworms in the world and about 500 in India alone. The earthworm is an aerator by making tunnels and by crushing and mixing of soil. By its enzymatic and biological activity, the earthworms stimulate soil microbiological activity. Vermicomposting began in Ontario, Canada in 1970 and today the USA, Japan and Phillippines are leading in vermicomposting both quantitatively as well as qualitatively.

A moist compost heap of 2.4m × 1.2m × 0.6m deep can support 50,000 worms. A shallow well drained heap is ideal for worms to feed and multiply. *Lumbricus rubellus* (the red worm) and *Eisenia foetida* are thermotolerant and can stand the heat generated in the composting process. There are many other suitable species of worms available with breeders which can be had from agricultural universities and centres of agriculture research. The bedding for multiplying worms comprises of any moistened organic residue such as saw dust, cereal straw, rice husks, bagasse, card board and so on. This bedding material kept in boxes is covered with damp sack and left for 4 weeks followed by the addition of chicken manure, green matter and water hyacinth. The pH must be around 7.0 and temperature about 20–27°C. The breeder earthworms are allowed to multiply in these boxes, taking care to avoid predators such as birds and frogs.

A series of compost pits of dimensions 3m × 4m × 1m deep are dug with sloping sides. Bamboo poles are placed at the bottom of the pits and lined with gunny sack to avoid the escape of worms (Fig. 81D). The pits are filled with moist farm wastes, animal manure and leaves that are well chopped. The worms are picked by hand from the boxes and placed into the pit. By incubating the compost pit in the shade and keeping it moist but not water logged for 2 months, a good organic matter rich vermicompost can be prepared for use on the farm.

Green Manure

The practice of green manuring of soil is as old as agriculture itself. Many leguminous and non-leguminous crops are grown and turned into the soil while they are still green to enrich soil nitrogen. When organic matter is decomposed, the nitrogen bound in the organic matter is released first as ammonia. The ammonia may be absorbed by the plant or converted to nitrate. Apart from enrichment of soil nitrogen, green manuring enriches the phosphorus, calcium, sulphur and other mineral content of soil. The foliage of following crops are used in India for green leaf manuring: *Gliricidia maculata, Pongamia glabra, Calotropis gigantea, Tephrosia purpurea, T. candida, Indigofera teysmanni, Cassia tora, Sesbania speciosa* and *Ipomoea carnea*. Many species of the following genera hold promise as potential green manure crops for rice cultivation: *Aeschynomene, Cassia, Crotalaria, Cyamopsis, Desmodium, Indigofera, Lathyrus, Melilotus, Stizolobium, Phaseolus, Sesbania* and *Vigna*.

The importance of stem nodulating *Sesbania rostrata* as a green manure plant in rice cultivation has already been mentioned in the chapter on *Rhizobium* and root nodulation.

Anaerobic Decomposition of Organic Matter

Under anaerobic conditions, decomposition of organic residues takes place by the activity of both mesophilic and thermophilic microorganisms resulting in the production of carbondioxide, hydrogen, ethyl alcohol and organic acids such as acetic, formic, lactic, succinic and butyric acids. Among the mesophilic flora, bacteria are more active than fungi or actinomycetes in cellulolytic activity. They belong to the genus *Clostridium* and are numerous in peaty soils and manure pits but rarely encountered in cultivated arable soils. In compost heaps, both mesophilic and thermophilic microorganisms (bacteria and actinomycetes) are important in the break down of cellulose substrates.

As stated above, the primary microbial colonizers initially break down the complex carbohydrates and proteins into organic acids and alcohols. At a later stage, the methane bacteria which are strict anaerobes begin to act upon the secondary substrates chiefly lactic, acetic and butyric acids and ferment them into CH_2 and CO_2 whose ratio is variable depending on the nature of reactions. It is not easy to isolate pure cultures of methane bacteria. However, by enrichment culture technique, methane bacteria have been isolated and grouped under four genera: *Methanobacterium*, *Methanobacillus, Methanosarcina* and *Methanococcus.* Experiments with pure cultures as well as mixed cultures of methane bacteria have shown that among the several types of reactions which can produce CH_4, the following typical ones are important:

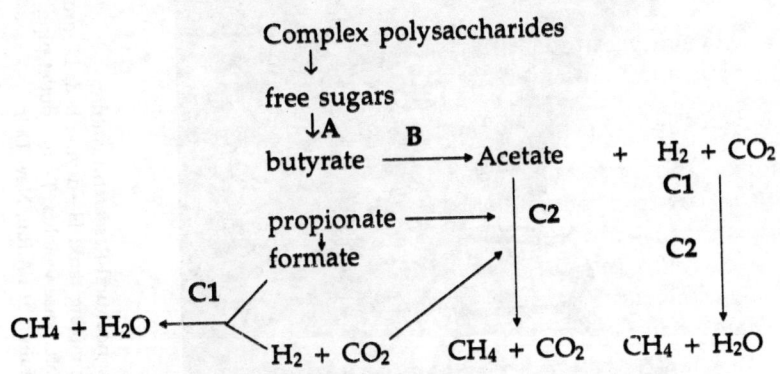

I Non-methanogens

A. *Clostridium acetobutylicum*
   Eubacterium limosum
   Clostridium propinicum
   coliforms

B. Volatile fatty acid oxidizers
   Syntrophomonas wolfeii
   Syntrophobacter wolinii

II Methanogens

C1 H_2 oxidizing methanogens
   Methanobacterium,
   Methanobrevibacter,
   Methanospirillum

C2 Aceticlastic methanogens
   Methanosorcina

(1) $CO_2 + 4H_2 \longrightarrow CH_4 + 2H_2O$

(2) $4HCOOH \longrightarrow CH_4 + 3CO_2 + 2H_2O$

(3) $CH_3COOH \longrightarrow CH_4 + CO_2$

(4) $2CH_3CH_2H \longrightarrow 3CH_4 + CO_2$

Biogas

Cattle-dung is a good substrate for anaerobic fermentation by methane bacteria. The fermentation of cattle-dung in specially designed and inexpensively fabricated plants (Fig. 82) not only provides cooking gas for the farming community by way of methane, but also leaves behind a slurry which can be used as an organic manure for growing crops. From laboratory experiments, it is now known that digestion of cattle-dung for 4 weeks at 30°C is necessary to achieve good gas production of about 1.5 cu ft/lb of fresh bullock-dung. The composition of the gas is approximately 55–60% methane, 5–10% hydrogen and 30–35% carbon dioxide. Dilution of the dung (1 to 1½ times with water) to bring the level of solids to about

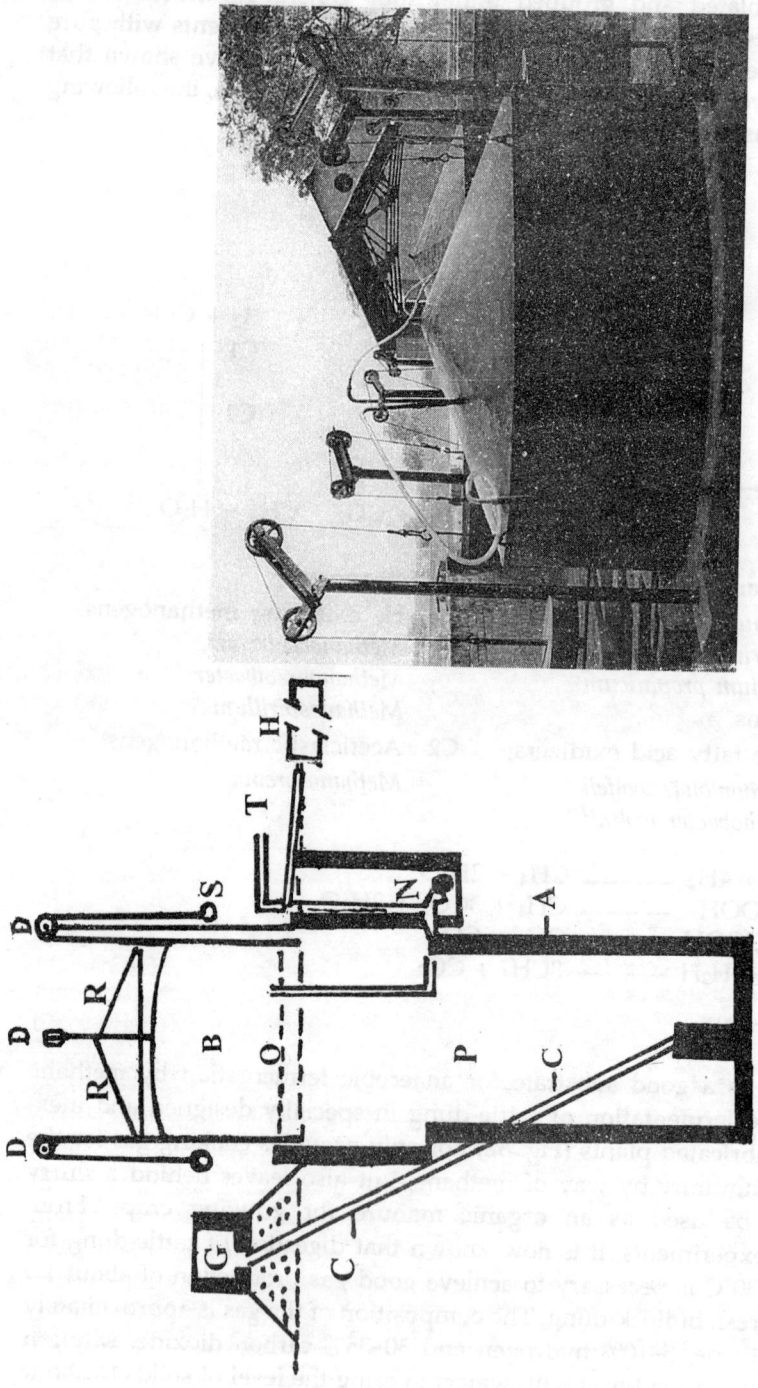

Fig. 82 On the left is a sketch of the biogas plant designed by Acharya at the Indian Agricultural Research Insitute; A—brickwall; B—gas holder; C—cowdung inlet pipe; D—pulley; G—cowdung mixing tank; H—drying bed; L—ground level; N—gas moisture exit trap; O—slurry level; P—fermentation tank; R—iron rods; S—counterpoise weights; T—gas outlet pipe On the right is a photograph of the biogas plant designed by Acharya still operating at I.A.R.I., New Delhi.

7 to 9% has been found to increase the gas production. Addition of nutrients like nitrogen, phosphoric acid, potash or minor elements will not augment gas production but supplementation of the dung with organic materials rich in cellulose and protein (sunnhemp, dhaincha, groundnut shells and sugarcane bagasse) increases gas production. On the other hand, organic materials rich in fermentable carbohydrates like jaggery, molasses, potato and maize do not contribute to increased gas production. The rate of gas production falls during winter in several parts of North India since temperatures below 15°C are not conducive for the activity of methane bacteria (Table 34). The sluggish rate of fermentation in a biogas plant in winter could be overcome by providing a casing to the fermentor so as to conserve heat. Alternatively, temperature insensitive methane bacteria may have to be evolved to allow fermentation at a wide range of temperatures.

Table 34 Influence of atmospheric temperature on the fermentation liquor at different depths (from Idnani and Varadarajan, 1974)

Atmospheric temp. (°C)	Temperature in °C at depths of cm						
	0	30	90	150	210	270	330
15.5	14.0	15.0	17.5	19.0	19.0	18.5	17.5
18.0	16.5	17.5	19.5	21.0	22.0	21.5	20.0
19.5	18.0	20.0	21.5	22.5	23.0	22.5	21.0
22.5	20.5	22.0	24.0	24.5	25.0	25.0	24.0
25.0	23.0	24.5	27.0	28.0	28.5	28.0	27.0
28.5	26.5	28.0	29.0	31.0	31.5	31.5	30.0
31.0	28.5	30.0	32.0	33.0	34.0	33.0	32.0

Degradation of Hydrocarbons

The natural sources of hydrocarbons in soil are waxes and other constitutents of tissues, hydrocarbon-like synthetic molecules produced by soil microflora and the oils used as base material for suspending insecticides which are sprayed for the control of insect pests of crops. In pure culture studies, it has been demonstrated that when the regular carbon source in synthetic media (such as sucrose, glucose etc.) are replaced by hydrocarbons such as paraffin, kerosene, gasoline and lubricating oils, growth of certain microorganisms takes place indicating the ability of certain microorganisms to utilize hydrocarbons. It is known that both saturated and unsaturated molecules are decomposed by certain microorganisms, the latter being more vulnerable to attack than the former. Many fungi, bacteria and actinomycetes are known to degrade hydrocarbons. For instance, ethane (C_2H_6), a short chain, low molecular weight paraffin hydrocarbon is metabolized by *Mycobacterium, Nocardia, Streptomyces, Pseu-*

domonas, Flavobacterium and several fungi. High molecular weight hydrocarbons are also degraded by a variety of microorganisms of the genera *Nocardia, Pseudomonas, Streptomyces* etc.

Selected References

Acharya, C.N.. 1961. *Preparation of Fuel Gas and Manure by Anaerobic Fermentation of Organic Material*. I.C.A.R. Series, No. 15, Krishi Bhavan, New Delhi.

Allison, F.E. 1973. *Soil Organic Matter and Its Role in Crop Production*. Elsevier Scientific Publishing Co., Amsterdam, London and New York.

FAO, 1977 Soils Bulletin No. 40. China: *Recycling of Organic Wastes*. FAO, Rome.

Flaig, W. 1966. *The Use of Isotopes in Soil Organic Matter Studies*. Pergamon Press, New York.

Gaur, A.C., Sadasivam, K.V., Vimal, O.P. and Mathur, R.S. 1971. A study of the decomposition of organic matter in an alluvial soil: CO_2 evolution, microbiological and chemical transformations. *Pl. Soil.*, 34, 17–28.

Hughes, D.E. and Rose, A.H. 1971. Eds. *Microbes and Biological Productivity*. Cambridge, Univ. Press, London.

Idnani, M.A. and Varadarajan, S. 1974. *Preparation of Fuel Gas and Manure by anaeorbic fermentation of organic material*. ICAR Tech. Bull. (Agric.), No. 46, Krishi Bhavan, New Delhi.

McLaren, A.D. and Peterson, G.H. 1967. *Soil Biochemistry*. Marcel Dekkar, Inc., New York.

National Academy of Sciences 1977. *Methane Generation from Human, Animal and Agricultural Wastes*. NAS., USA. Washington, DC.

Patrik, Z.A., Toussoun, T.A. and Koch, L.W. 1964. Effect of crop residue decomposition products on plant roots. *Ann. Rev. Phytopath.*, 2, 267–292.

Varshney, T.N. and Gaur, A.C. 1974. Effect of spraying sodium humate and hydroquinone on *Glycine max* var. Bragg and *Solanum lycopersicum* var. Heiz 1370, *Curr. Sci.*, 43, 95–96.

Wolfe, R.S. 1971. Microbial fermentation of methane. *Adv. microb. Physiol.* 6, 107–146.

11. Nitrification and Denitrification

Nitrification

As stated earlier, mineralization of nitrogen in organic materials results in the formation of ammonium which is the most reduced form of inorganic nitrogen. This ammonium in soils is the starting point for a series of reactions resulting in the formation of nitrites and nitrates mediated by specialised bacteria.

The biological oxidation of ammonium salts in soil to nitrites and the subsequent oxidation of nitrites to nitrates or to put it in general terms, the biological conversion of nitrogen in soil from a reduced to a more oxidized state may be taken as definitions for nitrification processes in soil.

The classical work of the Russian soil microbiologist, Winogradsky showed that certain chemoautotrophic bacteria, *Nitrosomonas* and *Nitrobacter* (Fig. 83) and other less important ones (*Nitrosococcus*, *Nitrosospira*, *Nitrosocystis*, *Nitrosogloea* and *Nitrocystis*) take part in nitrification. These bacteria are strict obligate autotrophs and are capable of synthesizing all their enzyme requirements from inorganic salts. *Nitrosomonas* obtains its energy by oxidizing ammonia to nitrite and *Nitrobacter* by oxidation of nitrite to nitrate, as follows:

$$HN_4^+ + 1\frac{1}{2} O_2 \rightarrow NO_2^- + 2H^+ + H_2O + 66Kcal \text{ —mediated by}$$

$$\text{Nitrosomonas}$$

$$NaNO_2 + \frac{1}{2} O_2 \rightarrow NaNO_3 + 18Kcal \text{ —mediated by } Nitrobacter$$

The bacteria are rod-shaped and are very difficult to isolate even on sophisticated media because even small amounts of chemicals in artificial culture media inhibit their growth. Notwithstanding this behaviour in pure cultures, it is known that all types of soils are congenial for the growth of nitrifying bacteria.

Several factors influence the growth of nitrifying bacteria in soil. The number of such bacteria in soil is dependent on the levels of ammonia and nitrite, aeration, moisture, temperature, pH and organic matter. Soils receiving good amount of organic matter appear to be congenial for the growth and proliferation of nitrifying bacteria although no factor or factors related to the products of organic matter

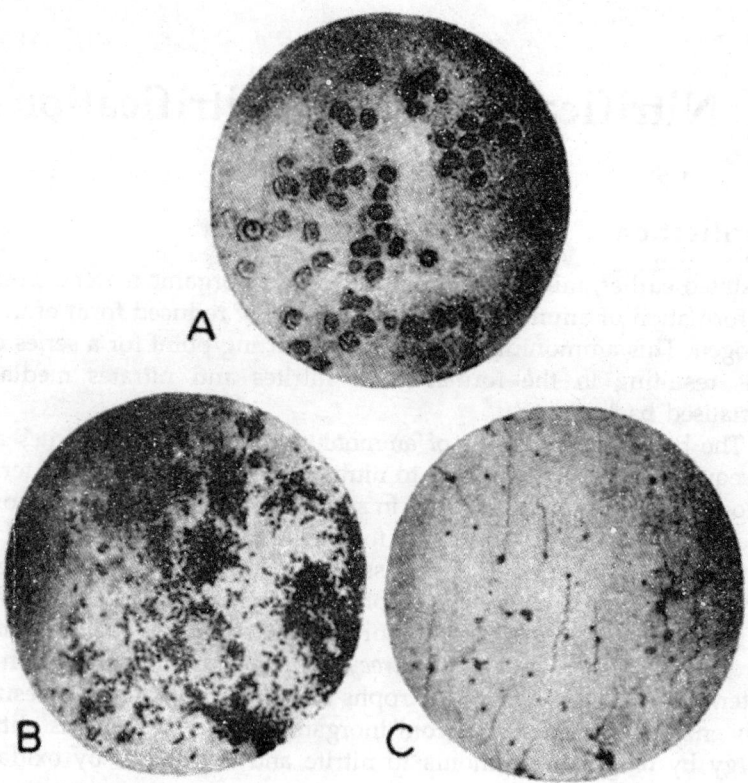

Fig. 83 Nitrifying bacteria: (A) Photomicrograph of *Nitrosomonas javanensis*; (B) Photomicrograph of *Nitrobacter* (Duito); (C) *Nitrosomonas* (Zurich). Surface colonies on silica gel. (from the textbook, *The Life of Bacteria* by K.V. Thimann, The Macmillan Company, New York.)

degradation in soil can be linked with the observed stimulatory effect. In acid soils, nitrification is poor due to a decrease in the population of nitrifying bacteria which can be counteracted by limiting and raising the pH level of soil to 6.0. Waterlogged soils deficient in oxygen are not congenial for nitrification. Similarly, either too low (below 5°C) or too high soil temperature (above 40°C) are not conducive for the optimum functioning of these organisms.

Several microorganisms belonging to the genera *Pseudomonas, Corynebacterium, Nocardia, Aspergillus, Streptomyces. Mycobacterium, Bacillus* and *Vibrio* have also been shown to produce either nitrite or nitrate from ammonia or other reduced forms of nitrogen. The reduced forms of nitrogen which have been used by several investigators in laboratory experiments are nitrophenols, nitrobenzoates, and oximes of a number of organic acids such as pyruvic, oxalacetic and α-ketoglutaric acids. The reactions mediated by these microorganisms may at best produce not more than 5 ppm equivalents of nitrite-nitrogen as against the 2000 ppm transformed by *Nitrosomonas* and *Nitrobacter*. Experiments with *Aspergillus flavus* to find out its significance as a soil nitrifier have shown that addition of such cultures to sterilzied soils and quartz resulted in no nitrate production. Therefore, the role of such heterotrophic microorganisms in nitrification processes in soil is still rather uncertain.

Denitrification

Nitrogen transformations in soil result in the loss of molecular nitrogen. The conversion of nitrate and nitrite into molecular nitrogen or nitrous oxide through microbial processes is known as denitrification. The escape of molecular nitrogen into the atmosphere, also known as volatilization of nitrogen is a drain on the availability of this vital element in soil for crop growth. Denitrification of bound nitrogen to gaseous nitrogen is mediated by numerous species of bacteria which normally use oxygen of the air as hydrogen acceptor (aerobically) but also possess the ability to use nitrates and nitrites in the place of oxygen (anaerobically). Thus these bacteria have the faculty to grow aerobically in the absence of nitrate but anaerobically in the presence of nitrate. The anaerobic conversion of nitrate into molecular nitrogen is also known as nitrate respiration. In fact such organisms capable of denitrification are isolated by enrichment cultures in anaerobic media containing excess of potassium nitrate. The bacterial genera which bring about denitrification are *Pseudomonas, Achromobacter, Bacillus* and *Micrococcus*. Of these, *Pseudomonas* and *Achromobacter* are the predominant ones in soil. The biochemical reactions can be summed up as:

$2NO_3^-$ (nitrate) $+ 10H \longrightarrow N_2 + 4H_2O + 2OH^-$ or
$2NO_2^-$ (nitrite) $+ 6H \longrightarrow N_2 + 2H_2O + 2OH^-$ or
N_2O (nitrous oxide) $+ 2 H \rightarrow N_2 + H_2O$.

In short, the denitrification pathway can be depicted as NO_3^- (nitrate $\rightarrow NO_2^-$ (nitrite) $\rightarrow NO$ (nitric oxide) $\rightarrow N_2O$ (nitrous oxide) $\rightarrow N_2$ (dinitrogen).

Many soil bacteria like *Thiobacillus denitrificans* which are known to oxidize sulphur chemoautotrophically also reduce nitrate to nitrogen. The source of energy is sulphur or thiosulphate and this energy is used to convert nitrate into molecular nitrogen. The biochemical reactions can be summarised as:

$5S + 6KNO_3 + 2H_2O \longrightarrow 3N_2 + K_2SO_4 + 4KHSO_4$ or
$5K_2S_2O_3 + 8KNO_3 + H_2O \rightarrow 4N_2 + 9K_2SO_4 + H_2SO_4$.

Fallow soils flooded with water are more congenial for denitrification than well drained and continuously cropped soils. In fact, the practice of continuous cropping which provides the much needed competition between plants and microorganisms for nitrate substrates minimizes the hazards of denitrification.

The denitrification process, besides depleting the fertility of soils, is also a cause for eutrophication of coastal waters because nitrate is a pollutant especially of drinking water. The process also influences the chemistry of the atmosphere by producing nitrous oxide, a 'green house gas' that destroys the ozone (O_3) of the stratosphere.

Losses of Nitrogen by Non-biological Ways

Leaching is one of the major causes of nitrogen losses extending to 20 to 50% of the fertilizer nitrogen supplied to cultivated oils. The most striking loss of nitrogen is in rice soils where more than half of the fertilizer nitrogen applied get lost through leaching. Another factor is the volatilization of ammonia in soil, often estimated at 5 to 20% of the fertilizer nitrogen applied to soil. Fixation of ammonium in soils is a minor contributory factor to the overall loss of nitrogen available for plant growth. Such losses of nitrogen by physical causes and by nitrification and denitrification processes can be controlled by the application of certain man-made chemicals. Some of these chemicals have been designed to control the rate of release of nutrient from nitrogenous fertilizers while others retard nitrification in soil by controlling the activity of nitrifying bacteria.

Controlled Release Fertilizers and Nitrification Inhibitors

The release of nutrient from nitrogenous fertilizers can be controlled by the use of 'controlled release fertilzers' which have been formulated and used successfully in Japan and U.S.A. Examples of this kind of new fertilizers are ureaform, isobutyledene diurea, crotonilidene diurea and sulphurcoated urea. They are sparingly soluble in water and by virtue of this property can regulate the release of nitrogen from fertilizers. Under experimental conditions in northern India, these fertilizers have been tried in rice fields and found to get mineralized at a slower rate than ordinary urea and also provide residual nitrogen for augmenting the yield of subsequent wheat crop (Table 35). The prohibitive cost of these materials, at present, prevents their immediate utility on the farm.

Some of the chemicals which are known to act as nitrification inhibitors are produced in U.S.A. and Japan, and are expensive. These chemicals are substituted pyridines, pyrimidines, acetanilides, anilines and isothiocyanates. The two major compounds which have been commercially

Table 35 Grain yield (q/ha)* of wheat as influenced by residual N from rice and rates of N applied to wheat. (from Prasad, 1975)

Source of N applied (10 kg/ha) to rice	Rates of N (kg/ha) applied to wheat		
	0	50	100
Urea	9.2	31.2	35.9
Sulphur-coated urea (TVA)	14.9	38.2	44.2
Lac-coated urea	12.7	34.8	41.9
Neem cake treated urea	14.9	38.8	50.7
Sulphathiazole treated urea	15.4	40.0	49.6
Coaltar treated urea	13.5	31.4	43.9
Control (no nitrogen)	3.9	17.0	33.0
C.D. 5%	–	1.7	–

*—lq = 100 kg.

produced by the Dow Chemical Co., U.S.A. and the Toyo Koatsu Co., of Japan are 2-chloro-6-(trichloromethyl)-pyridine, commonly known as N-serve and 2-amino-4-chloro-6-methyl pyridine commonly known as AM. At level of 1.0 ppm, N-serve inhibits the growth of *Nitrosomonas europea* and *N. agilis*. Oxidation of ammonium by fresh cell suspensions of *Nitrosomonas* was completely suppressed at 1.0 ppm of the chemical. In experiments done at the Indian Agricultural Research Institute, it was seen that N-serve and AM effectively retarded the nitrification of ammonium sulphate and reduced the losses of nitrogen from soil under water-logged conditions. Increase in yields of rice due to application of ammonium sulphate in the presence of N-serve has also been observed (Table 36). The beneficial effect of N-serve has also been recorded in experiments carried out by independent workers with grasses, tomato and spinach.

Table 36 Influence of nitrification inhibitor treatment and variety on grain yield and nitrogen uptake by rice. (from Prasad, 1974)

Treatment (Averaged over 40, 80 and 120 kg N/ha)	1966		1967	
	Taichung Native-1	NP 130	Taichung Native-1	NP 130
	Grain yield (q/ha)*			
Ammonium sulphate	44.1	41.3	63.4	48.1
Ammonium sulphate + 'N-Serve'	48.4	45.1	63.7	50.8
Ammonium sulphate + 'AM'	50.1	43.9	67.7	49.3
C.D. 5%	2.5		2.4	
	Nitrogen uptake (kg N/ha)			
Ammonium sulphate	87	81	103	103
Ammonium sulphate + 'N-Serve'	100	89	128	111
Ammonium sulphate + 'AM'	104	89	131	110
C.D. 5%	6		3	

*lq = 100 kg.

The seeds of neem tree (*Azadirachta indica*) contain certain lipid associates which act as nitrification inhibitors and thereby increase the efficiency of urea fertilizer. Increased yields due to the application of neem cake with urea fertilizers have been recorded and further work in this direction may open up the possibility of providing suitable and inexpensive substitute to N-serve and AM.

Selected References

Alexander, M. 1961. *Introduction to Soil Microbiology*. John Wiley & Sons, Inc., New York and London.

Alexander, M. 1965. Nitrification. In *Soil Nitrogen*, pp. 307–343, Eds. M.V. Bartholomew and F.E. Clark, American Society of Agronomy, Madison, Wis., U.S.A.

Allison, F.E. 1955. The enigma of soil nitrogen balance sheets. *Adv. Agron.*, 7, 213–250.

Bartholomew, M.V. 1965. Mineralization and immobilization of nitrogen in the decomposition of plant and animal residues. In *Soil Nitrogen*, pp. 285–306, Eds. M.V. Bartholomew and F.E. Clark, American Society of Agronomy, Madison, Wis., U.S.A.

Belser, L.W. 1982. Inhibition of nitrification. pp. 267–293. In *Advances in Agricultural Microbiology*, Ed. N.S. Subba Rao, Oxford and IBH Publishing Co., New Delhi.

Broadbent, F.E. and Clark, F.E. 1965. Denitrification. In *Soil Nitrogen*, pp. 344–359, Eds. M.V. Bartholomew and F.E. Clark, American Society of Agronomy, Madison, Wis., U.S.A.

Groffman, P.M. 1998. Denitrification and its impact on soil fertility and environment quality, pp. 163–191. In *Microbial Interactions in Agriculture and Forestry*, Vol I. Eds. N.S. Subba Rao and Y.R. Dommergues, Oxford and IBH Publishing Co. New Delhi.

Knowles, R. 1981. Denitrification. In *Soil Biochemistry*, Vol. 5 Eds. E.A. Paul and J. Ladd, Marcel Dekker, Inc., New York.

Knowles, R. 1982. Denitrification in soils, pp. 243–266. In *Advances in Agricultural Microbiology*, Ed. N.S. Subba Rao, Oxford and IBH Publishing Co., New Delhi.

Prasad, R. 1968. Dry matter production and recovery of fertilizer nitrogen by rice as affected by nitrification retarders 'N-serve' and 'AM'. *Pl. Soil*, 29, 327–330.

Prasad, R. 1998. Fertilizer, water, food security, health and the environment. *Curr. Sci. (India) 75*, 677–683.

Prasad R., Rajale, G.B. and Lakhdive, B.A. 1971. Nitrification retarders and slow release nitrogen fertilzers. *Adv. Agron,.*, 23, 337–376.

Prasad, R. 1974. Research on nitrification inhibitors and slow release nitrogen fertilizers in India—Prospects for their use and problems. Proc. FAI-FAO Seminar on *Optimising Agriculture Production under Limited Availability of Fertilizers*, pp. 167–185, Fertilizer Association of India, New Delhi.

Tiedje, J.M. 1982. Denitrification, pp. 323–369. In *Methods of Soil Analysis*, Vol. 2, Eds. R.H. Miller and others, Amer. Soc. Agronomy, Madison.

12. Microbial Products Influencing Plant Growth

Indole Acetic Acid

Soil microorganisms produce a variety of substances which directly or indirectly affect plant growth. Many species of bacteria and fungi are known to produce indole acetic acid (IAA) in small amounts, especially when the growth medium is supplemented with tryptophane, a precursor to IAA (Fig. 84). For example, *Agrobacterium tumefaciens, Ustilago maydis, Synchytrium endobioticum, Gymnosporangium juniperi-virginianae, Nectria galligena, Endophyllum sempervivi, Rhizobium* spp., *Rhizopus suinus* and *Pseudomonas fluorescens* produce IAA in pure cultures or in association with higher plants. Some of the important morphogenetic effects of IAA on plant growth are elongation of the stem and gall formation representing the reaction of the host to the presence of the auxin.

A simple example of interaction of microbially produced IAA with the host plant is the root hair curling phenomenon observed in leguminous plants inoculated with *Rhizobium*. An instance of microbially produced IAA causing hypertropny of plant cells is the crown-gall tissue caused by *Agrobacterium tumefaciens*. The gall-like structures seen in various plants have been primarily attributed to IAA acting together with an unidentified tumour-inducing principle (TIP) produced by the causative organism. Several cases of hyperauxiny (accumulation of auxin in host tissues as a result of interaction between pathogenic microorganisms and the host plants) have been cited in literature. Examples of hyperauxiny are wilt diseases of plants caused by *Verticillium* resulting in the death of plants due to water shortage and maize smut caused by *Ustilago maydis* which produces large rounded galls on leaves, stems and inflorescence. However, it should be borne in mind that IAA produced by these parasitic microorganisms acts in conjunction with other substances produced in the host plant due to host-parasite interactions.

Gibberellins

As early as 1926, Japanese investigators, while studying a disease of rice caused by *Gibberella fujikuroi* (Saw) Wr. (imperfect state of *Fusarium*

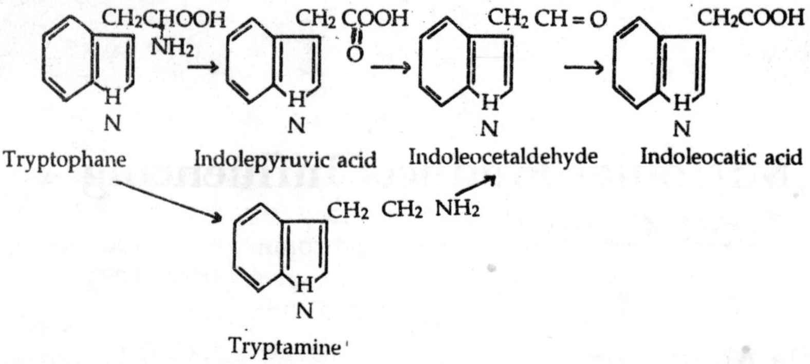

Fig. 84 Biochemical pathway from tryptophane to indole acetic acid.

moniliforme Sheld), discovered that leaves and stems of infected plants showed abnormal growth. The infected plants were usually taller than the healthy ones. They called the infection as 'Bakanae' disease, literally meaning foolish seedlings. Subsequently, a purified preparation containing highly active crystalline material was obtained from culture filtrates of the fungus which was named as gibberellin. When applied to healthy plants, the crystalline material reproduced the symptoms of the disease. These initial findings led to the subsequent purification of several fractions of the original gibberellin and today fourteen gibberellins have been identified and many of them are available on the market for experimental and commercial purposes. Some of the gibberellins are known to be natural components of plants controlling their growth activities, dormancy, flowing and responses to light and temperature. Among the several gibberellins, GA_1 was probably the first to be identified both from the fungal medium and also from higher plants. Subsequently, GA_2, GA_3, GA_4, GA_7, and GA_9 were identified from the fungal medium and GA_5 GA_6 and GA_8 from flowering plants

Several bacteria, actinomycetes and fungi are known to produce gibberellins or gibberellin-like substances. The bacterial genera are *Arthrobacter, Azospirillum, Azotobacter, Bacillus, Brevibacterium, Flavobacterium, Pseudomonas* and *Rhizobium*. The actinomycetes come under the genera *Actinomyces* and *Nocardia*. The fungal genera capable of producing the plant growth regulator are *Alternaria, Aspergillus, Fusarium, Gibberella, Penicillium, Rhizopogon, Rhizopus, Sphaceloma* and *Suillus*.

Some of the roles attributed to gibberellins are: (1) they overcome dormancy and dwarfism in plants; (2) they induce flowering of some photoperiodically sensitive and other low temperature dependent plants; (3) they alter the sex of flowers and contribute to fruit setting; and (4) they stimulate stem growth and at the same time suppress the growth of lateral branches.

Cytokinins

Under favourable conditions, *Azotobacter* and *Azospirillum* are known to produce cytokinins in the rhizosphere along with other plant growth regulators. Other bacteria that produce cytokinins or cytokinin-like substances come under the genera *Agrobacterium, Arthrobacter, Bacillus, Corynebacterium, Escherichia, Pseudomonas* and *Rhizobium*.

Ethylene

Ethylene production has been detected in soil as well as in pure cultures of microorganisms. A variety of substrates in soil including methionine, other aminocids, organic acids, carbohydrates are known to be the sources of ethylene. There appears to be a direct relation between the rate of C_2H_4 production and the organic matter content of soil. Small amounts of exogenous application of ethytlene can influence plant response and ethylene has a role in ripening of fruits. Bacterial genera capable of producing ethylene are *Aeromonas, Arthrobacter, Citrobacter, Clostridium, Enterobacter, Erwinia, Escherichia, Klebsiella, Pseudomonas, Serratia* and *Streptomyces*. Ethylene producing fungi come under the genera *Acremonium, Agaricus, Alternaria, Ascochyta, Aspergillus, Blastomyces, Botrytis, Candida, Mucor, Fusarium, Laccaria, Neurospora, Penicillium, Pythium* and *Rhizoctonia*.

Antibiotics

Antagonism among microorganisms is a common phenomenon in soil resulting from the production of antibiotics. In human medicine, antibiotics (Table 37) have proved to be of great therapeutic value. Some of these antibiotics have also a potential role in controlling plants diseases.

As examples of antifungal antibiotics useful in the control of plant diseases, two prominent products can be cited: griseofulvin, a metabolic product of *Penicillium griseofulvum* and aureofungin, a metabolic product of *Streptoverticillium cinnammomeum* var. *terricolum*. Application of griseofulvin has proved successful in minimizing infection of plants by *Botrytis*. Extensive trials in Japan have shown excellent control of monilia disease of apple caused by *Sclerotinia*. Similar results were obtained in

Table 37 Some examples of antibiotics produced by actinomycetes, fungi and bacteria

Antibiotic	Isolated from	Active against
(1)	(2)	(3)
Amphomycin	*Streptomyces canus*	Gram-positive bacteria
Amphotericin B	*Streptomyces nodosus*	Yeast, fungi
Aterrimin	*Bacillus subtilis*	Gram-positive bacteria
Bacitracin	*Bacillus subtilis*	Gram-positive bacteria
Blasticidin S	*Streptomyces griseochromogenes*	Fungi
Candicidin B	*Streptomyces griseus*	Yeast, fungi
Cephalosporins and chemical derivatives	*Cephalosporium acremonium*	Staphylococcus, Steptococcus, E. coli, K. pneumoniae, Serratia
Chloramphenicol	*Streptomyces venezuelae*	Gram-positive and Gram-negative bacteria; Rickettsiae
Colistin	*Bacilius colistinus*	Gram-negative bacteria
Cycloheximide	*Streptomyces griseus*	Fungi
Cycloserine	*Streptomyces orchidaceus*	Gram-positive and TB bacteria
Dactinomycin (Actinomycin D)	*Streptomyces antibioticus*	Gram-positive bacteria; anti-tumor principle
Erythromycin	*Streptomyces erythreus*	Gram-positive bacteria
Fusidic acid	*Fusidium coccineum*	Gram-positive bacteria
Gentamycin	*Micromonospora purpurea*	Gram-positive bacteria
Gramicidin	*Bacillus brevis*	Gram-positive bacteria
Griseofulvin	*Penicillium griseofulvum*	Fungi
Hygromycin	*Streptomyces hygroscopicus*	Gram-positive and Gram-negative bacteria; Helminths
Kanamycin	*Streptomyces kanamyceticus*	Gram-positive, Gram-negative and TB bacteria
Leucomycin	*Streptomyces kitasoensis*	Gram-positive bacteria
Lincomycin	*Streptomyces lincolnensis*	Gram-positive bacteria
Neomycins	*Streptomyces fradiae*	Gram-positive, Gram-negative and TB bacteria
Novobiocin	*Streptomyces niveus*	Gram-positive bacteria
Nystatin	*Streptomyces noursei*	Fungi and yeast
Oleandomycin	*Streptomyces antibioticus*	Gram-positive bacteria
Paromomycin	*Streptomyces rimosus*	Gram-positive, Gram-negative and TB bacteria; Protozoa
Penicillin and its chemical derivatives	*Penicilliium chrysogenum*	Gram-positive bacteria
Polymyxin B	*Aerobacillus polymyxa*	Gram-negative bacteria
Pristinamycin	*Streptomyces* sp.	Gram-positive bacteria
Rifomycin SV	*Streptomyces mediterranei*	Gram-positive and TB bacteria
Ristocetin	*Nocardia lurida*	Gram-positive bacteria

(Contd.)

Table 37 (Contd.)

(1)	(2)	(3)
Spiramycin	*Streptomyces ambofaciens*	Gram-positive and Gram-negative bacteria; Rickettsiae
Staphylomycin	*Streptomyces virginiae*	Gram-positive bacteria
Stendomycin	*Streptomyces endus*	Gram-positive and Gram-negative bacteria
Streptomycin and chemical derivatives	*Streptomyces griseus*	Gram-positive, Gram-negative and TB bacteria
Tetracycline and chemical derivatives	*Streptomyces aureofaciens*	Gram-positive and Gram negative bacteria; Rickettsiae
5-Hydroxytetracycline	*Streptomyces rimosus*	Gram-positive and Gram-negative bacteria; Rickettsiae
Thiostrepton	*Streptomyces azureus*	Gram-positive bacteria
Trichomycin	*Streptomyces hachijoensis*	Fungi and yeast
Tylosin	*Streptomyces fradiae*	Gram-positive bacteria
Tyrothricin	*Bacillus brevis*	Gram-positive and Gram-negative bacteria
Vancomycin	*Streptomyces orientalis*	Gram-positive and TB bacteria
Variotin	*Paecilomyces varioti*	Fungi and yeast
Viomycin	*Streptomyces floridae*	Gram-positive, Gram-negative and TB bacteria

melon canker caused by *Mycosphaerella* and powdery mildews caused by *Erysiphe*. Griseofulvin is a systemic fungicide since the antibiotic permeates plant tissues uniformly and forms a barrier to the penetration by pathogenic fungi.

Aureofungin is a broad spectrum antifungal antibiotic with the unique property of inhibiting growth of a large number of phytopathogens. It was once commercially produced in India at the Hindustan Antibiotics, Pimpri, Poona. The antifungal activity of the antibiotic has been demonstrated in pure cultures on *Pyricularia oryzae, Helminthosporium oryzae, H. turcicum, H. nodulosum, Alternaria tenuis, Curvularia lunata, Verticillium alboatrum, Phytophthora citrophthora, Aspergillus niger, A. Fumigatus, Candida albicans, Cryptococcus neoformans, Trichophyton mentagrophytes* and *T. rubrum*. The minimum inhibitory concentration of the antibiotic ranges from 0.005 to 1 µg/ml. The antibiotic is insoluble in water but soluble in alcohols and can be made soluble in water at alkaline pH. Its effectiveness against citrus gummosis incited by *Phytophthora* sp., powdery mildew of apple caused by *Podosphaera leucotricha* and diseases of grapes such as powdery mildew, downy mildew and the anthrachnose is well established. It has also been reported that aureofungin has potentialities in the control of seed-borne infection and seedling blight caused by *Helminthosporium oryzae* and *Pyricularia oryzae* disease of ragi, and many other disease including post-harvest and storage diseases.

Some antibiotics are of limited application such as cycloheximide (actidione), blasticidin-S, kasugamycin and streptomycin. Cycloheximide has been shown to be active against leaf spots, powdery mildews and blister rust of pine while blasticidin-S has shown promise against *P. oryzae*. Streptomycin has proved effective against diseases caused by *Erwinia, Xanthomonas, Pseudomonas, Corynebacterium, Agrobacterium, Pseudoperonospora, Peronospora* and *Sphaerotheca*.

In Japan, agriculturally useful antibiotics have been widely used to protect plants against diseases and pests. Some of the examples are cited below: (1) Cycloheximide (from *Streptomyces griseus*) as a wettable powder against onion downy mildew and shoot blight of Japanese larch, (2) Kasugamycin (from *Streptomyces kasugaensis*) as dust against rice blast, (3) Polyoxins (produced by *Streptomyces cacaoi* var *asoensis*) as dust, wettable powder and emulsion against rice sheath blight and fungal diseases of fruits and vegetables, (4) Validamycin A (from *Streptomyces hygroscopicus* var. *limoneus*) as dust against rice sheath blight, (5) Streptomycin (from *Streptomyces griseus*) as wettable powder or liquid against bacterial diseases of fruits and vegetables, (6) Oxytetracycline (from *Streptomyces viridifaciens*) as wettable powder against citrus canker and peach bacterial leaf spot and (7) Tetranactin as emulsion against carmine mite of fruits and tea. The primary sites of action of these antibiotics relate to chitin synthesis of cell wall, cation leakage from mitochondria, biosynthesis of inositol and protein and DNA synthesis. Limitations in the use of antibiotics in agriculture are difficulties in analysis because they are used in small amounts and there is a likelihood of development of plant pathogens resistant to antibiotics. Therefore, the use of chemicals and antibiotics in alternate years has been recommended to overcome such a possibility.

Antibiotics are also widely used as growth stimulants in poultry and livestock feeds. The use of aureomycin, terramycin and penicillin at the rate of 5–20 g/ton of wheat increases the rate of growth of farm animals from 5 to 50%. Although the mechanism behind such stimulating effect of antibiotics is not known, it is believed that they may kill unwanted bacteria in the intestine and also increase the availability of vitamin B_{12}.

Mechanism of Antibiotic Action

The penicillins inhibit cell wall synthesis of bacteria. The mechanism is as follows: The cell wall of both Gram positive and Gram negative bacteria is composed of peptidoglycan and murein. Petptidoglycan is made up of polysaccharide chains containing alternating residues of N-acetylmuramic acid (M) and N-acetylglucosamine (G). Peptide units cross-link extending from the N-acetylmuramic acid and during bacterial wall synthesis these linkages are caused by precursors which are catalysed by specific enzymes (ex: transpepidases and carboxypeptidases). These enzymes are regulatory proteins which bind to penicillin and hence are known as penicillin binding proteins (PBPs). When bacteria are exposed to penicillin, the antibiotic binds to the PBPs in the cell membrane followed by the release of autolytic enzymes that degrade the preformed cell wall and arrest further cell wall synthesis leading to death of bacteria (Fig. 85).

The principal mode of action of tetracyclines is the inhibition of protein synthesis. The antibiotics bind to the 30S ribosomal sub-unit and prevent the attachment of aminoacyl-tRNA to be acceptor site on the messenger RNA (mRNA)-ribosome complex (Fig. 86).

Resistance to Antibiotics

Resistance to antibiotics may be intrinsic (inherent) or acquired. Prior to the use of antibiotics, inherent or intrinsic resistance was common whereas acquired resistance is a result of exposure to these agents. Acquired resistance may be the result of spontaneous chromosomal mutations or due to entry of plasmids with resistance factors (R factors) and transposons. The resistance genes may be transferred to plasmids from chromosomes and *vice versa* by recombination (exchanges of DNA pieces) in the presence of a *recA* gene showing genetic homology. Transposons are also pieces of DNA that carry resistance genes. They are known as 'jumping genes' and can get inserted into chromosomes or plasmids even if homology does not exist and this process is independent of a *recA* gene.

In general, the biochemical mechanisms of bacterial resistance to antibiotics can be classified into 4 types—(1) alteration of the target site of the antibioitic causing decreased affinity of the antibiotic for the binding site or the inability of the antimicrobial agent to bind, (2) inactivation of the antibiotic by the production of enzymes that are able to convert an active drug into an inactive derivative, (3) decreased permeability of the membranes and entry or accumulation of the antibiotic and (4) a combination of the above three mechanisms.

Aflatoxin

Aflatoxin is a metabolic product of *Aspergillus flavus* although there are claims of production of aflatoxin-like substances by other *Aspergillus* and *Penicillium* spp.

Aflatoxin attracted the attention of microbiologists in the United Kingdom in 1960 when countless turkeys and ducks died due to contamination of their feed of groundnut cakes with *Aspergillus flavus*. The disease manifestation in infected turkeys was due to the damage caused to the liver and kidneys of the animals by aflatoxin. Since this initial outbreak of the turkey-disease, considerable work has been done on the biochemical and microbiological aspects of aflatoxin and it is now known

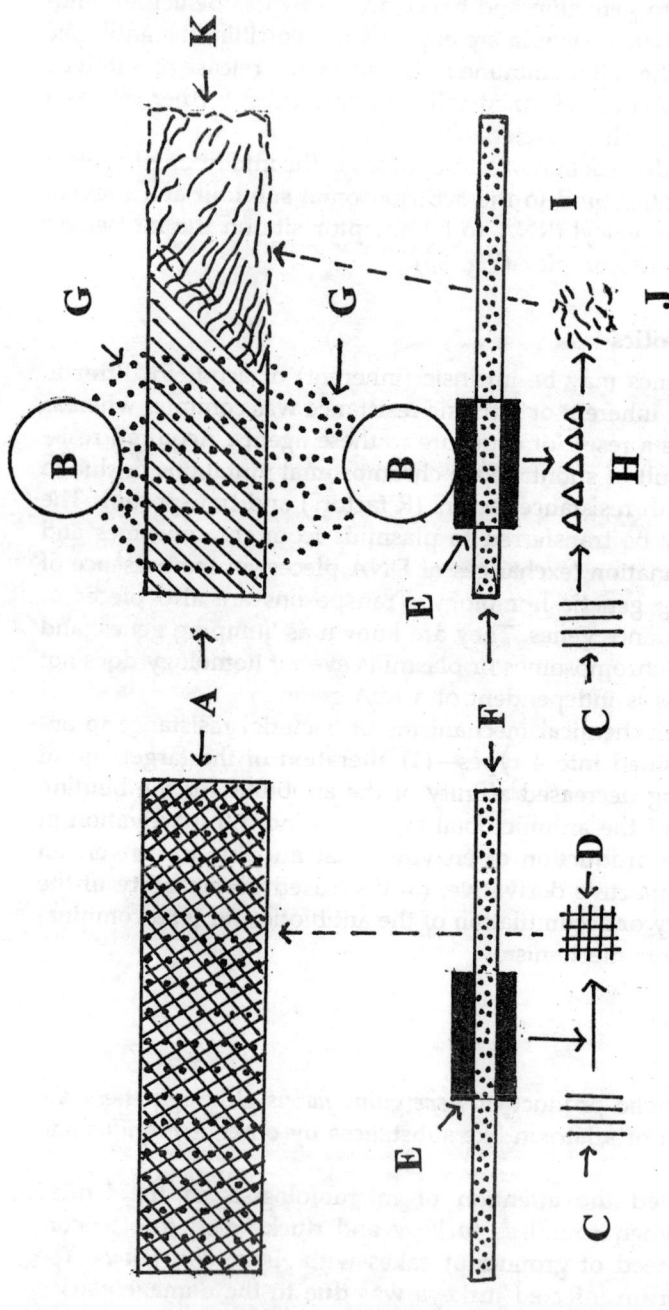

Fig. 85 The inhibition of cell wall synthesis by penicillin: Normally, as shown in the left side of the figure, cell wall precursors (C) are cross-linked (D) by specific enzymes, transpeptidases and carboxypeptidases, also known as penicillin binding proteins (E) in the cell membranes (F) and then added on to the growing bacterial cell wall (A). However, when penicillin (B) gets into the cell through porins (G) as shown in the right side of the figure, the antibiotic binds to the specific enzymes (E). This leads to the release of autolysins (H) by the cell membrane (F) into the cytoplasm (I). The autolysins (H) prevent the cell wall precursors (C) to form cross linkages leading to their disruption (J) and lack of participation in cell wall growth which ultimately leads to a loss in cell wall turgidity (K) and death of the bacterial cell.

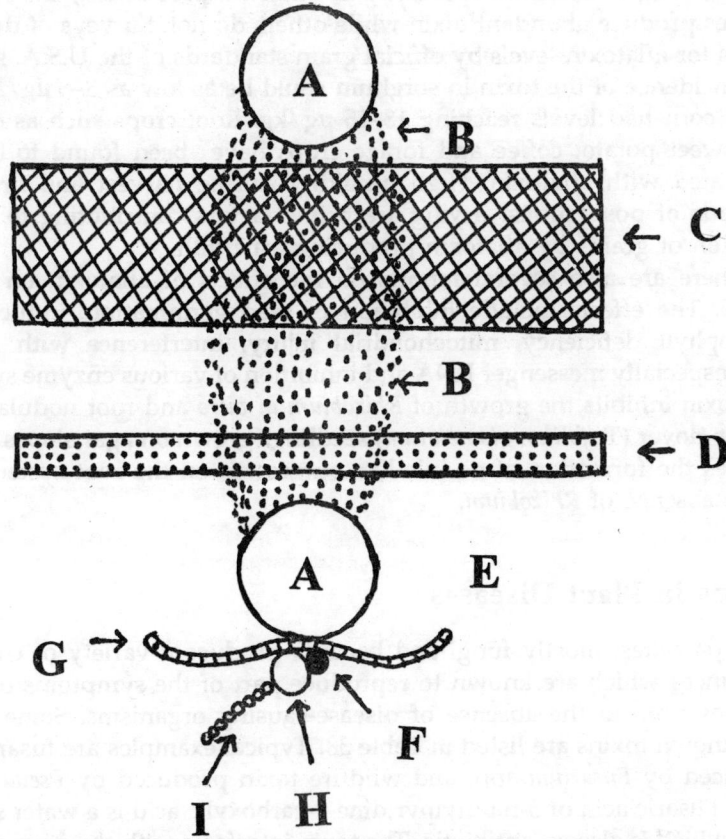

Fig. 86 Inhibition of steps involved in protein synthesis by tetracyclines: The antibiotic tetracycline (A) enters bacteria through porins (B) in the cell wall (C) and is actively transported across cytoplasmic membrane (D). In the cytoplasm (E), the antibiotic binds to 30S ribosomal sub-unit (F) of mRNA (G) causing misreading of tRNA (H) leading to deformed protein (I).

that aflatoxin consists of several chromatographically distinguishable components like aflatoxin B_1, B_2, G_1, G_2, M_1, M_2, etc.

Aflatoxins are known to possess carcinogenic properties and mainly affect animals such as birds, fish, cattle, swine, sheep, goats, dogs and monkeys. Human beings are rarely affected by aflatoxin unless large quantities of groundnut infested by *A. flavus* are consumed. Reports are available which imply the accumulation of considerable amounts of aflatoxin in the breast milk of mothers fed on preparations made from moldy grains.

Experimental production of aflatoxins by inoculation with *A. flavus* has been reported in several agricultural commodities such as copra, wheat, rice, cotton, oats, groundnut, corn and clovers. There appears to be some degree of strain specificity in aflatoxin production, since certain isolates produce abundant toxin while others do not. Surveys of domestic grains for aflatoxin levels by official grain standards of the U.S.A. showed that incidence of the toxin in sorghum could be as low as 3–6 µg/kg seed while corn had levels reaching 13–15 µg/kg. Root crops such as cassava and sweet potato, coffee and forage crops have been found to be contaminated with aflatoxin. High humidity during harvest and improper methods of post-harvest drying are prime factors contributing to the infestation of grains by aflatoxin producing aspergilli.

There are also several reports on the effects of aflatoxin on higher plants. The effects include inhibition of seed germination, induction of chlorophyll deficiency, mitochondrial injury, interference with nucleic acids especially messenger RNA and inhibition of various enzyme systems. The toxin inhibits the growth of *Rhizobium in vitro* and root nodulation of cluster clover (*Trifolium glomeratum*) seedlings grown on agar slopes. It also induces the formation of nodule-like outgrowth on the root system even in the absence of *Rhizobium*.

Toxins in Plant Diseases

Plant parasites, mostly fungi and bacteria produce a variety of chemical substances which are known to reproduce part of the symptoms of plant diseases even in the absence of disease-causing organisms. Some of the well-known toxins are listed in Table 38. Typical examples are fusaric acid produced by *Fusarium* spp. and wildfire toxin produced by *Pseudomonas tabaci*. Fusaric acid or 5-n butylpyridine-2-carboxylic acid is a water soluble toxin which is also an antibiotic. The toxin interferes with the permeability of the plasma membrane and finally affects the water economy of plants. The impediment caused to water movement in plants leads to an irreversible pathological wilting resulting in the death of plants as in the case of wilt diseases of cotton and tomato caused by *Fusarium* spp. The wildfire toxin, on the other hand, is an amino acid derivative which is highly specific and localized in action. The toxin is biologically very active and even 0.05 mg of the substance gives rise to typical chlorotic lesion when introduced into the tobacco leaf.

Bacterial and Fungal Insecticides

There are about 90 species of bacteria pathogenic to insect pests and they serve as new tools in biological control of plant pests. Among them, *Bacillus thuringiensis* which was first discovered in 1902 by a Japanese bacteriologist, Ishiwata from infected silkworms stands out prominent. The bacteria form a protein crystal inclusion body (molecular weight 800–900)

Table 38 Examples of known toxins involved in plant diseases (Kalyanasundaram, 1963)

Causal agent	Host of choice	Toxin	Chemical nature	Molecular weight
Fusarium lycopersici	Tomato	Lycomarasmine	Dipeptide	277
Fusarium lycopersici	Tomato	Fusaric acid	Pyridine-carboxylic acid	179
Fusarium oxysporum	Potato	Enniatin 'A'	Polypeptide	455
Penicillium patulum	Potato	Patulin	Lactone	154
Alternaria solani	Potato	Alternaric acid	Dibasic acid	410
Endothia parasitica	Chestnut	Diaporthin	Bianthraquinone	250
Pseudomonas tabaci	Tobacco	Wildfire toxin	A new type of amino acid	206

which is an endotoxin active in inhibiting the growth of about 130 species of insects and larvae. The protein crystal synthesis and spore formation in the bacterium proceed simultaneously and the two processes are in many ways interlinked. The organisms can be grown in naturally available cheap media (such as bran) and spores harvested to produce a mixture of spores and endotoxin crystals. Commercial preparations containing B. *thuringiensis* have been produced in many countries notably in U.S.A. where they are used on several agricultural crops, trees and ornamental shrubs.

A strain of B. *thuringiensis* which showed high toxicity to mosquito larvae (*Anopheles, Culex, Aedes*) described as subspecies *isaraelensis*, has been isolated in Israel and its efficacy proved against mosquito larvae. The preparation from this subspecies is not toxic to lepidopterous larvae and hence specific to mosquitoes. It has potentialities in the control of malaria in man.

Other bacterial agents used against insect pathogens are *Bacillus popilliae, Coccobacillus acridorum* and *Serratia marcescens*.

Fungi and protozoa are also efficient in controlling insect pests on plants. Some of the species of fungi which are currently being used in the U.S.A. in controlling insect pests are *Entomophthora* spp.., *Beauveria* spp., *Metarrhizium anisopliae* and *Aeschersonia* spp. Examples of protozoans which hold promise as insect pathogens are *Thelohania hyphantriae, Mattesia grandis* and *Malameba locustiae*. (See also chapter on Biotechnology in Agriculture).

Virus Insecticides

There are more than 300 viruses with are known to rapidly infect susceptible species of insects. Unlike plant and other animal viruses, insect viruses are encased in protein crystals either singly or in groups. The protein crystals are insoluble in water and they are produced abundantly inside insect tissues and released when insects die. The crystals can re-infect live insects and retain their infectivity even after long storage outside the living tissues of insects.

Insect viruses are classified as polyhedroses and granuloses. Polyhedroses are those which contain many virus particles in the protein crystal while granuloses contain only one virus particle in each crystal. Yet another minor group is recognised as nuclear polyhedrosis viruses which develop on the nuclei of host cells. Such viruses are largely used in the biological control of insect pests on a commercial scale. In polyhedrosis type of viruses, the infective unit is embedded in protein polyhedron which protects the viruses from damage during its shelf life. Their host specificity is confined to a species or a group of species.

When artificially introduced in pest-ridden plant populations, the viruses multiply and are dispersed by air currents and rain water thus becoming potent insecticides. For instance, the European spruce sawfly, once recognised as a serious plant pest in Canada and U.S.A. in the early part of this century, was controlled by the chance introduction of a nuclear polyhedrosis virus. Since then, numerous successful attempts have been made to use insect viruses in biological control of plant pests either by collecting naturally occurring virus-infected insects and applying their extracts or by spraying preparations from laboratory grown virus infested insects. Many virus insecticides have been developed on an industrial scale in U.S.A. by artificial rearing of infected insects. Among them, the nuclear polyhedrosis virus of cotton bollworm (*Heliothis zea*) and that of the cabbage worm (*Trichoplusiani*) hold promise as revealed by results of extensive field trials.

The commercial aspects of microbial insecticides have been included in the chapter on "Biotechnology in Agriculture".

Microbial Herbicides

The concept of herbicides of microbial origin stems from the efforts of plant pathologists to use endemic or exotic pathogens to kill weeds. Some of the successes with exotic pathogens have been: 1) the use of *Puccinia chondrillina* from Southern Europe to control skeleton weed (*Chondrilla juncea*) in Australia, 2) the use of *Cercosporella riparia* to control *Ageratina riparia* introduced in Hawaii from Jamaica, and the use of introduced rust *Phragmidium violaceum* to control wild blackberry (*Rubus* spp.). Similarly, some of the successes with endemic pathogens are: 1) the use of *Cercospora rodmanii* to control water hyacinth (*Eichhornia crassipes*) which chokes waterways in Florida and Louisiana of U.S.A., 2) the use of *Colletotrichum gloeosporioides* to control jointveth (*Aeschynomene virginica*) in rice fields of Arkansas in U.S.A., and 3) the use of host-specific pathotype *Phytophthora citrophthora* to control milkweed vine (*Morrenia odorata*) in Florida, U.S.A.

It should , however, be stated that exploitation of the exact microbial herbicides is dependent on close international cooperation whereas the use of endemic microbial herbicides would need national and local cooperation from various agencies.

Biological Control of Plant Diseases

Several microorganisms have shown potentialities as biological control agents against important plant diseases caused by soil-borne pathogens. They are *Pseudomonas fluorescens* against the take-all disease of wheat caused by *Gaeumannomyces graminis, Erwinia caratovora* infection of wheat, *Thielaviopsis basicola* infection of tobacco and damping off caused by *Pythium* in cotton; *Bacillus subtilis* against *Fusarium roseum* wilt of corn: *Trichoderma harzianum* against *Alternaria* spp. infection of radish; *Penicillium oxalicum* against root rot of peas; non-pathogenic *Fusarium oxysporum* against *Fusarium* wilt of cucumber; *Trichoderma viride* against *Verticillium* wilt of tomatoes; *Cytophaga* sp. against damping off of conifer seedlings; *Bacillus* sp., *Penicillum* sp., and *Alcaligenes* sp. against crown gall disease of cherry seedlings caused by *Agrobacterium tumefaciens; Chaetomium globosum* against damping off of sugar beets; *Pseudomonas putida* against *Fusarium solani* wilt of beans; *Bacillus* sp. and *Pseudomonas* sp. against *Fusarium oxysporum* wilt of carnations.

Although research on biological control of plant diseases has been carried out for a number of years, it is relevant to mention some examples of success in this regard.

One example is the use of *Peniophora gigantea* to control *Heterobasidium annosum* infection of pine trees. In the U.K. and the U.S.A., felled pine tree stumps are coated with *P. gigantea* which prevents the colonization of stumps by *H. annosum*, thereby restricting the survival and multiplication of the pathogen. The second example is the use of *Agrobacterium radiobacter* in the control of crown-gall disease caused by *A. tumefaciens*. Considerable damage to stone fruit, pome fruit, other rosaceous plants, walnut, grapevine, chrysanthemum and dahlia is caused by crown-gall infection (plant cancer). This disease has been successfully controlled in Australia by using the strain 84 of the non-pathogenic *A. radiobacter*, which produces an unusual antibiotic called "Agrocin 84" which belongs to a new group of antibiotics known as nucleotide bacteriocins. A bacteriocin is a substance produced by certain strains of bacteria and active on other strains of the same or closely related species.

A large conjugative plasmid called Ti plasmid (tumour-inducing circular DNA molecule) controls crown-gall infection. This plasmid can be transferred to another cell by conjugation and in this process a non-pathogen can become a pathogen. The plasmid directs the synthesis of chemicals called opines—nopaline, agrocinopine A, octopine and agropine. Strains of *A. tumefaciens* which produce the first two opines are known as 'nopaline strains' and those which produce the other opines are known as 'Octopine strains'. All octopine strains are not controlled by agrocin 84 but nopaline strains are susceptible to the bacteriocin and fortunately crown-gall on stone fruit is caused by nopaline strains of *A. tumefaciens* and hence can be controlled.

The role of rhizobacteria in the control of soil-borne diseases of plants has been discussed in the chapter on "The Rhizosphere."

Selected References

Angus, T.A. 1965. Bacterial pathogens as microbial insecticides. *Bact. Rev.* 29, 364–372.

Armbrocht, B.H. 1972. Aflatoxin residues in food and feed derived from plant and animal sources. *Residue Rev.*, 41, 13–54.

Braun, A.C. 1955. A study on the mode of action of the wildfire toxin. *Phytopathology*, 45, 659–664.

Brooks, W.M. 1980. Production and efficacy of protozoa. *Biotechnology and Bioengineering*, 22, 1415–1440.

Bruehl, G.W. Ed. 1975. *Biology and Control of Soil-borne Plant Pathogens*. Amer. Phytopath. Society, St. Paul, Minnesota.

Charudattan, R. 1978. *Biological Control Projects in Plant Pathology. A Directory*. Plant Pathology Miscellaneous Publications, University of Florida, Gainsville, Florida, U.S.A.

Dekker, J. 1963. Antibiotics in the control of plant diseases. *Ann. Rev Microbiol.*, 17, 243–262.

Detroy, R.W. Lillehoj, E.B. and Ciegler, A. 1971. Aflatoxin and related compounds. In *Microbial Toxins*, pp. 4–155. Eds. A. Ciegler, S. Kadis and S.J. Ajil, Academic Press, New York.

Freeman, T.E. 1982. Microbial herbicides, pp. 419–428. In *Advances in Agricultural Microbiology*, Ed. N.S. Subba Rao, Oxford and IBH Publishing Co., New Delhi.

Ferron, P. 1978. Biological control of insect pests by entomogenous fungi. *Ann. Rev. of Entomology*, 23, 409–442.

Gaüman, E. 1957. Fusaric acid as a wilt toxin. *Phytopathology*, 37, 342–357.

Kalyanasundaram, R. and Venkataram., C.S. 1956. Production and systemic translocation of fusaric acid in *Fusarium* infected cotton plants. *J. Indian bot. Soc.* 35, 7–10.

Kerr, A. 1982. Biological control of soil-borne microbial pathogens and nematodes, pp. 429–463.. In *Advances in Agricultural Microbiology*, Ed. N.S. Subba Rao Oxford and IBH Publishing Co., New Delhi.

Luthy, P. 1980. Insecticidal toxins of *Bacillus thuringiensis*. *FEMS Microbiology Letters*, 8, 1–17.

Misato, T. and Yoneyama, K. 1982. Agricultural antibiotics, pp. 465–506. In *Advances in Agricultural Microbiology*. Ed. N.S. Subba Rao, Oxford and IBH Publishing Co., New Delhi.

Moore, L.W. and Warren, G. 1979. *Agrobacterium radiobacter* strain 84 and biological control of crown gall. *Ann. Rev. Phytopathol.*, 17, 163–179.

Nickle, W.R. 1980. Possible commercial formulations of insect parasitic nematodes. *Biotechnology and Bioengineering*, 22, 1407–1414.

Norris, J.R. 1971. Microbes as biological control agents. In *Microbes and Biological Productivity*, pp. 197–229, Eds. D.E. Hughes and A.H. Rose, Cambridge Univ. Press, London.

Thirumalachar, M.J. 1968. Antibiotics in the control of plant pathogens. *Adv. appl. Microbiol.*, 10, 313–337.

Vijayalakshmi, K. and Subba Rao, N.S. 1971. Effect of aflatoxins on *Rhizobium* spp. *Indian Phytopath.*, 24, 791.

Moore, L. W. and Warren, G. 1979. *Agrobacterium radiobacter* strain 84 and biological control of crown gall. *Ann. Rev. Phytopathol.*, 17: 163-179.

Mickle, W. R. 1980. Possible commercial formulations of insect parasitic nematodes. *Protozoology and Biotechnology*, 22: 360-3414.

Pramer, J. S. 1971. Microbes as biological control agents. In *Microbial Products Influencing Plant Growth*. Inter. Sci. Press, London.

Dhanraj, M. 1948. Antibiotics. *Advances in Microbiology*, 10: 211-229.

Raj, Sashi, K. and Subba Rao, N.S. 1977. Effect of Rhizoxin on Rhizobacteria. *Indian Fertilizers*, 23, 291.

13. Sulphur, Phosphorus and Trace Element Nutrition

Sulphur

Besides nitrogen and carbon which are important constituents of plants, microorganisms are also known to influence the availability of sulphur, phosphorus and certain trace elements in soil for absorption by plants. The inorganic component of soil sulphur is in the form of sulphate and constitutes only a minor portion of the total sulphur content of soils. The bulk of soil sulphur is in the organic form which is metabolized by soil microorganisms to make it available in an inorganic state for plant nutrition. Sulphur is bound in organic state in proteins of vegetable and animal origin and in the protoplasm of microorganisms in the form of sulphur containing amino acids (cystine and methionine) and B-vitamins. The conversion of organically bound sulphur to the inorganic state is termed as mineralization of sulphur and is medicated through microorganisms. The sulphur thus released is either absorbed by plants or escapes to the atmosphere in the form of oxides. In the absence of oxygen, certain microorganisms produce hydrogen sulphide from organic sulphur substrates especially in waterlogged soils.

Bacteria capable of oxidizing inorganic sulphur compounds could be either aerobic or anaerobic. Their morphology varies from non-filamentous (*Thiobacillus*) to filamentous forms (*Beggiatoa, Thiothrix* and *Thioploca*). Several fungi and actinomycetes have also been reported to be sulphur oxidizers. (*Aspergillus, Penicillium, Microsporeum*). Among these microorganisms. *Thiobacillus* deserves special mention as it produces sulphuric acid when elemental sulphur is added to soil with the result that the pH of soil may fall as low as 2.0 after prolonged incubation with the bacterium. The possible role of *Thiobacillus* in controlling plant diseases in sulphur amended soils has been demonstrated with regard to potato scab caused by *Streptomyces scabies* and the rot of sweet potatoes caused by *S. ipomoea*. Under acidic soil conditions (below pH 5.0), inoculation of soil with thiobacilli after addition of sulphur effectively minimizes losses due to these pathogens. The application of sulphur coupled with thiobacilli inoculation has also the potentiality of rendering alkali soils fit for cultivation of crops. The formation of H_2SO_4 in soil following additions of elemental sulphur augments nutrient mobilization by increasing the level of soluble phosphate, potassium, calcium, manganese, aluminium and magnesium. In fact, manganese deficiency in soils can be corrected by sulphur applications.

Thiobacilli can also be used in the manufacture of 'Biosuper' a form of organic fertilizer once favoured in Australia. In biosuper, a mixture of rock phosphate and sulphur is inoculated with *Thiobacillus thiooxidans*. The H_2SO_4 produced in the mixture dissolves the phosphate and thereby enhances phosphorus nutrition of plants. This is somewhat similar to the 'Lipman process' developed earlier in U.S.A. where a compost was made of soil, manure, elemental sulphur and rock phosphate to improve plant nutrition.

Sulphate-reducing bacteria, i.e., those bacteria which reduce inorganic sulphate into hydrogen sulphide may diminish the availability of sulphur for plant nutrition and thus influence agricultural production. *Desulfovibrio desulfuricans* is a species belonging to this class of bacteria which is an obligate anaerobe capable of producing hydrogen sulphide at a rapid rate. Other species of *Desulfovibrio* are also active in inorganic sulphate reduction but the exact pathway is not yet clearly understood.

The various transformations of sulpur in the biosphere can be summed up as a cyclic reaction involving: (1) decomposition of organic sulphur compounds into subunits which are, in turn, converted into inorganic compounds through a process of mineralization, (2) assimilation of sulphur into the protoplasm of micro-organisms, a process referred to as immobilization, (3) oxidation of inorganic sulphur compounds into elemental sulphur, and (4) reduction of sulphate. The overall sequence of events in the transformation of sulphur is summarized in Fig. 87.

Phosphorus

Both inorganic and organic phosphates occur in soil. The inorganic forms are compounds of Ca, Fe, Al, and P. The bulk of phosphorus on earth exists as apatites, with the basic formula $M_{10} (PO_4)_6 X_2$. The M refers to calcium and the X is the anion fluorine but it can also be Cl^- OH^- or CO_3^- which means that phosphorus can exist as flour, chloro, hydroxy and carbonate apatites. Different substitutions and combinations of M and X may result in 200 different types of phosphorus occurring in nature. Rock phosphates high in carbonate apatite are commonly mined for fertilizer. The organic phosphorus containing compounds are derived from plants and microorganisms and are composed of nucleic acids, phospholipids and phytin. Organic matter derived from dead and decaying plant debris is rich in organic sources of phosphorus.

The deficiency of phosphorus may occur in crop plants growing in soils containing adequate phosphates. This may be partly due to the fact that plants are able to absorb phosphorus only in an available form. Soil

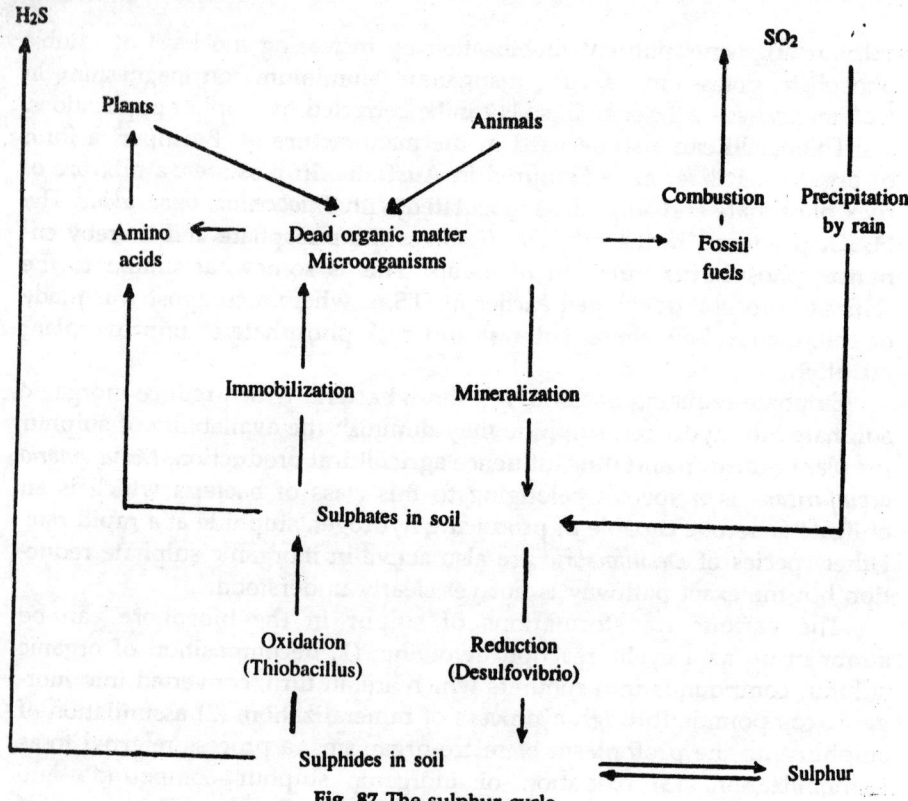

Fig. 87 The sulphur cycle.

phosphates are rendered available either by plant roots or by soil microorganisms through secretion of organic acids. Therefore, phosphate-dissolving soil microorganisms play some part in correcting phosphorus deficiency of crop plants. They may also release soluble inorganic phosphate (H_2PO_4) into soil through decomposition of phosphate-rich organic compounds. On the other hand, certain microorganisms, through assimilation, may immobilize available phosphates in their cellular material. Such immobilization processes in soil may also contribute to phosphorus deficiency of crop plants (Fig. 88).

Sulubilization of phosphates by plant roots and microorganisms is dependent on soil pH. In neutral or alkaline soils having a high content of calcium, precipitation of calcium phosphates takes place. Microorganisms and plant roots readily dissolve such phosphates and render them easily available to plants. On the contrary, acid soils are generally poor in calcium ions and, therefore, phosphates are precipitated in the form of ferric or aluminium compounds which are not so easily amenable to solubilization by plant roots or by soil microorganisms. If such conditions prevail in acid soils, deficiency of phosphorus in plants may also occur. One of the ways

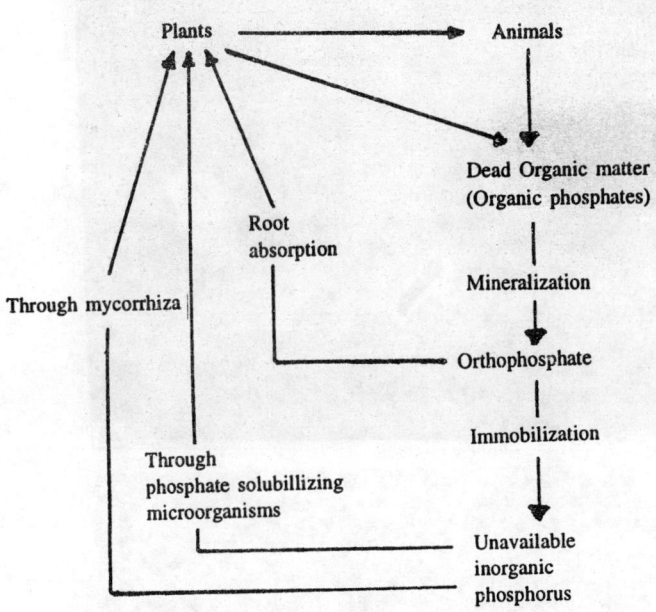

Fig. 88 The phosphorus cycle.

to correct deficiency of phosphorus in plants is to inoculate seed or soil with phosphate-dissolving microorganisms along with phospatic fertilizers.

Many fungi and bacteria (for example, *Aspergillus, Penicillium, Bacillus* and *Pseudomonas*) are potential solubilizers of bound phosphates as revealed by experiments in pure culture (Fig. 89A). Although bacteria have been used in the commercial preparation of phosphate-dissolving cultures to improve the growth of plants, fungi seem to be better agents in the dissolution of phosphates (Table 39). Phosphate-dissolving bacteria are known to reduce the pH of the substrate by secretion of a number of organic acids such as formic, acetic, propionic, lactic, glycolic, fumaric and succinic acids. Some of these acids (hydroxy acids) may form chelates with cations such as Ca and Fe and such chelation results in effective solubilization of phosphates. Although organic acid production is invariably associated with phosphate solubilization in pure cultures of microorganisms, many workers have not been able to correlate changes in the pH of the medium after growth with the amount of phosphates solubilized.

Labelled phosphate (^{32}P) has been used to test the phosphate solubilizing property of soil microorganisms and also to find out the extent of phosphorus uptake by the plant. Such experiments have been done in sterilized and unsterilized soils using tricalcium and rock phosphates,

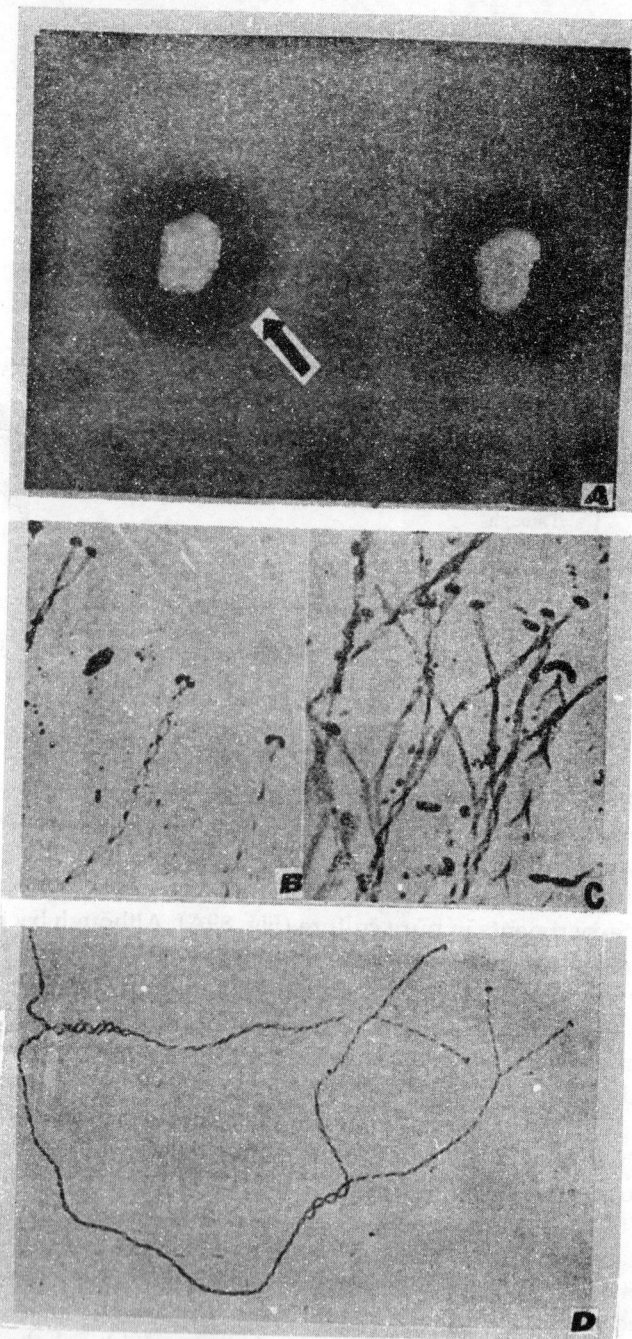

Fig. 89   (A)—Microbial solubilization of phosphate by *Pseudomonas striata* on agar. Arrow shows the zone of solubilization. (Courtesy: Dr. A.C. Gaur.) B, C and D—Photomicrographs of iron bacteria: (B) *Gallionella minor*, (C) *G. major*, (D) *G. ferruginea.* (from the text book, *The Life of Bacteria* by K.V. Thimann, The Macmillan Company, New York).

Table 39 Solubilization of tricalcium phosphate or calcium phytate by isolates of soil fungi from Delhi and Ludhiana. (from Sethi and Subba Rao, 1968)
The figures in parentheses refer to calcium phytate (average of four replicates)

Fungi	Number tested/Number significant in solubilization	No. of isolates showing significance	P_2O_5 in mg/ 50 ml	pH of culture filtrate
Control (uninoculated)	–	–	2.59 (2.50)	6.6 (9.6)
Acrothecium	1/1	1	2.48 (3.62)*	8.0 (8.4)
Aspergillus	10/4	1	3.30* (2.10)	7.5 (8.1)
		2	4.25* (3.07)*	6.1 (8.1)
		3	5.75* (1.73)	5.8 (7.6)
		4	11.32* (11.27)*	4.0 (4.6)
Cephalosporium	1/0	–	–	–
Chaetomium	1/0	–	–	–
Cladosporium	1/1	1	4.80* (3.67)*	5.5 (5.6)
Fusarium	9/2	1	2.83 (7.33)*	7.0 (6.1)
		2	9.67* (2.63)	5.1 (6.0)
Mortierella	1/0	–	–	–
Paecilomyces	3/2	1	6.94* (2.38)	5.8 (7.5)
		2	3.89* (2.66)	6.5 (8.5)
Penicillium	10/9	1	3.78* (2.75)	6.1 (7.5)
		2	6.56* (2.60)	5.0 (8.5)
		3	6.51* (3.23)*	5.5 (8.0)
		4	11.45* (1.75)	6.3 (5.6)
		5	9.07* (8.07)*	5.7 (6.8)
		6	4.55* (2.75)	5.8 (6.8)
		7	5.98* (3.17)*	7.0 (6.7)
		8	3.59* (7.21)*	5.0 (5.0)
		9	11.83* (6.19)*	6.4 (5.4)
Phoma	1/1	1	2.07 (4.00)*	7.3 (6.2)
Rhizoctonia	1/1	1	2.47 (3.84)*	6.0 (7.1)
Rhizopus	2/0	–	–	–
Scolecobasidium	1/0	–	–	–
Sordaria	1/0	–	–	–
Thielavia	3/0	–	–	–
Trichoderma	2/0	–	–	–

C.D. at 5% = 0.584 (0.480).
*Significantly different from controls. Note that accumulation of P_2O_5 in the medium does not always accompany a fall in pH.

apatite and bone meal. The results seem to be inconclusive since several contradictory reports have been made on this aspect.

Agronomical Aspects

Vegetables respond better than cereal crops to the application of phosphate-dissolving microorganisms. It was in USSR that a commercial preparation under the name 'phosphobacterin' containing bacterial cells of *Bacillus megatherium* was widely used for the first time. The usefulness of such bacterization of soils was rather overemphasised in spite of the fact that increase in grain yield was of the order of 5 to 10%. Several field trials have been conducted by the Indian Agricultural Research Institute with wheat, berseem, maize, arhar and rice to test the efficiency of phosphate-solubilizing bacteria on the yield of crop plants. The results have shown that significant increases were obtained in 13 out of 38 experiments conducted in different agro-climatic conditions (Table 40). These results point out that consistent response in increasing yield was not always seen.

Table 40 Effect of phosphate-solubilizing microorganisms on various crops in some field experiments in India (experimental results of W.V.B. Sundara Rao and A.C. Gaur)

Name of the crop	Oganism	No. of field experiments	No. of experiments showing significant increase	% yield increase over control
Berseem (*Trifolium alexandrinum*)		6	4	10–20
Maize (*Zea mays*)		3	2	0–14
Wheat (*Triticum aestivum*)		6	1	0–37
Rice (*Oryza sativa*)	Phosphobacterin	12	2	12–31
Gram (*Cicer arietinum*)	(*Bacillus megatherium*)	1	0	–
Arhar (*Cajanus cajan*)	var. *phosphaticum*	1	0	–
Urad (*Phaseolus aureus*)		3	1	16–19
Soybean (*Glycine max*)		1	0	–
Groundnut (*Arachis hypogea*)		1	0	–
Potato (*Solanum tuberosum*)		1	0	–
Wheat (*Triticum aestivum*)	*Pseudomonas striata*	1	1	10
Rice (*Oryza sativa*)	*Bacillus polymyxa*	1	1	9.5
Potato (*Solanum tuberosum*)	*Pseudomonas striata*	1	1	25.0

Trace Elements

Plants as well as microorganisms require traces of iron, manganese, copper, zinc, molybdenum and cobalt. Non-availability of these trace metals in soil may result in the manifestation of specific symptoms on plant parts.

Some of these trace elements participate in celluar enzyme activities. Trace elements deficiency of plants may be due to: (1) poor supply of one or more elements in a given soil, (2) inimical soil conditions which prevent the absorption and uptake of trace elements even when soils are not deficient with respect to some of the elements, (3) non-availability of trace elements in soil due to fixation in the presence of high amounts of organic matter, (4) antagonism between trace elements which may render them unavailable to plants and (5) under certain circumstances, microorganisms capable of converting the elements form a bound state to an easily assimilable form may be absent from soils. It is proposed here to briefly highlight some influences of microorganisms on the trace element nutrition of plants.

Manganese

Manganese is liable to oxidation in soil depending on pH, oxygen supply and organic matter content of soil. In acid soils it is present in a bivalent state (Mn^{++}) in which state it is easily available for absorption by plants. In neutral and alkaline soils, on the other hand, manganese occurs in trivalent or tetravalent state (Mn^{++++}) when the element is not easily available for absorption by plants. Therefore, the oxidation state of the element is the primary factor affecting its absorption by plants. The conversion of manganous (Mn^{++}) to manganic ions (Mn^{++++}) may be a microbiological process involving bacteria such as *Azotobacter chroococcum, Pseudomonas fluorescens, P. trifolii, Leptothrix* sp., *Aerobacter* sp., *Proteus* sp., *Corynebacterium* sp., *Flavobacterium* sp., *Chromobacterium* sp., *Metallogenium* sp., *Hypomicrobium* and several other unidentified ones. Some workers have also found that yeasts such as *Cryptococcus albidus* are involved in the process. Other fungal genera reported to be involved in the oxidation of the manganous ion are *Cladosporium, Curvularia, Helminthosporium* and *Cephalosporium*. A significant observation in this regard has been that mixed cultures of bacteria and fungi are more effective in the conversion of manganous to manganic ions than pure cultures.

Soil perfusion techniques which facilitate continuous reactions in soil have been extensively used in studies on microbiological transformation of manganese. When $MnSO_4$ is perfused in unsterilized soil oxidation of Mn^{++} occurs with an accumulation of insoluble oxidized manganese compounds such as MnO_2. An indirect proof of the participation of microorganisms in manganese nutrition of crop plants comes from the work on the rhizosphere microflora of oat plants (*Avena sativa*) in relation to grey-speck disease caused by manganese deficiency. A variety of oats susceptible to the disease had greater number of Mn oxidizers in its rhizosphere than the resistant varieties.

Copper

Deficiency of copper in higher plants, especially in cereals is dependent on the extent of fixation of the element in soil. The amount of copper fixed in soil depends on pH and such fixation is high in acidic peaty soil. The black organic matter of soil is largely responsible for the fixation of copper, a fact borne out by several independent observations.

Copper is precipitated by H_2S producing soil microorganisms. Apart from *Desulfovibrio desulfuricans* which reduces sulphate, bacteria such as *Clostridium lentoputrescens*. *Proteus vulgaris* and *Escherichia coli* may produce H_2S from sulphur containing amino compounds such as cystine, methionine and glutathione under anaerobic conditions. That H_2S producing bacteria are involved in the precipitation of copper rendering the element insoluble has been clearly demonstrated in experiments with barley and oats. In such experiments, the precipitate of copper formed by *D. desulfuricans*. and *E. coli* were fed to copper-deficient plants and the plants showed no recovery. A second set of plants treated with copper sulphate along with sterilized bacterial medium recovered from copper-deficiency symptoms. Apparently, in the later case, there was no bacteria-mediated precipitation of copper.

Iron

Certain bacteria oxidize ferrous iron to ferric state which precipitate as ferric hydroxide around cells. These bacteria commonly known as iron bacteria (Fig. 89 B, C, D) are usually non-filamentous and spherical or rod-shaped (*Gallionella, Siderophacus, Siderocapsa, Siderosphaera, Ferribacterium, Naumanniela, Ochrobium, Sideromonas, Sideronema, Ferro-bacillus, Siderobacter* and *Siderococcus*). Filamentous forms resembling algae are also encountered (*Leptothrix, Sphaerotilus, Toxothrix, Crenothrix* and *Colnothrix*). In addition to these bacteria, certain algae belonging to Cyanophycea, Volvocales, Chlorococcales, Euglenineae, Conjugales and Ulotrichales also transform ferrous salts to ferric state and deposit the precipitation around the filaments. The ferric hydroxide deposits give a brown or rust-red colour to these organisms.

The iron bacteria can be grouped into: (1) obligate chemoautotrophs, capable of utilizing energy released in the process of ferric hydroxide formation (*Gallionella ferruginea, Thiobacillus ferroxidans* and *Ferrobacillus ferroxidans*), (2) facultative chemoautotrophs, utilizing energy derived in the process of ferric hydroxide formation or alternatively, from organic matter (*Leptothrix ochraceae*) and (3) heterotrophs represented by most other iron bacteria which do not derive energy from iron oxidation but depend on organic matter for their nutrition. The iron bacteria play no significant role in cultivable soils and hence their importance in plant-microorganisms interaction, barring indirect effects, is less noteworthy.

Bioaccumulation and Mineral Leaching

It is now well known that in sewage treatment processes, metals such as copper, nickel and cadmium are precipitated by bacteria, thereby purifying the waste water. Bacterial leaching of metals has ben exploited and nearly 5 per cent metal production in the world comes from this 'low technology' source applicable to low grade mineral ores such as pyrite and metal sulphides. Somewhat similar technology is applicable to enrichment of uranium and gold by microbial leaching.

Thiobacillus ferrooxidans, T. thiooxidans and consortium of other bacteria are involved in metal sulphide leaching and iron oxidation. The *Leptospirillum* group (iron bacteria) also contribute to recovery of iron. The economics involved in microbial processes by mechanization and scaling up of such processes need careful review. However, the science of biometallurgy remains a fascinating area of study for microbiologists and metallurgists alike.

Selected References

Alexander, M. 1961. *Introduction to Soil Microbiology.* John Wiley & Sons, Inc., New York and London.

Bromfield, S.M. 1956. Oxidation of manganese by soil micro-organisms. *Austr. J. biol. Sci.*, 9, 238–252.

Bromfield, S.M. and Skerman, V.B.D. 1950. Biological oxidation of manganese in soil. *Soil Sci.*, 69, 337–348.

Chhonkar, P.K. and Subba Rao, N.S. 1967. Phosphate solubilization by fungi associated with legume root nodules. *Can. J. Microbiol*, 13, 749–753.

Gaur, A.C. 1990. *Phosphate Solubilizing Microorganisms as Biofertilizer.* Omega Scientific Publishers New Delhi.

Gerretsen, F.C. 1937. Manganese deficiency of oats in its relation to soil bacteria. *Ann. Bot.*, 1, 207–230.

Gerretsen, F.C. 1948. The influence of microorganisms on the phosphate intake by the plant. *Pl. Soil*, 1, 51–81.

Heintze, S.G. 1957. Studies on soil manganese. *J. Soil Sci.*, 8, 287–300.

I.C.A.R. 1964. *Hand Book of Manures and Fertilizers.* Indian Council of Agricultural Research, New Delhi.

Ishimoto, M., Koyama, J. and Nagai, Y. 1954. Biochemical studies of sulfate-reducing bacteria. IV. The cytochrome system of sulfate-reducing bacteria. *J. Biochem.* (Tokyo) 41, 760–770.

Jones, L.H.P. and Leeper, G.W. 1951a. The availability of various manganese oxides of plants. *Pl. Soil*, 3, 141–153.

Jones, L.H.P. and Leeper, G.W. 1951b. Available manganese oxides in neutral alkaline soils. *Pl. Soil*, 3, 154–159.

Louw, H.A. and Webley, D.M. 1959. A study of soil bacteria dissolving mineral phosphate fertilizers and related compounds. *J. appl. Bact.*, 22, 227–233.

Mulder, D.G. 1964. Iron bacteria, particularly those of the *Sphaerotilus-Leptothrix* group and industrial problems. *J. appl. Bact.*, 27, 151–173.

Mulder, E.G. and Lie, T.A. 1963. Investigations, on the *Sphaerotilus-Leptothrix* group. *Antonie van Leeuwenhoek*, 29, 121–153.

Mulder, E.G., Lie., T.A. and Woldendrop, J.W. 1969. Biology and soil fertility. In *Soil Biology, Reviews of Research*, pp. 163–208, UNESCO, Paris.

Murr, A.E., Torma, A.E. and Brierley, J.A. 1978 (Ed.) *Metallurgical Applications of Bacterial Leaching and Related Microbiological Phenomenon.* Academic Press, New York.

Page, E.R. 1962. Studies in soil and plant managanese II. The relationship of soil pH to manganese availability. *Pl. Soil*, 16, 247–257.

Postgate. J.R. 1949. Competitive inhibition of sulphate reduction by selenate. *Nature, Lond.*, 164, 670–671.

Quastel, J.H. 1955. Soil metabolism. *Proc. R. Soc.*, 143B, 159–179.

Sethi, R.P. and Subba Rao, N.S. 1968. Solubilization of tricalcium phosphate and calcium phytate by soil fungi. *J. gen. appl. Microbiol.*, 14, 329–331.

Sperber, J.I. 1957. Solution of mineral phosphates by soil bacteria. *Nature, Lond.*, 180, 994–995.

Starkey, R.L. 1950. Relations of microorganisms to transformations of sulphur in soils. *Soil Sci.*, 70, 55–65.

Subba Rao, N.S. and Bajpai, P.D. 1965. Fungi on the surface of legume root nodules and phosphate solubilization. *Experientia*, 29, 386–387.

Sundara Rao, W.V.B. 1968. Phosphorus solubilization by microorganisms, *Proc. All India Sympositum on Agricultural Microbiology*, Univ. Agric. Sci., Bangalore, pp. 21–29.

Timonin, M.I. 1947. Microflora of the rhizosphere in relation to the manganese-deficiency disease of oats. *Soil Sci., Soc. Amer. Proc.*, 11, 284–292.

Vishniac, W. and Santer, M. 1957. The Thiobacilli. *Bact. Rev.*, 21, 195–213.

Zavarzin, G.A. 1962. Symbiotic oxidation of manganese by two species of *Pseudomonas. Mikrobiologiya*, 31, 586–588. (English Translation: Microbiology, 31, 481–482, 1962/1963).

14. Biodegradation of Pesticides and Pollutants

Pesticides in General

The chemicals used to control all kinds of pests are known as pesticides. The word pest is an all embracing general term but in the current context includes insects, fungi, bacteria, viruses, nematodes that damage crops or incite diseases on crops; weeds in cultivators' plots that rob the nutrients from soils; snails, birds and rodents that destroy seedlings at sowing time and grains at harvest time; and finally insects, microorganisms and rodents that consume stored grains and fruits in the post harvest stage. The word pesticides is again a general term that includes insecticides, fungicides, nematocides, rodenticides, herbicides and algicides. There are other chemicals which do not directly kill or mitigate pests but are nevertheless classed as pesticides. They are growth regulators, chemosterilants (sterilize insects, birds, rodents), defoliants (remove leaves), pheromones (attract insects) and repellents (repel insects, dogs, rabbits, birds). These chemicals, however, do not form part of the writing here which confines only to insecticides, herbicides and fungicides.

Pesticides include one or more elements such as arsenic (As), boron, (B), bromine (Br), cadmium (cd), carbon (C), chlorine (Cl), copper (Cu), fluorine (F), hydrogen (H), iron (Fe), lead, (Pb), magnesium (Mg), manganese (Mn), nitrogen (N), oxygen (O), phosphorus (P), sodium (Na), sulphur (S), tin (Sn) and zinc (Zn). The benzene ring is quite frequently found in pesticide publications. It is a six carbon ring with six hydrogen atoms. For ease of presentation it is usually indicated as a hexagon with double bonds. When other groups have replaced one or more of the hydrogens, the ring is referred to as the phenyl radical rather than a benzene radical. For example, DDT contains the words diphenyl which refers to the two benzene rings with other groups attached (see under DDT).

The original pure form of a pesticide is formulated to technical grade materials that can be directly used effectively and safely and are amenable to storage, handling and application. They come in the form of sprays that are emulsions, water soluble concentrates, wettable powders, water soluble powders, oil solutions, soluble pellets, suspensions, dusts with active or

Benzene   hexagon with double bonds
                           (represents benzene)

inert diluents, aerosols; granulars; fumigants; impregnates either volatile or non-volatile; fertilizer combinations; baits and slow-release materials.

Apart from pesticides, oil-spills from many sources constitute another aspect of hazard for plant growth and this chapter also focuses problems and solutions faced by such pollutants.

Residues of pesticides persist in soil water and food and have posed problems all over the world, especially in the U.S.A. where in 1988, global sales of pesticides produced by some top 10 companies of the world amounted to 20 million US dollars, of which herbicides alone constituted 5 billion dollars. During the subsequent decade, bewildering amounts of a variety of organic compounds have been used in diverse human activities. An approximate estimate made in 1992 shows that the following organic chemicals were produced in the U.S.A the quantities of the chemicals indicated in parenthesis as $Kg \times 10^9$: ethylene (18.33), propylene (10.25), ethylene dichloride (7.23), vinyl chloride (6.00), benzene (5.45), ethylbenzene (4.99), methyl t-butyl ether (4.93). Styrene (4.06), methanol (3.96), Formaldehyde (3.17), Xylene (2.89), Toluene (2.74), p-xylene (2.57), terephthalic acid (2.56), ethylene oxide (2.52), ethylene glycol (2.32), cumene (2.07), phenol (1.68), butadiene (1.44), acrylonitrile (1.28), propylene oxide (1.22), vinyl acetate (1.21), acetone (1.08), cyclo-hexane (1.00), totalling to $9495 \text{ kg} \times 10^9$. The residues of these chemicals remain hazardous to living beings. In recent years, oil spills have been considered as prime pollutants to the detriment of life in aquatic habitats.

Insecticides

There are several classes of insecticides but only few examples have been outlined here:

DDT

This well known organochlorine type of insecticide is in existence for about 100 years and won the Nobel prize for medicine in favour of Dr. Paul Muller of Germany in 1948. The insecticide is known as diphenyl aliphatic with two phenyl rings attached. The DDT is converted to DDE through dehalogenation by an enzyme dehydrochlorinase as shown below by several bacteria such as *Achromobacter, Aerobacter, Agrobacterium, Bacillus, Clostridium, Corynebacterium, Escherichia, Erwinia, Kurthia, Pseudomonas* and *Streptococcus.*

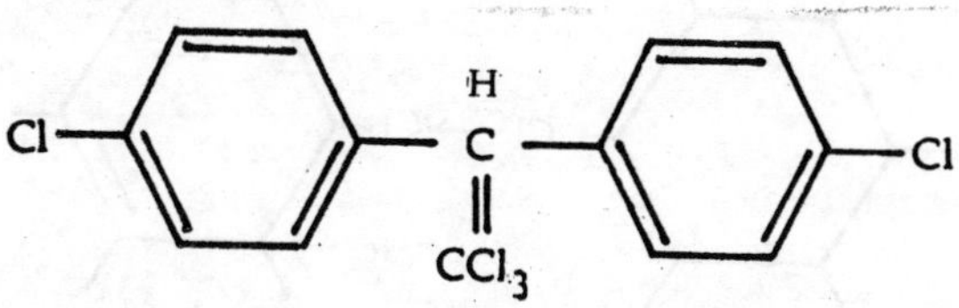

DDT or [1,1,1-trichloro-2,2-bis(p-chlorophenyl)]

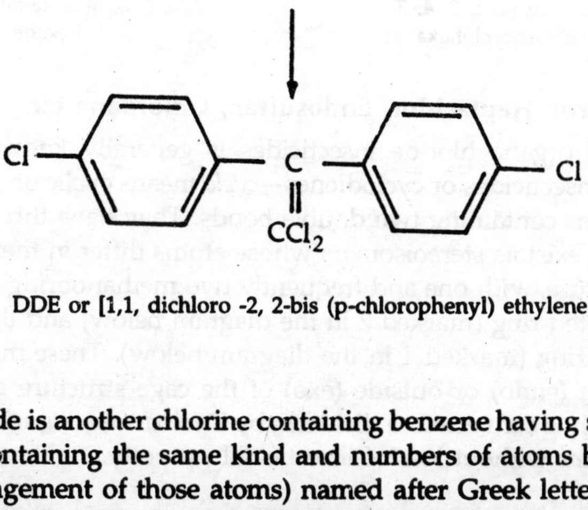

DDE or [1,1, dichloro -2, 2-bis (p-chlorophenyl) ethylene]

BHC

This insecticide is another chlorine containing benzene having several isomers (molecules containing the same kind and numbers of atoms but differing in internal arrangement of those atoms) named after Greek letters, alpha, beta, gamma, delta and epsilon but only the gamma isomer has insecticidal properties and the remaining serve as inactive filler ingredients.

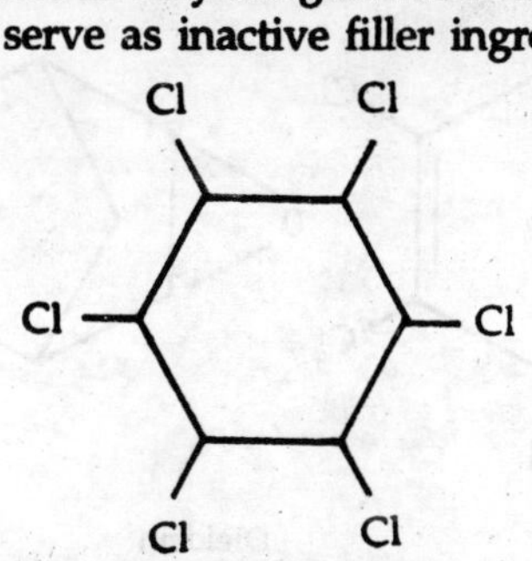

BHC or 1,2,3,4,5,6-hexachlorocyclohexane

Lindane

The product containing 99 per cent gamma isomer of BHC is known as lindane which is most active against several insects. The degradation of lindane is mediated by *Clostridium* and *Escherichia*. The dehalogenation (removal of chlorine) process converts lindane to 2, 3, 4, -pentachloro-1 cyclo-hexane as shown below:

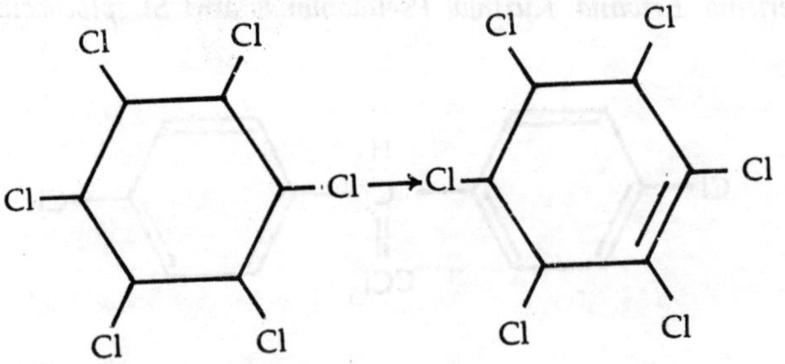

| Lindane *or* [1, 2, 3, 4, 5, 6-hexachlorocyclohexane] | 2, 3, 4, 5, 6-pentachloro-1cyclo-hexane |

Aldrin, Dieldrin, Heptachlor, Endosulfan, Chlordane etc.

This group of organochlorine insecticides is generally known as dieneorganochlorine insecticides or cyclodienes—*cyclo* means cyclic or ring structures and *diene* means containing two double bonds. They have three dimensional structures and exist as stereoisomers whose atoms differ in their spatial location and structure, with one and frequently two methanobridges, one located in the chlorinated ring (marked 2 in the diagram below) and the other in the unchlorinated ring (marked 1 in the diagram below). These methanobridges are bent inside (endo) or outside (exo) of the cage structure as exemplified by aldrin which is converted to dielderin by *Trichoderma, Fusarium, Penicillium* and *Pseudomonas* without loss of insecticidal property.

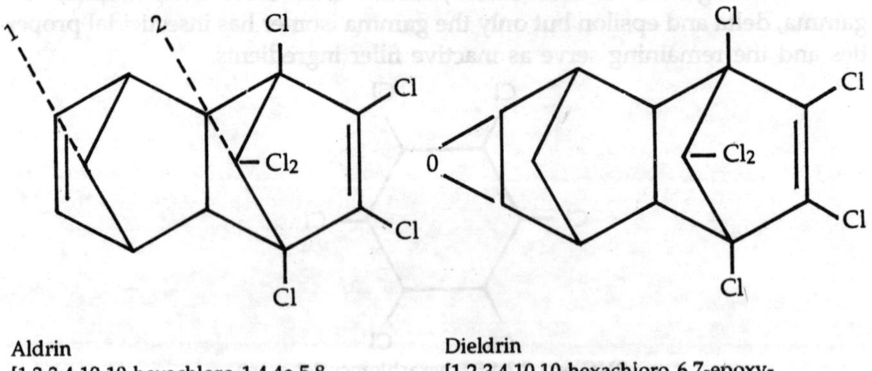

Aldrin
[1,2,3,4,10,10-hexachloro 1,4,4a,5,8, 8a-hexahydro-1, 4-endo-exo-5, 8-dimethanonaphthalene]

Dieldrin
[1,2,3,4,10,10-hexachloro 6,7-epoxy-1,4,4a,5,6,7,8,8a-octahydro-1,4-endo-exo-5,8-dimethanonaphthalene

The names aldrin and dieldrin go after the Nobel Prize winning German chemists Otto P.H. Diels and Kurt Alder. These insecticides are persistent and stable in soil and are very effective against termites.

Malathion

This insecticide is an example of organophosphates, all of which are derived from phosphoric acid in combination with alcohols described as esters. These esters of phosphorus have varying combinations of oxygen, carbon, sulphur and nitrogen attached to phosphorus. They are divided into three classes, the aliphatic, phenyl and heterocyclic derivatives. They are degraded by *Torulopsis, Chlorella, Pseudomonas, Thiobacillus* and *Trichoderma*. Malathion is an aliphatic derivative with linear arrangement of carbon atoms.

$$
\begin{array}{c}
\quad\quad\quad\quad\quad\quad\quad\quad\quad\quad\quad\quad O \\
\quad\quad\quad\quad\quad\quad\quad\quad\quad\quad\quad\quad \| \\
\quad S \quad\quad\quad\quad CH_2 - C - OC_2H_5 \\
\quad \| \quad\quad\quad\quad\quad\quad | \\
(CH_3O)_2\, P - S - CH - C - OC_2H_5 \\
\quad\quad\quad\quad\quad\quad\quad\quad \| \\
\quad\quad\quad\quad\quad\quad\quad\quad O
\end{array}
$$

Malathion
(O, O-dimethyl-S-1, 2-di (carboethoxy) ethylphosphorodithioate)

Parathion

This insecticide is an example of organophosphates belonging to phenyl derivatives. It has a benzene ring with one of the ring hydrogens displaced by attachment to the phosphorus moiety.

Ethylparathion *or* (O, O-diethyl
O-p-nitrophenyl phosphorothioate)

Methyl parathion *or* (O, O-dimethyl
O-p-nitrophenyl phosphorothioate)

Diazinon

This insecticide comes under the class heterocyclic derivatives. The term heterocylic denotes that the ring structures are composed of unlike atoms, one or more of the carbon atoms being displaced by oxygen, nitrogen or sulphur and the ring may have 3, 5, or 6 atoms. In the case of diazinon, the 6 membered ring contains 2 nitrogen atoms.

$$(C_2H_5O)_2 \; P - O \underset{}{\overset{S}{\parallel}} \text{---} \quad CH(CH_3)_2$$

Diazinon
[O, O-diethyl O-(2-isopropyl-4-methyl-6-pyrimidyl) phosphorothioate]

Herbicides

Herbicides or weed killers are classified as selective when they kill weeds without detriment to the main crop and non-selective when the objective is to eradicate all vegetation. Contact herbicides kill parts of the plant (mostly annual weeds) to which the chemical is applied whereas translocated herbicides are absorbed by roots or above ground parts of plants (mostly perennials) and then translocated to different tissues. Application of herbicides may be done at pre-planting stage, pre-emergence and post-ermergence stages by banding, broadcasting, spot treatments or direct spraying. The two major classes or herbicides are inorganic and organic ones. Examples of inorganic ones, used prior to the development of organic herbicides, are copper sulphate, arsenic trioxide, ammonium sulphate, ammonium thiocyanate, ammonium sulphate, iron sulphate, sodium tetraborate, sodium chlorate etc., but many of them have been discontinued because of their persistence in soil. Some examples of organic herbicides are 2,4-D; 2-4, 5-T; 2,4-DB; and MCPA [2,4-dichlorophenoxyacetic acid]; [2,4,5-trichloro-phenoxyacetic acid]; [4-(2,4-dichloro-phenoxy) butyric acid]; and [4-chloro-2-methylphenoxy acetic acid].

2,4-D

Known as phenoxyalkanoic acids or phenoxy herbicides, these herbicides are extensively investigated and well known. They are degraded by microorganisms of the genera *Pseudomonas, Achromobacter, Flavobacterium, Cornebacterium, Arthrobacter* and *Sporocytophaga.* Known as hormone weed killers, they are active against broad leaved weeds.

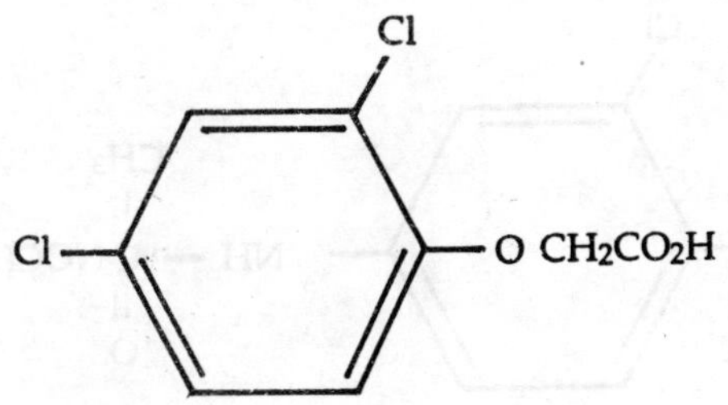

$$Cl$$

Cl —⟨ ⟩— O CH₂CO₂H

2,4-D
[(2,4-dichlorophenoxy) acetic acid)]

Simazine and Atrazine

These are examples of widely used S-triazine herbicides. The triazines are 6-member rings containing 3 nitrogens (tri = 3 and azine = a nitrogen containing ring). The triazines are applied to soil primarily for their post-emergence action through inhibition of photosynthesis. The microorganisms capable of degrading the two triazines belong to the genera *Aspergillus, Rhizopus, Fusarium, Penicillium* and *Trichoderma*.

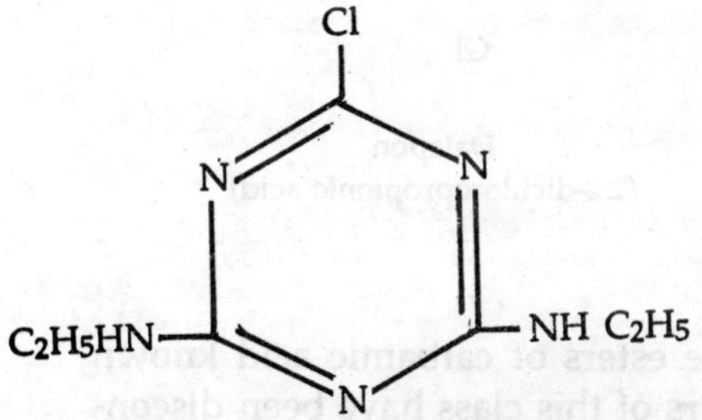

Simazine
[2-chloro-4,6-bis
(ethylamino)-s-triazine]

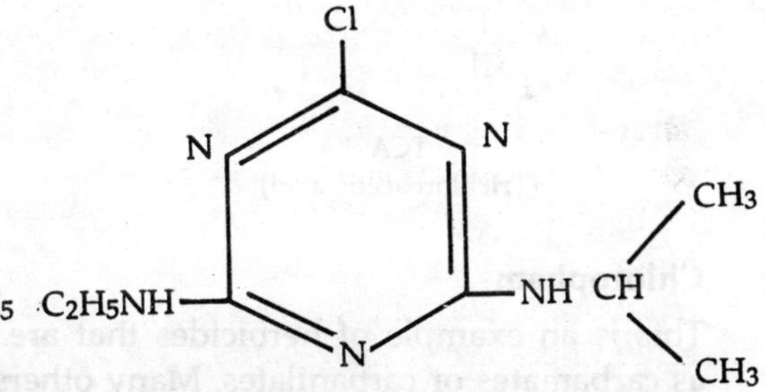

Atrazine
[2-chloro-4- (ethylamino)-
6-(isopropylamino)-s-triazine]

Linuron

This is an example of phenylureas that have the hydrogen atoms replaced by various carbon chain and ring structures to form compounds that are primarily used as selective pre-emergence herbicides. Phenylureas act by inhibiting photosynthesis. Several bacteria and fungi are capable of degrading the herbicide, especially *Aspergillus nidulans*.

Linuron
[3,-(3,4,-dichlorophenyl)-1-methoxy-1-methylurea]

TCA and Dalapon

These are two most widely used herbicides, known as chlorinated aliphatic acids to get rid of grasses. Both are degraded by bacteria such as *Bacillus*, *Pseudomonas*, *Agrobacterium*, *Arthrobacter*, *Micrococcus*, *Alcaligenes* and *Flavobacterium*.

TCA
(Trichloroacetic acid)

Dalapon
(2,2-dichloropropionic acid)

Chloropham

This is an example of herbicides that are esters of carbamic acid known as carbamates or carbanilates. Many others of this class have been discontinued but chloropham is used as an inhibitor of sprouts in potatoes. The microorganisms that degrade chloropham are *Pseudomonas*, *Flavobacterium*, *Agrobacterium* and *Achromobacter*.

Chloropham
(Isopropyl m-chlorocarbanilate)

Dicamba

This is an example of aliphatic acids or carbon chain acids, whose action may be similar to 2,4-D which attacks the nucleic acid metabolism of weeds.

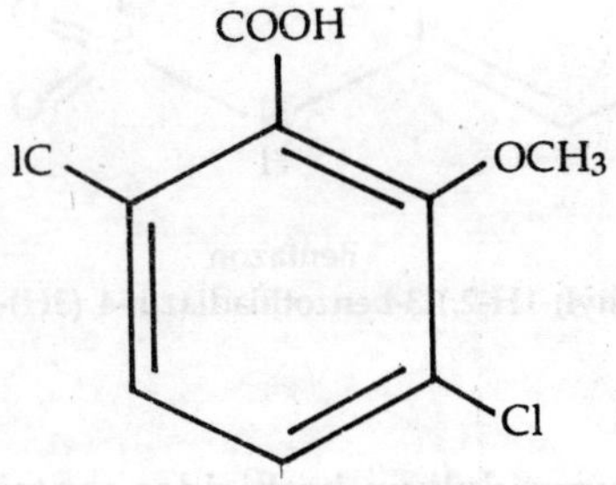

$$COOH$$
$$IC \qquad OCH_3$$
$$Cl$$

Dicamba
(2-methoxy-3, 6-dichlorobenzoic acid)

Dichlobenil (Casoron), Bromoxynil (Brominal)

These are benzonitriles with a benzene ring containing the C≡N or cyanide group acting as rapid contact herbicides and used in the cultivation of corn, grain sorghum and turf grass.

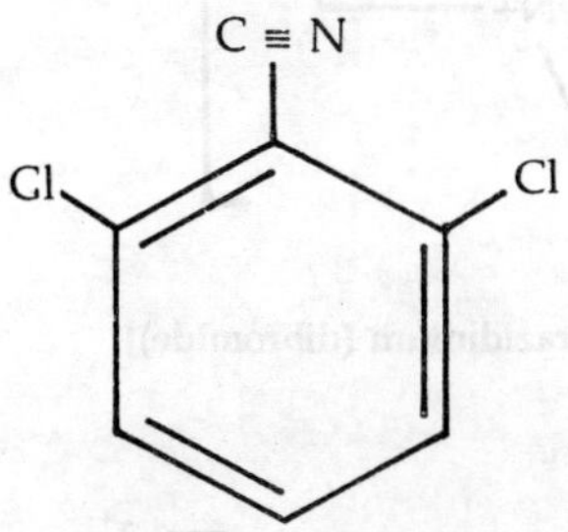

C ≡ N	C ≡ N
Cl Cl	Br Br
	OH
Dichlobenil	Bromoxynil
(2,6-dichlorobenzonitrile)	(3,5-dibromo-4-hydroxy-benzonitrile)

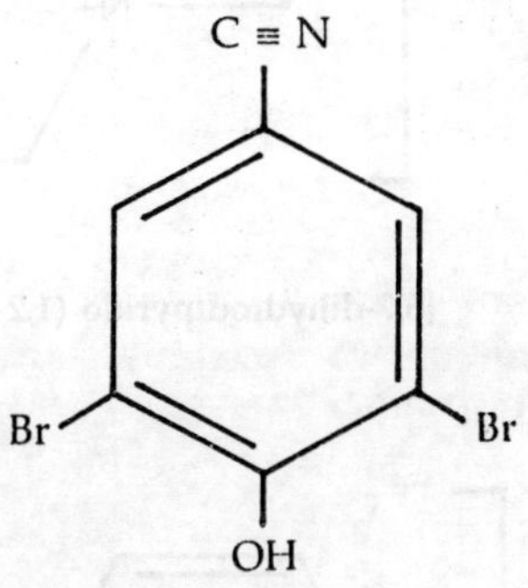

Bentazon

This is an example of photosyntheis inhibiting benzothiadiazoles post emergence herbicides used in the cultivation of soybeans, rice, corn, groundnut, beans and peas.

Bentazon
[3-(1-methylethyl)-1H-2,1,3-benzothiadiazin-4 (3H)-one 2,2-dioxide].

Diquat and Paraquat

These are known as bipyridylium herbicides containing two pyridyl rings. They act as contact herbicides but are also used as preharvest dessicants in cotton, sugarcane, sunflower and aquatic weeds.

Diquat
[6,7-dihydrodipyrido (1,2-α 2,1C) pyrazidinium (dibromide)]

Paraquat
[1,1-dimethyl-4,4'-bipyridylium ion (dichloride)]

Fungicides

Chemicals used to control fungi that cause plant disease are generally regarded as fungicides. Historically these compounds are based on sulphur, copper and mercury compounds. These inorganic fungicides, have been largely replaced in recent years by organic fungicides. The first organic fungicide was discovered in 1931 under the name thiram followed by zineb and captan in 1943 and 1949, respectively. One generally accepted theory of the fungicides action of metal containing organic fungicides (copper, mercury, cadmium etc.) is the formation of chelates in fungal cells followed by disruption of protein synthesis. Some examples of fungicides have been outlined here:

Thiram, Maneb, Ferbam, Nabam and Zineb

These are known as dithiocarbamates which are believed to inactivate the sulphydryl groups in amino acids and enzymes of fungal cells.

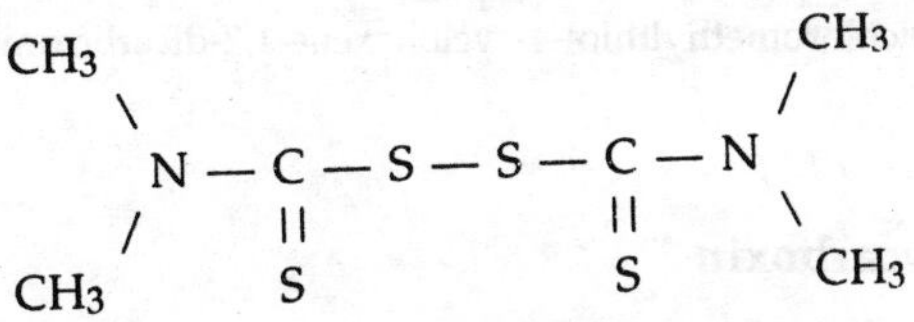

Thiram
[bis(dimethylthio-carbomoyl) disulfide]

PCP, PCNB, Chloroneb etc.

These are known as substituted aromatics or simple benzene derivatives and act on fungi by combining with —NH2 or —SH groups of metabolites in fungi.

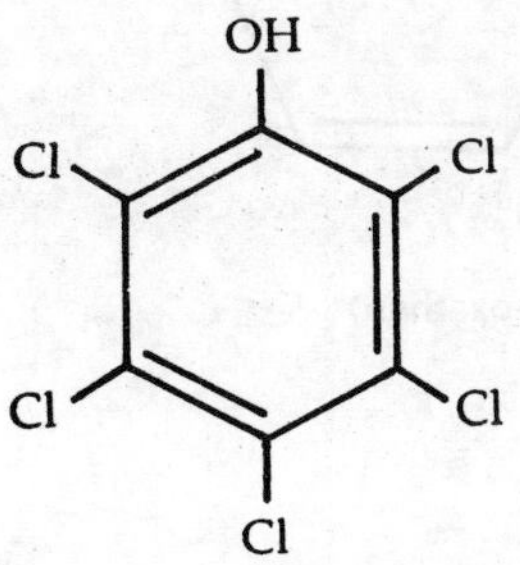

PCP
(Pentachlorophenol)

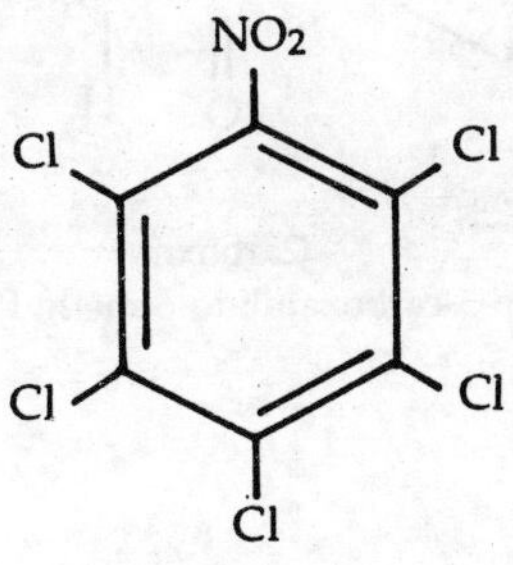

PCNB
(Pentachloronitrobenzene)

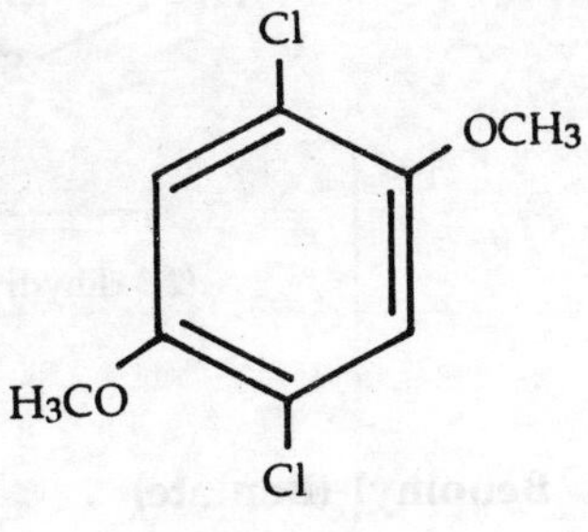

Chloroneb
(1,4-dichloro-2,
5-dimethoxybenzene)

Captan

This is an example of dicarboximides or sulphanimides used as foliage dusts and sprays on fruits, vegetables and ornamentals and the fungicidal action is believed to be the inhibition of amino compounds and enzyme synthesis.

Captan
[N-(trichloromethylthio)-4-cyclohexene-1,2-dicarboximide]

Carboxin and Oxycarboxin

These are systemic fungicides that are absorbed by the plant and carried by translocation to growing points. Known as oxathins, they affect mitochondrial systems of fungal cells.

Carboxin
(2,3-dihydro-5-carboxanilido-6-methyl-1-1,4-oxathiin)

Benomyl (Benlate)

This is an example of benzimidazoles that are systemic in action capable of inhibiting fungal spore germination and growth by interfering in mitosis.

Streptomycin, Cycloheximide, Blasticidin-S, Terramycin

These are antibiotics which are active against bacterial and fungal diseases and many of them are metabolic products of the actinomycete, *Streptomyces*.

Fate of Pesticides in Soil (Biodegradation)

Several factors are known to influence the fate of pesticides in soils—(1) chemical decomposition, (2) photochemical decomposition, (3) volatilization, (4) movement in soil (5) plant uptake, (6) adsorption and (7) microbial decomposition. Biodegradation of organic compounds by soil microorganisms involves a process known as mineralization whereby microorganisms convert the organic molecules to obtain carbon and energy for growth and multiplication, releasing the inorganic forms of N, P, S or other elements. In this process, the parent molecule becomes detoxicated or rendered harmless to life by enzymatic reaction. There is a lag period or acclimation period during which microorganisms adjust to the presence of the organic compound which may extend to several days or weeks. After this lag phase, depending upon the nature of the compound, detoxication or partial detoxication can proceed by any of the several processes such as hydrolysis, dehalogenation, demethylation, methylation, nitro reduction, deamination, cleavage of ether linkages, conversion of nitrile to amide and conjugation. Some examples are cited which give an insight into some of the types of biodegradation involved.

Pentachloronitrobenzene (PCNB) is a common fungicide used in the control of *Rhizoctonia* and other soil-borne pathogens of plants. This fungicide is gradually converted to pentachloroaniline (PCA) in moist soil and the conversion is greatly enhanced by submergence of soil in water. In sterilized soil, PCNB remains unchanged while in unsterilized soils, loss of PCNB is accompanied by the increase in PCA (Fig. 90) indicating the role of microorganisms in the degradation of the fungicide. The metabolic pathway of microbial degradation is as follows:-

PCNB   PCA

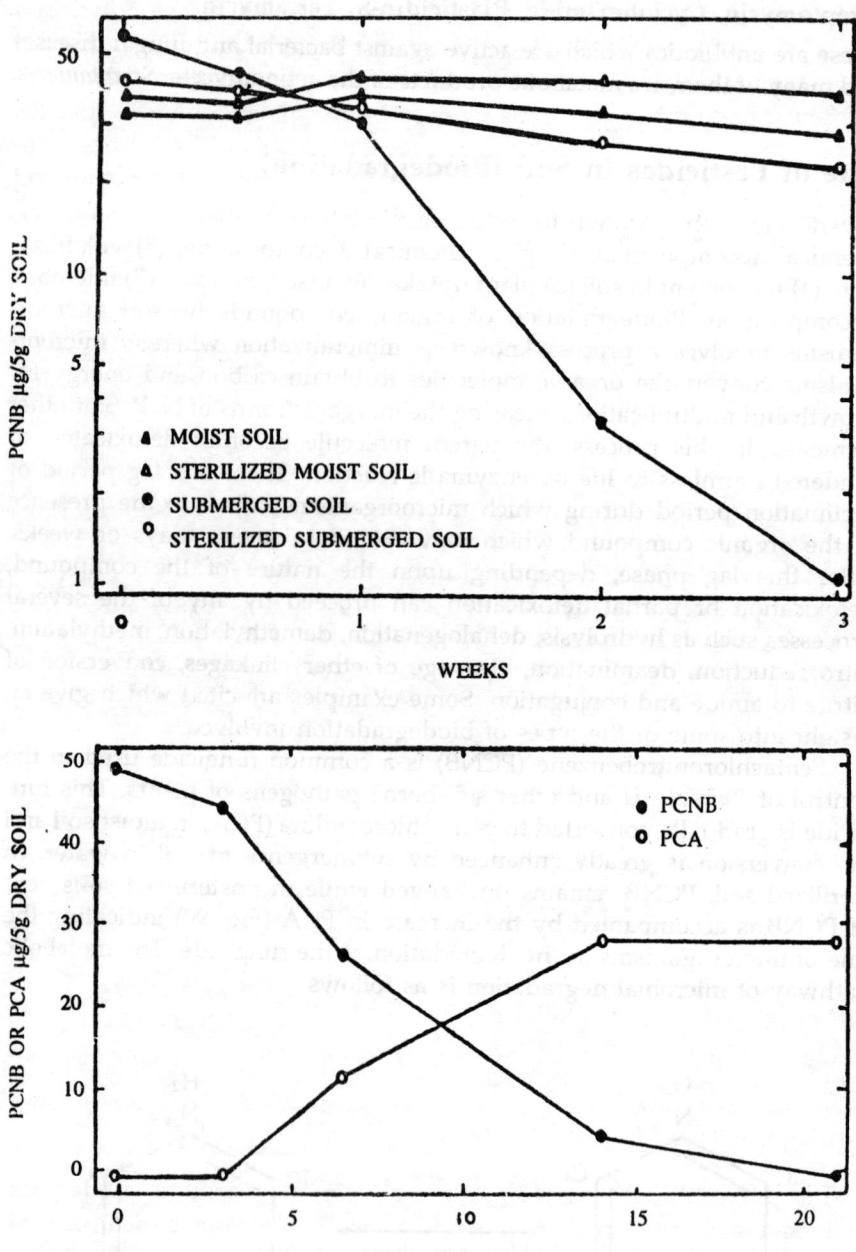

Fig. 90 *Top:* Loss of pentachloronitrobenzene (PCNB) from nonsterilized or sterilized, moist or submerged soil. *Bottom:* Relation between disappearance of PCNB and appearance of pentachloroaniline (PCA) in nonsterilized submerged soil. (from KO and Farley, 1969).

The herbicide atrazine undergoes degradation in soil by two routes. One route involves replacement of the 2-chloro substituent by a hydroxyl group. A second route is mediated by soil fungi (*Aspergillus fumigatus*) which involves dealkylation to give rise to 2-chloro-4-amino-6-isopropylamino-s-triazine and 2-chloro-4-ethylamino-6-amino-s-triazine. Dealkylation need not necessarily insure detoxication. The former compound is almost as toxic to oats as atrazine whereas the latter compound is toxic only at high concentrations (Fig. 91).

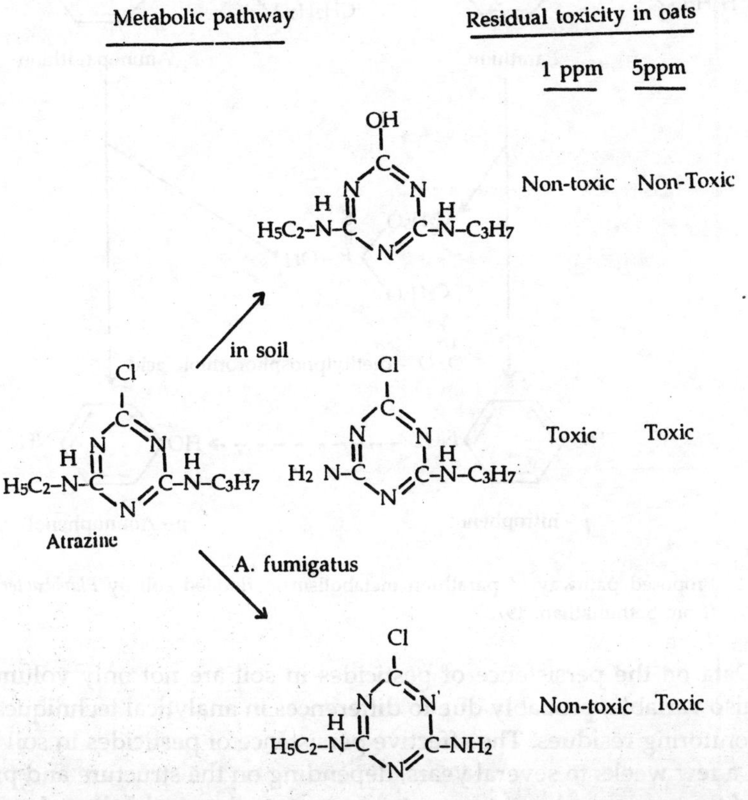

Fig. 91 Atrazine metabolism to toxic and non-toxic metabolites. For details see text. (from Kaufman and Blake, 1969).

In paddy fields under standing water, parathion undergoes biodegradation after repeated applications due to the enrichment of *Flavobacterium* sp. in soil. The standing water in fields turns yellow after the third addition of the insecticide due to the formation of p-nitrophenol. Earlier work, however, indicates that parathion breaks down in aerobic microbial cultures resulting in the formation of aminoparathion by the reduction of nitro group in the parent molecule of the insecticide. In *Flavobacterium* sp. (a facultative anaerobe) inoculated culture, p-nitrophenol is detected when the cultures are grown either aerobically or anaerobically but the secondary product is resistant to further degradation. Based on work in flooded soil and in pure culture containing *Flavobacterium* sp., a scheme for parathion metabolism has been proposed (Fig. 92).

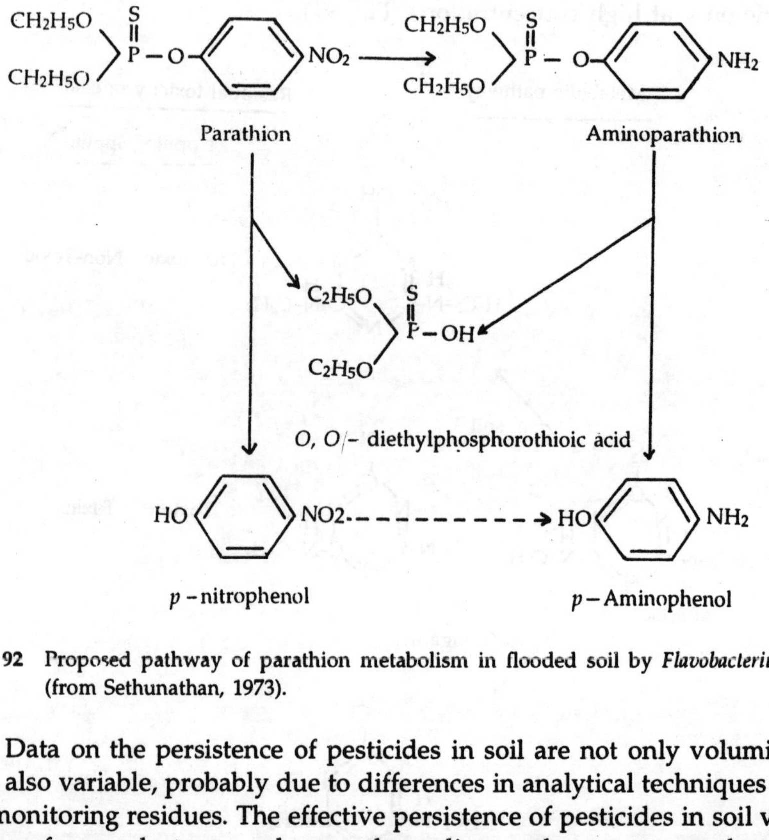

Fig. 92 Proposed pathway of parathion metabolism in flooded soil by *Flavobacterium* sp. (from Sethunathan, 1973).

Data on the persistence of pesticides in soil are not only voluminous but also variable, probably due to differences in analytical techniques used in monitoring residues. The effective persistence of pesticides in soil varies from a few weeks to several years, depending on the structure and properties of the compound and to a certain extent on the availability of moisture in soil. The variability in persistence is best illustrated in the case of insecticides—for instance, the highly toxic phosphates do not persist for more than three months in contrast to some of the chlorinated hydrocarbon insecticides which are known to persist for extended periods (4 to 5 years) at normal rates of application. From the agricultural point of view, accumulation of residues in soil may lead to increased absorption of such chemicals by plants to a level at which the consumption of plant products may prove deleterious to livestock and human beings.

Persistence is a relative term and, therefore, 75 to 100% loss of the pesticide in normal arable soil under recommended practices of application has been taken by several workers as the persistence value of a pesticide. Based on this parameter, the persistence of some chlorinated hydrocarbon insecticides is as follows: chlordane—5 years, DDT—21 years, BHC—3 years and heptachlor and aldrin which are metabolized in soil to their epoxides, heptachlor epoxide and dieldrin—16 years. On the other hand, the persistence of phosphate insecticides in soil is very low—diazinon (3 months), disulfoton (4 weeks), phorate (2 weeks) and malathion and parathion (up to 2 weeks).

The persistence of herbicides in soil may also vary from few weeks to 18 months, depending on the nature of the herbicide. Among the urea, triazine and picloram group of herbicides, propazine and picloran persist for 18 months followed by simazine (12 months), atrazine and monuron (10 months), fenron and diuron (8 months), linuron (4 months) and prometryne (3 months). In the group of benzoic acid and amide herbicides, 2,3,6-trichlorobenzoic acid persists for 12 months followed by bensulide (10 months) diphenamide (8 months), amiben (3 months) and others (less than 3 months). Among the phenoxy, toluidine and nitrile group of herbicides, the order of persistence is as follows: trifluralin (6 months) 2,4,5-T (5 months), dichlobenil (4 months) MCPA (3 months) and 2,4-D (one month). The carbamate and aliphatic acid group of herbicides are comparatively less persistent among all types of herbicides with TCA having a persistence life of 12 weeks followed by dalapon (8 weeks) and barban (2 weeks).

The disappearance of a pesticide depends on the initial concentration of the chemical in soil although factors such as volatilization, photodecomposition and erosion by water and wind contribute to the loss of pesticide from soil. When a biodegradable pesticide is applied to a newly cultivated soil, it disappears fairly rapidly after an initial lag phase but periodic application of the same chemical may lead to an accumulation of the substance depending on the period of persistence of the pesticide. Similarly, periodic application of heavy metal containing pesticides such as arsenic and mercurial pesticides results in a progressive accumulation of the heavy metals with every application even though a portion of the molecule of the pesticide gets degraded or lost. For instance, metallic mercury vapour and trace amount of phenyl mercuric acetate (PMA) have been detected in air surrounding PMA-treated soils. However, the problem of arsenic and mercury contamination in food chain of human beings is more important than their persistence in soil.

Mercury enters the food chain from seeds or crops treated with mercury containing fungicides. A survey of mercury levels of plants in Britain, Canada, New Zealand and Scandinavia has revealed that translocation of mercury does occur into tissues of plants raised from seeds treated with mercury containing fungicides. One example of mercury containing seed protectants is methylmercury dicyanodiamide (panogen). It has been demonstrated through experiments that hens fed on grains obtained from crops raised from panogen-treated seeds concentrated mercury in their liver and eggs. The food chain of eggs and chicken from such contaminated hens may transfer mercury to man and the indications at preset are that mercury tends to accumulate in the tissues of brain. Fortunately, methyl mercury has a biological half-life of 60 to 74 days in human beings but it should be alarming to note that as little as 6 ppm in brain cells is sufficient to cause an irreversible brain damage.

Of all the pesticides investigated so far for residues in food chains, DDT has received universal attention and therefore, a reference to the persistence of DDT may prove useful in understanding the hazards faced by man through intensive use of chemicals to combat pests. DDT has been in use since World War II and has permeated our environment and contaminated human beings and wild life of the earth including such remote areas like the Antarctic. Apart from wind and water, migrating fish and birds can transport the chemical to long distances. DDT is sparingly soluble in water, has the property of entering into fatty substances of living beings, is extremely stable and is very slowly broken down in the environment. These attributes provide enough scope for the chemical to persist in living matter for extended periods and render DDT a potent source of contamination in food chains.

Several compounds by themselves are non-toxic but may be converted to toxic products by a process known as 'activation' which is also microbially mediated. This leads to the formation of carcinogens, teratogens, neurotoxins, phytotoxins, and insecticidal or fungicidal chemicals. There are several examples of this kind, some of which are as follows: Many bacteria convert trichloroethylene (TCE) to vinyl chloride which is a potent carcinogen. TCE can also be converted to chloral hydrate (2,2, 2-trichloroacetaldehyde), a mutagen as well as a toxin, by methanotrophic bacteria. Nitrosamines are potential carcinogens, teratogens and mutagens, even at low concentrations. These compounds are products of microbial action of naturally formed NO_2 ions from NO_3 ions present in soil and water with synthetic chemicals such as secondary amines that are constituents of many pesticides. These nitrosamines contaminate river waters and become part of sewage and other effluents. Some insecticides such as aldrin and parathion are liable to conversion into more toxic compounds which are known to persist in soil microorganisms. Aldrin is converted to dieldrin and the latter compound is known to persist in soil for more than 15 years. Similarly parathion gets transformed to paraoxon. There have been instances, when some harmless precursors get converted by microorganisms to potent toxic chemicals as exemplified hereunder:

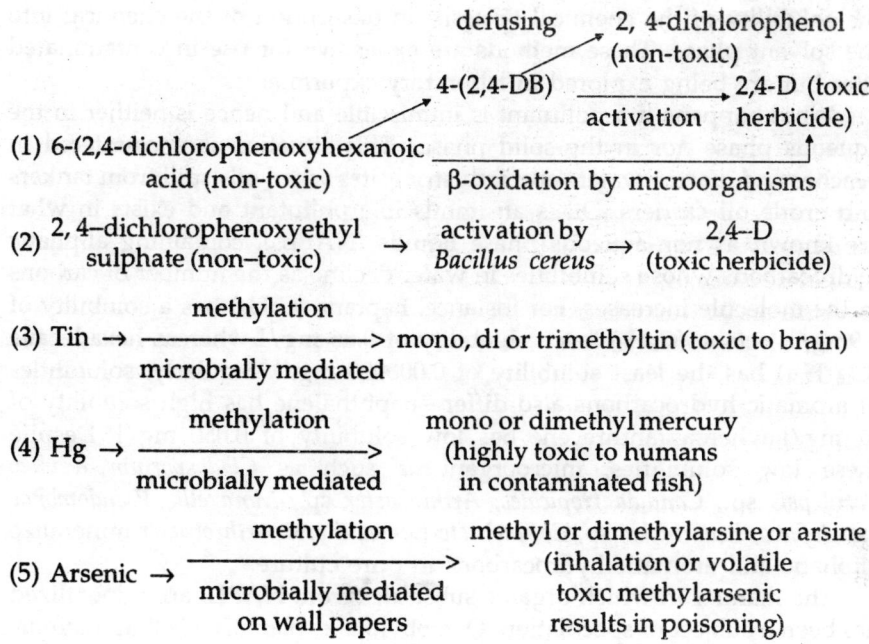

(1) 6-(2,4-dichlorophenoxyhexanoic acid (non-toxic)

defusing → 2, 4-dichlorophenol (non-toxic)

4-(2,4-DB) → 2,4-D (toxic activation herbicide)

β-oxidation by microorganisms

(2) 2, 4–dichlorophenoxyethyl sulphate (non–toxic) → activation by *Bacillus cereus* → 2,4–D (toxic herbicide)

(3) Tin → methylation (microbially mediated) → mono, di or trimethyltin (toxic to brain)

(4) Hg → methylation (microbially mediated) → mono or dimethyl mercury (highly toxic to humans in contaminated fish)

(5) Arsenic → methylation (microbially mediated on wall papers) → methyl or dimethylarsine or arsine (inhalation or volatile toxic methylarsenic results in poisoning)

Bioavailability of Pesticides/Pollutants

The readiness with which a solute in soil is in the reach of a microorganism has been referred to as bioavailability which is dependent on sorption (adsorption as well as absorption) and retention of the solute in the soil matrix and aquatic sediments composed of clay and organic matter. Bioavailability is also dependent on the nature of the pesticide/pollutant. For instance, the organic fraction of soils and sediments have great affinity for hydrophobic compounds such as polycyclic aromatic hydrocarbons and other non-polar chemicals. Herbicides like simazine and atrazine have great affinity for organic matter and hence biodegrade slowly because of low bioavailability. The silicon and aluminium constituents of montmorillonite type of clays are arranged tightly in 2:1 ratio thereby allowing facility to expand and sorp microbial substrates in the matrix. The herbicide diquat when sorbed to montmorillonite clays becomes recalcitrant for degradation because of low bioavailability caused by the high cation exchange capacity of the clays. The extent of degradation also gets minimized since microorganisms often remain trapped and functionless in the matrix. Therefore, the crux of biodegradation of such bound chemicals lies in the ways and means by which chemicals can be desorbed into the aqueous phase from the solid matrix. This limiting factor of low bioavailability in the aqueous phase can be overcome by physical operations such as pulverisation of soil by slurry systems, soil heating and application of detergents to increase the solubility of the chemical or assist in desorption of the chemical into the solvent phase. These methods are expensive for use in contaminated sites but are being explored in laboratory experiments.

What happens if a pollutant is immiscible and hence is neither in the aqueous phase nor in the solid phase? This situation is encountered in beaches and marine waters and industrial sites when oils spill from tankers and crude oil carriers. Oil is an immiscible pollutant and exists in what are known as non-aqueous phase liquids (NAPLs), containing aliphatic hydrocarbons whose solubilities in water decline as the number of carbons in the molecule increases. For instance, heptane (C_7H_6) has a solubility of 2.9mg/l, octane (C_8H_8) has a solubility of 0.66 mg/l whereas hexadecane ($C_{16} H_{34}$) has the least solubility of 0.000020 mg/l. Similarly, solubilities of aromatic hydrocarbons also differ—naphthalene has high solubility of 31 mg/l whereas anthracene has low solubility of 0.050 mg/l. Despite these low solubilities, microorganisms such as *Cladosporium resinae*, *Torulopsis* sp., *Candida tropicales*, *Arthrobacter* sp., *Moraxella*, *Pseudomonas pseudoflava*, *Beijerinckia* sp, *Corynebacterium equi* and *Arthrobacter* mineralize aliphatic and aromatic hydrocarbons in pure culture.

The manner in which organic substrates from NAPLs are mineralized has been a subject of speculation. One obvious explanation is that microorganisms use the substrates as and when they are solubilized and enter the water phase as exemplified by the dissolution and utilization of phenanthrene by *Flavobacterium* sp. and *Beijerinckia* sp. A second hypothesis is that bacteria attach to substrates by fimbriae and pili and directly dissolve the substrate from NAPL as has been shown in the case of *Acinetobacter calcoaceticus* in a NAPL composed of pure alkane.

An interesting pseudosolubilization hypothesis involves the bacterial excretion of emulsifiers composed mainly of polysaccharides, polysaccharide protein complexes or glycolipids that serve not only to increase the surface of the interphase between the NAPL (pure alkanes or oils) and water but also act as surfactants to reduce tensions between the different phases. This dual role of emulsifying agents results in the conversion of the NAPL to droplets or particles of 0.1 to 1.0 μm dimensions. These molecules have hydrophobic (water repellant) and hydrophilic (water loving) ends that aggregate to form 'micelles'. In a micelle, the hydrophilic ends tend to face outwards to the waterphase while the hydrophobic ends get clustered in the centre. An analogy may make the layout of the micelle easy to understand. If several roundshaped small rubber balloons, each tied with a thread are arranged in a circle by facing the balloons (the hydrophilic ends) to the circumference and the ends of threads (hydrophobic ends) coalesing to the centre of the circle, we can then imagine how micelle appears in water. It is presumed that the hydrophobic substrate contained in the NAPL would be entrapped in the inner region of the micelle in a semi soluble form and hence the usage of the word 'pseudosolubilization' in this context. One can surmise that a continuity may exist from hydrophobic to hydrophilic portions of the surfactant molecules thereby rendering the substrate optimally available for microbial action. There are other hypotheses put forward to explain how immiscible substrates get mineralized by microorganisms. One possibility can be the occurrence of an emulsifier in the form of a thin layer on the bacterial surface. Another proposition has been the existence of a lipophilic layer on the bacterial surface that facilitates the binding of the substrate to the cell.

The terms 'aged residue' and 'bound residue' have been used in discussions on bioavailability of pesticides for microbially mediated biodegradation. When pollutants get incorporated into humic compounds via oxidative coupling reactions by changing or altering the original chemical/biological activity and remain non extractable even with vigorous solvents, they are referred to as bound residues. On the otherhand, if the residue are amenable to extraction by solvents, they are referred to as aged residues. This can be illustrated in the case of the insecticide aldrin which is converted to dieldrin by soil microorganisms. When monitored at yearly intervals, it was observed that the residues in the form of both the compounds were traceable even after 5 years that get integrated in organic matter within compact deeper layers of soils. Such soils become deep polluted soils with limited access to degrading bacteria for interaction. These aged residues were released however by solvent extraction.

Acceleration of Biodegradation

Three methods by which biodegradation can be augmented have been explored. They are (1) addition of surfactants, (2) supplementation with inorganic nutrients and (3) inoculation with biomass of enriched bacterial species known to degrade a specific compound, the method referred to as 'bioaugmentation'. The enzymes responsible for biodegradation are genetically encoded in bacterial plasmids. These plasmids possess broad host range and can be transferred within the same species or genera. They have been identified in species of *Pseudomonas, Alcaligenes, Acinetobacter, Flavobacterium, Beijerinkia, Klebisella, Moraxella* and *Arthrobacter*. Most of the novel strains have been genetically engineered from the genus *Pseudomonas* and have been patented for cleaning up oil spills.

Some surfactants such as 'Surgee 2' and 'Corexit' in combination with bacteria have been marketed to clean up crude oil spills. Since the volume of carbon to be degraded in crude oil is large, there is obviously, a need for replenishing or reinforcing the surroundings of NAPLs with nitrogen and phosphorus for faster biodegradation. This strategy has met with success in the case of phenanthrene degradation in an hexane NAPL.

Where normal microflora cannot take care of the degradation processes, inoculation with specific microorganisms becomes necessary and infact reduces the acclimation period. Success appears to be the rule in closed bioreactors designed for specific pollutants when bioaugmented with biomass of known microorganisms capable of degrading the pollutant in question. Industrial waste treatment systems use immobilized microbial cells (cells trapped in an inert matrix) or biofilms carrying the desired microbes in bioreactors. The substrates used for immobilization of desired bacteria have been alginate diatomaceous earth, activated carbon and nylon membrane, powder or tubes. However, success as well as failures have been encountered in *in situ* clean ups of aquifers, surface water and soils contaminated with pollutants. Some instances of successes have been listed in Table 41.

Bioremediation

In many instances, indigenous microflora have been enriched or bioaugmented by adding nitrogenous and phosphatic fertilizers to soil or liming and aeration (bioventing) for degradation of petroleum wastes or oil spills. This method of enhancing the activity of native soil microflora currently in vogue using sophisticated engineering devises was earlier known under the name 'land farming'. Not all bioremediation experiments have been successful but in one instance in U.S.A, success was achieved in a field contaminated with 1.9 million litres of kerosene by bioventing and addition of fertilizers. In another situation with an aerated lagoon, the benzene levels came down from 64.4 to 1.9 mg/kg and likewise naphthalene content decreased from 290.0 to zero mg/kg. Bioremediation has been used to clean up leakages of petroleum hydrocarbons in underground storage tanks, ground water near gasolene outlets and marine oil spills.

Herbicides and Plant Disease

The chemotherapeutic effects of fungicides on plant diseases are well known and they will not be considered here. On the other hand, from the point of view of interaction between soil microorganisms and plants, any consideration on herbicidal effects will be relevant since herbicides are directly applied to soil in all cases. When herbicides are applied to soil, they are bound to influence the growth and activity of various organisms in the surrounding ecosystem besides mitigating the growth of weeds. Such changes in the ecosystem may often increase the incidence of a plant disease in more than one way—by stimulating the pathogen, by increasing the virulence of the pathogen, through increased susceptibility of the host and by the suppression of microorganisms antagonistic to the pathogen. The following are examples of enhanced disease incidence due to the ap-

Table 41 Some examples of pesticides/pollutants degradation

The pesticide/pollutant/intermediaries/final product	Specific microorganisms used for mineralisation/degradation	Comments, if any
A. Insecticide Parathion (O, O-diethyl O-(p-nitrophenyl phosphorothioate → 4-nitrophenol → full degradation	Stepwise, *Pseudomonas stutzeri* and *P. aeroginosa*	90 per cent degradation in soil.
B. The herbicide IPC (Isopropyl N-phenyl carbamate)	*Arthrobacter* Sp.	Completely degraded in soil.
C. The herbicide chloropropham (Isopropyl-N-3-chlorophenyl carbamate)	*Pseudomonas* spp.	in soil.
D. The fungicide pentachlorophenol (PCP)	*Rhodococcus Chlorophenolicus, Arthrobacter* sp. and *Flavobacterium* sp.	in soil.
E. Crude oil	*Candida guillermondii*	Nil
F. The herbicide 2,4-D(2,4-dichlorophenoxy acetic acid)	Suspension of microorganisms	in soil
G. PCB (Polychlorinated biphenyl)	*Acinetobacter* sp.	Nil
H. The herbicide pyrazon (5-amino-4-chloro-2-phenyl pyridazine-3 (2H)-one	Gram negative coccus	Nil
I. The herbicide 2,4,5-T (2,4,5-trichlorophenoxy acetic acid)	*Pseudomonas cepacia* and *Phanerochaete chrysosporium*	*P. cepacia* in soil and *P. chrysosporium* in ground corn cob.
J. The insecticides Lindane (1,2,3,4,5,6-hexachlorocyclohexane) and chlorodane (1,2,4,5,6,7,8,8-octachloro-2,3,3a,4,7, 7a-hexahydro-4,7-methanoidane)	*Phanerochaete chrysosporium*	in sterile soil amended with corncobs.
K. The herbicide dicamba (3-6-dichloro-2-methoxybenzoic acid) and nitrophenols (2-3, and 4-nitrophenol and 2,4-dinitrophenol)	microbial mixture	Decreased in flooded soil.
L. 4-nitrophenol	*Pseudomonas* sp.	Enhances in sewage system.
M. Wood preservative pentachlorophenol (PCP)	*Phanerochaete* sp. and a mixture of microorganisms	Under field condition 90 per cent degradation in 7 weeks.

plication of herbicides (shown in parenthesis)—*Alternaria solani* on tomato (2,4-D), *Helminthosporium sativum* on barley (MH), and *Botrytis fabae* on broad bean (simazine). Several *in vitro* studies have revealed the ability of several herbicides to enhance the growth of fungi. For instance, 2,4-D has been shown to enhance the growth of *H. sativum* and atrazine to augment the growth of *Fusarium* species in culture media as well as in sterilized soil.

On the other hand, the incidence of plant disease are known to be decreased by the application of herbicides as a result of direct toxic effects on the pathogen by creating resistance in the host and by alteration in the microbial equilibrium in favour of decreased population of the disease causative organism. The following are examples of decreased incidence of diseases due to herbicide application (shown in parenthesis)—*Puccinia graminis* on oats and wheat (2,4-D) and *Sclerotium rolfsii* on groundnut and *Cercospora arachidicola* on groundnut (dinoseb).

Selected References

Alexander, M. 1994. *Biodegradation and Bioremediation* Academic Press. San Diego.

Ashton, F.M. and Craft, A.S. 1981. *Mode of Action of Herbicides*, Wiley, New York.

Katan, J. and Eshel, Y. 1973. Interactions between herbicides and plant pathogens. *Residue Rev.*, 35, 145–177.

Kaufman, D.D. and Blake, J. 1969. Atrazine degradation by soil fungi. *Weed Sci. Soc., Amer., Abstr.* No. 230.

Leisinger, T., Cook, A.M., Hutter, R. and Nuesch Eds. 1981. *Microbial Degradation of Xenobiotics and Recalcitrant Compounds*. Academic Press, London.

Martin, J.P. 1963. Influence of pesticide residues on soil microbiological and chemical properties. *Residue Rev.*, 4, 96–129.

Miller, M.W. and Berg, G.G. 1969. Eds. *Chemical Fallout—Current Research on Persistent Pesticides*. Second Printing. Charles C. Thomas, Springfield, Illinois, U.S.A.

Schnoor, J.L. Ed. 1992. *Fate of Chemicals and Pesticides in the Environment*, Wiley (Interscience), New York.

Walker, N. 1982. Interactions of pesticides with soil microorganisms, pp. 377–395, In *Advances in Agricultural Microbiology*, Ed. N.S. Subba Rao, Oxford and IBH Publishing Co., New Delhi.

Ware, G.W. 1991. *Fundamentals of Pesticides: A Self-instruction Guide 3rd Edition.*, Thomson Publication, Fresno, California.

15. Mycorrhizae

Fungal Symbioses with Roots

The symbiotic associations between fungi and root systems of higher plants come under the general name, mycorrhiza (plural mycorrhizae) which literally means 'fungus roots'. Fungus roots were discovered by the German botanist Frank in the last century (1855) in forest trees such as pine but subsequent work has pointed out that such symbiotic associations with fungi exist under natural conditions in root systems of many other economically important crops (Fig. 93).

There are two main kinds of mycorrhizae—the ecto and the endomycorrhizae. In the ectomycorrhizae (also called ectotrophic mycorrhizae), the fungus completely encloses each feeder rootlet in a sheath or mantle of hyphae (Fig. 94). The hyphae penetrate only between the cells of the root cortex (intercellular). Ectomycorrhizae are known to occur in the following families: Pinaceae, Salicaceae, Betulaceae, Fagaceae, Juglandaceae, Ceasalpinioideae and Tiliaceae. Several genera such as *Pinus, Picea, Abies, Pseudotsuga, Cedrus, Larix, Quercus, Castanea, Fagus, Nothofagus, Betula, Alnus, Salix, Carya* and *Populus* have ectomycorrhizal infections. In endomycorrhizae, the fungus does not form an external sheath but lives within the cells of the root (intracellular) and establishes direct connections between the cells of the root and the surrounding soil. Endomycorrhizae are found in representative species of most of the families of angiosperms. They are also found in conifers except Pinaceae and in certain pteridophytes and bryopytes. Unmistakable evidences have been presented to indicate the beneficial effects of both ecto- and endomycorrhizae on plant growth.

Ectomycorrhizae

In general, the fungi involved in ectomycorrhizae come under Basidiomycetes from the families, Amanitaceae, Boletaceae, Cortinariaceae, Russulaceae, Tricholomataceae, Rhizopogonaceae and Sclerodermataceae. They are included in the genera—*Amanita, Boletus, Cantharellus, Cortinarius, Entoloma, Gomphidius, Hebeloma, Inocybe, Lactarius, Paxillus, Russula, Rhizopogon, Scleroderma* and *Cenococcum*. Many of these fungi show a wide host spectrum. Likewise, one and the same host may be infected by more

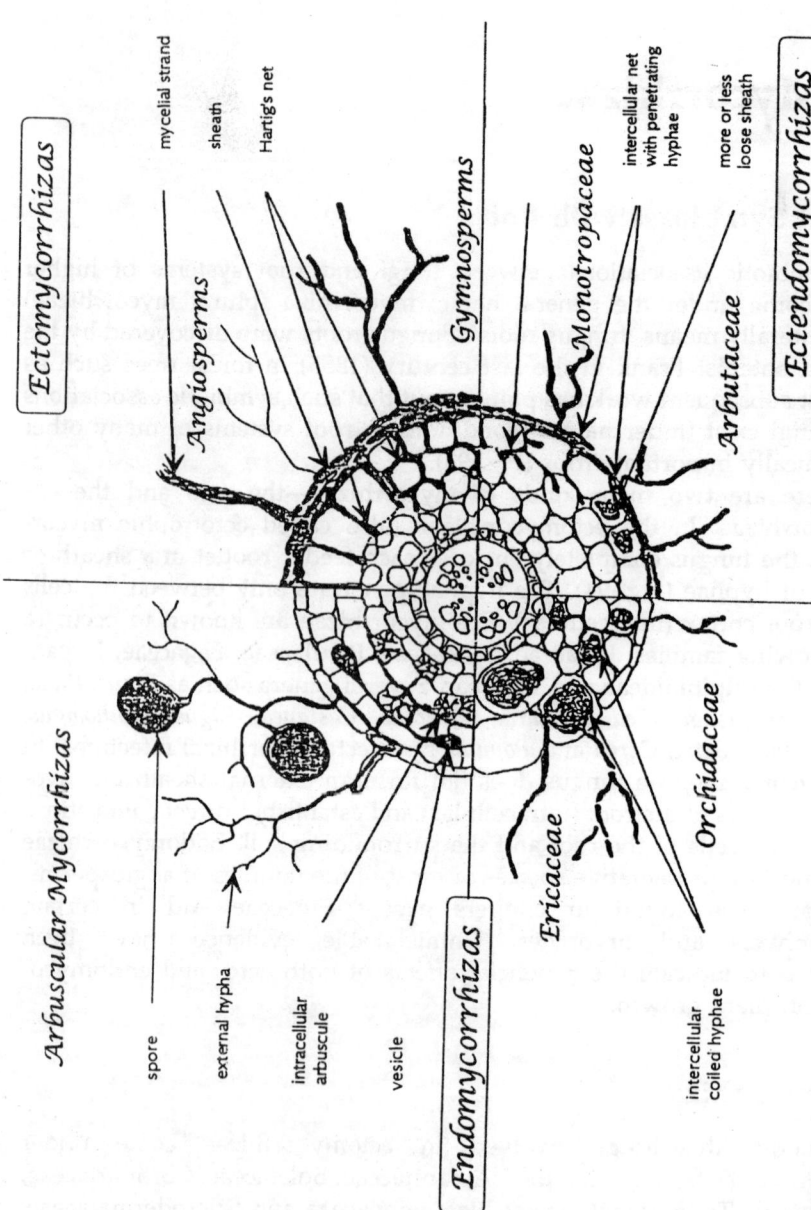

Fig. 93 Diagram representing different types of mycorrhizae (from Budi *et al* 1998).

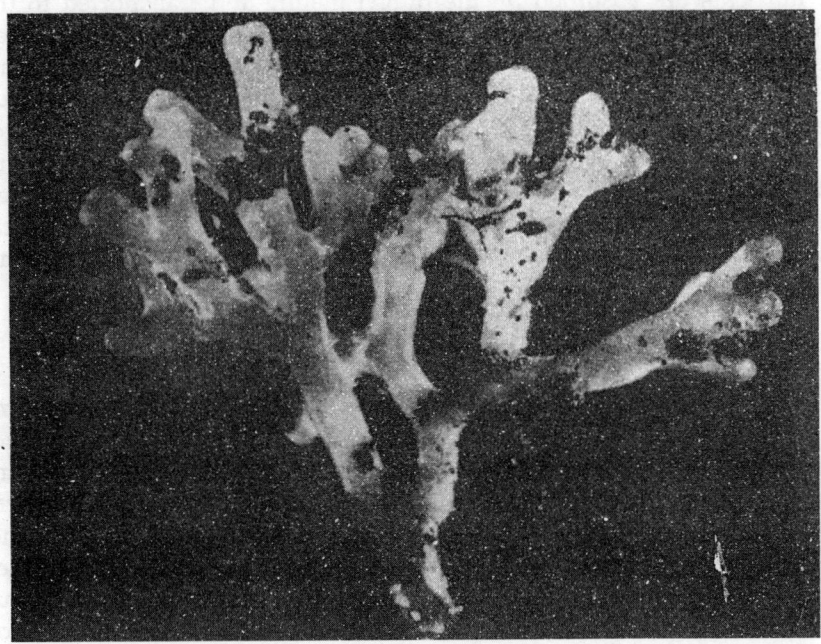

Fig. 94 Photograph showing dichotomously branching roots of *Pinus* sp., showing the ectotrophic fungal mantle around roots. (Courtesy: E. Hacskaylo, USDA).

than one fungus as exemplified by *Pinus sylvestris* from whose mycorrhizal roots as many as 40 fungal species have been isolated. When the defence reaction of the higher symbiont diminishes as it is likely to happen in senescent or diseased trees, the lower symbiont may become endotrophic. Such instances have been designated as 'ectendotrophic mycorrhizae' or 'ectendomycorrhizae'. A survey of literature on mycorrhizal associations also reveals descriptions known as 'Pseudomycorrhizae' which refer to the thin and long roots which are infected with fungi. The fungi are intracellular in such roots and it is difficult to define whether fungi exist in ectendomycorrhizae as parasites or symbionts. Mention must also be made of the mycorrhiza of *Monotropa* (which in Greek means living alone), a native of North America growing in forests under *Pinus* and *Quercus*. The plant does not contain chlorophyll and a common mycorrhizal fungus *Boletus* transports nutrients from the forest trees to *Monotropa*.

An ectomycorrhizal root is devoid of root hairs and is covered by a layer or sheath of fungal hyphae which almost looks like host tissue. This layer is called pseudoparenchymatous sheath. From this sheath, hyphae enter the cortex and remain only in the outer cortical cells to form a network called the 'Hartig net' (named after the German botanist Hartig). All the nutrients are absorbed by the fungal mantle and transported to the root through the Hartig net. One of the frequent questions that have been posed is whether such plants can grow in the absence of ectomycorrhizae. Researches on this subject have shown that if essential nutrients are made available, plants can still grow in the absence of its symbiont. The mycorrhizal habit is, however, indispensable in soils and seedlings in nursery beds deficient in major nutrients, like nitrogen and phosphorus. When trees are introduced to new regions, inoculation of soil with mycorrhizal fungi has been found to be a necessary prerequisite for the establishment of the mycorrhizal habit.

Cultural Characteristics of Ectomycorrhizal Fungi

The ectomycorrhizal fungi can easily be isolated in the vegetative form although the identification of such fungi becomes difficult since reproductive bodies are not readily formed in culture media. Fruiting bodies can, however, be seen on the soil surface near the trees, from which fungal cultures can easily be isolated. These fungi grow slowly in culture and require special nutrients such as thiamine, simple amino acids and other undefined constituents (collectively known as the M-factor) of root exudates. Melin and his associates in Sweden have shown that the M-factor is exuded by roots of plants susceptible to mycorrhizal infection and not by plants resistant to it. Experiments with *Boletus variegatus* have shown that low doses of M-factor are stimulatory to growth of the fungus whereas high doses become inhibitory. The inhibitory portions of M-factor appear to be excessive in old secondarily thickened axils or roots while the stimulatory portions appear to be so in the primary rootlets. Ectomycorrhizal fungi are generally not cellulolytic or lignolytic and therefore have to depend on carbohydrates from their host plants. Experiments with ^{14}C labelled sucrose, glucose, fructose and $^{14}CO_2$ have shown that most of the carbon requirements of the fungus come from the host plant by way of root excretions. Metabolites produced by the fungus influence the structure and morphology of the root system. These substances include auxins such as indole acetic acid and other unknown growth substances. They are partly responsible for the dichotomy of pine rootlets. In this way, the fungus and the host plant mutually control the morphological and physiological activities of the symbiotic system. The carbon nutrition of the fungus is dependent on the photosynthetic activity of the host which is balanced by the greater efficiency of absorption and storage of nutrients afforded by the fungal partner.

Absorption of Nutrients

Mycorrihizal roots lack root hairs. The fungal sheath together with the hyphae extending to the soil absorb nutrients. The Hartig net acts as a liaison tissue between the fungal sheath and the host cells. The ectmycorrhizal habit increases the surface area of the root system and hence affords better intake of nutrients such as nitrogen, phosphorus and potassium from the surrounding soil. Using intact plants as well as excised roots of pine (*Pinus radiata*) many workers have demonstrated the transfer of isotopically labelled phosphorus, nitrogen, calcium and sodium from the soil into the roots through the fungal mycelium. The labelled isotopes were detected in all parts of the plants including leaves.

The pattern of movement of ions has been investigated in ectomycorrhizae of *Fagus sylvatica*. The mycorrhizal association of this large rooted species can easily be distinguished into an outer fungal sheath and an inner host core. The host core consists of the Hartig net of the fungus between cells of the host. Experiments done by Harley and his associates in England with excised mycorrhizal roots have shown that 80–90% of the absorbed phosphate remains accumulated in the fungal sheath whereas experiments done by Melin and Nilsson in Sweden with intact as well decapitated plants have, however, shown that accumulation of phosphate in the sheath is variable depending on the rates of transpiration since decapitated plants accumulated more of phosphates in the sheath than the intact ones. It is likely that the sheath acts as a phosphate reservoir (Table 42) and releases the nutrient during certain deficiency conditions while under normal conditions, a steady uptake of phosphate to the plants is maintained by the fungal hyphae.

Table 42 Relative phosphorus absorption (Radioactivity per milligram of dry-weight) by mycorrhizas dissected before and after immersion in a buffer containing 0.16 mM. KH_2PO_4 labelled with ^{32}P (from Harley, 1969)

	Sheath removed after immersion			Sheath removed before immersion		
	Total	Sheath	Host	Total	Sheath	Host
Experiment 1	5.18	1200	85	1200	2370	428
Ratio $\frac{Stripped}{intact}$	–	–	–	2.32	1.98	5.04
Experiment 2	351	795	65.6	971	2153	265
	332	765	56.5	801	1671	228
Ratio $\frac{Stripped}{intact}$	–	–	–	2.60	2.47	4.04

Techniques of Ectomycorrhizal Inoculation

Since ectomycorrhizal habit in trees is the nature's way of scavenging scarce nutrients from the forest floor, questions have been posed whether inoculation with artificially grown specific fungal cultures can help in improving tree stands and growth in existing or newly afforested lands. Undoubtedly, inoculation requirement exists in freshly mined soil and grasslands where afforestation programmes are planned. Inoculation with identified fungi can be done at the nursery stage with the help of pure cultures of specific fungi isolated from fruiting bodies seen near the trees. Fast growing species of fungal symbionts are easily grown while fastidious slow growing ones need growth factors to generate enough inoculum biomass. Satisfactory results have been reported in experiments with *Pinus radiata* with isolates of *Rhizopogon luteolus, Suillus granulatus, S. luteolus, Cenococcum geophilum* and *Pisolithus tinctorius*. The fungus is grown for 3–4 months in 2 litre jars containing sterilized peatmoss-vermiculite substrate moistened with a suitable nutrient medium. The inoculum is washed well in tap water to get rid of the unutilized substrate and spread on nursery beds and mixed with soil for a depth of 8–10 cm. It is necessary that the soil for nursery beds must be fumigated with a methyl bromide-chloropicrin mix or any other sterilant to minimize the population of other microorganisms which may inhibit the growth of the ectomycorrhizal inoculum. The fumigation of soil is done prior to inoculation with the fungal symbiont and planting of seed or seedlings. Air-dried inocula have also proved successful and drying facilitates uniform mixing of fungal propagules in soil.

Seedlings may be raised in containers with a potting mix and these containorized seedlings may be inoculated with the fungus. One limitation of pure culture inoculation is the slow growth of the fungus in culture and hence low inoculum yield is not easy for rapid field application.

Other methods of inoculation are the transfer of soil from the root region of well established particular species of a tree to constitute a nursery bed so that spores and propagules from the already established tree stands serve as natural inoculum. Sporocarps, spores and other fruiting bodies of fungi near a particular tree stand may also serve as a natural source of inoculum. These are labour intensive practices and subject to transfer of other tree pathogens to nursery beds. Planting mycorrhizal "nurse" seedlings or incorporation of chopped roots of ectomycorrhizal roots of a given tree species have also proved useful in certain instances.

Resistance to Plant Diseases by Ectomycorrhiza

Feeder root pathogens such as *Phytophthora, Pythium, Rhizoctonia* and *Fusarium* infect immature and meristematic cortical tissues of roots and cause necrosis. However, one of the physiological benefits of ectomycorrhizae is the protection afforded by the fungal mantle against such root pathogens. Well formed mycorrhizal roots are resistant to infection and non-mycorrhizal feeder roots are prone to fungal necrosis even when adjacent roots have become mycorrhizal. The resistance is purely due to the mechanical barrier afforded by the mycorrhizal fungal mantle. However, species of certain fungal genera causing ectomycorrhizae such as *Lactarius, Cortinarius* and *Hygrophorus* produce antibiotic substances while species of

Russula produce none at all. Some of these antibiotics are antifungal on *Rhizoctonia salani, Pythium debaryanum* and *Fusarium oxyporum*. Nevertheless, it remains to be seen whether antibiotics are elaborated by ectomycorrhizal fungi *in vivo* in association with the higher symbionts.

Boletus variegatus is known to produce volatile fungistatic compounds in pure culture. They have been identified as isobutanol and isobutyric acid. Infection of roots of *Pinus sylvestris* with *B. variegatus* resulted in the production and accumulation of volatile and fungistatic terpenes and sesquiterpenes to the extent of eight times the concentrations of such compounds in non-mycorrhizal roots. In this connection, it is relevant to point out that tubers of several species of orchids produce orchinol, coumarin, hircinol and other phenolic compounds as a defense reaction to the presence of fungi such as *Rhizoctonia*. These substances were not detected in tubers of orchids free of fungal symbionts. Very likely, substances such as orchinol act as deterrents to pathogenic fungi by restricting the activity of certain fungi such as *Rhizoctonia* more to a symbiotic state than to a parasitic state.

Endomycorrhizae of Orchids

As the name indicates, in endomycorrhizae (also known as endotrophic mycorrhizae) the fungal hyphae enter the cells of the host plant and thus penetrate the host tissues. The fungi involved in endotrophic association belong either to the Phycomycetes (possessing aseptate hypae) or to the Basidiomycetes or Fungi Imperfecti (possessing septate hyphae).

All orchids have endomycorrhizae. However, a few species of some genera of orchids are capable of growing without a fungal partner. They are *Cephalanthera, Listera, Epipactis* and *Cypripedium*. Other orchid species of the genera *Neottia, Limodorum, Epipogon, coralliorhiza, Galeola, Vanilla, Gastrodila* and *Didymoplexis* depend entirely on the fungal partner since most of them have little or no chlorophyll. However, when orchid seeds germinate they become infected by hyphae from the soil and most of the orchids depend on fungi in seedling stages. After penetration, the fungi appear in the cortical cells in the shape of a coil, cause swelling and disorganisation of the cells and ultimately get disintegrated within the host cell. The disintegration of the hyphae within the cell has also been referred to as 'tolypophagy' and 'ptyophagy' which are different ways of digestion by the host. The same host cell may again get re-infected and the process may continue. The infection is restricted to the cortical cells of the root or to the portion of the plant acting as root system. One group of the orchid mycorrhizal fungi belong to the genera of Basidiomycetes such as *Armillaria, Fomes, Zerotus, Corticium* and *Marasmium*. The other group belongs to Fungi Imperfecti under the genus *Rhizoctonia*. Many of these fungi can break down lignin and cellulose and thus contribute to the decaying of organic matter. In this respect, they differ from ectomycorrhizal fungi which rely on the host for their carbon nutrition. The orchid is dependent on the fungus for its carbon requirements in the early stages of its establishment.

Arbuscular Mycorrhiza (AM)

Arbuscular mycorrhizae occur in roots of most angiosperms, pteridophytes and bryophytes, although absent in plants which form only ectomycorrhiza (Pinaceae, Betulaceae) or the two other specific types of endomycorrhiza of Ericales and Orchidales. Arbuscular mycorrhizae develop special characteristic structures called arbuscles and vesicles (Fig. 95). The arbuscles help in the transfer of nutrients (especially phosphates) from the soil into the root system (Fig. 96).

The taxonomy of AM fungi is in a state of flux and purely based on spore morphology five genera of AM fungi are recognized—*Glomus, Gigaspora, Acaulospora, Sclerocystis* and *Endogone*, the latter restricted to plants that form ecto or no mycorrhiza. These fungi are obligate symbionts and have not been isolated in pure culture and can be maintained only on live plants inoculated with spores of a species and collecting the pieces of roots with soil for experimental purposes. The root biomass heavily infected by a specific AM fungus serves as the inoculum for subsequent experimental plants. Specificity between the two partners of this symbiosis has not been well established and cross-inoculations with the any other susceptible hosts are fairly easy.

Isolation of AM Fungal Spores

Soil samples with rootlets, preferably close to the root system are collected from a depth of 10–15cm after the surface soil has been scraped and discarded. Such samples from several locations within a plot or a given area are pooled and subjected to wet sieving and decanting procedure as follows: The composite soil sample (say 50g) is placed in a beaker and mixed with water by frequent stirring. The suspension is passed through a 710 μm sieve placed in a funnel and the filtrate collected in a 1 litre measuring cylinder, taking care to wash the root-soil debris in fine jets of water until the filtrate reaches the high mark. The residue containing the rootlets in the sieve is set aside. The filtrate in the cylinder is stirred constantly and allowed to pass through a 250 μm sieve and the filtrate collected in a second 1 litre measuring cylinder until the filtrate reaches the high mark. The residue in the sieve is set aside. This procedure is again repeated twice, once with 105 μm sieve and the other with a 53 μm mesh sieve and the filtrates collected in separate 1 litre measuring cylinders. The objective of stirring, sieving and washing so many times is to facilitate quantitative

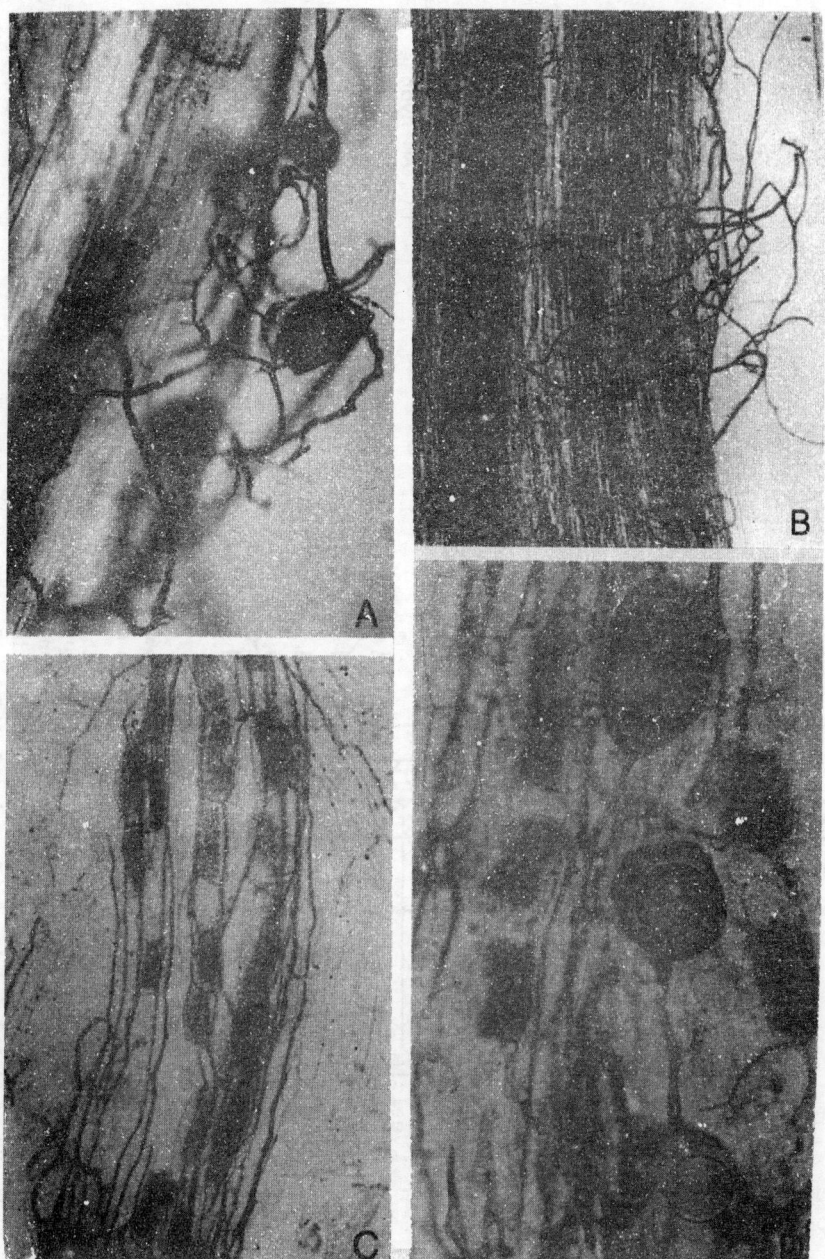

Fig. 95 AM fungi on roots: (A) spores and external mycelium attached to roots; (B) Part of root with clumps of external mycelium; (C) arbuscles showing branched form; (D) arbuscles showing granular appearance and vesicles with oil globules (Courtesy: Dr. D.S. Hayman).

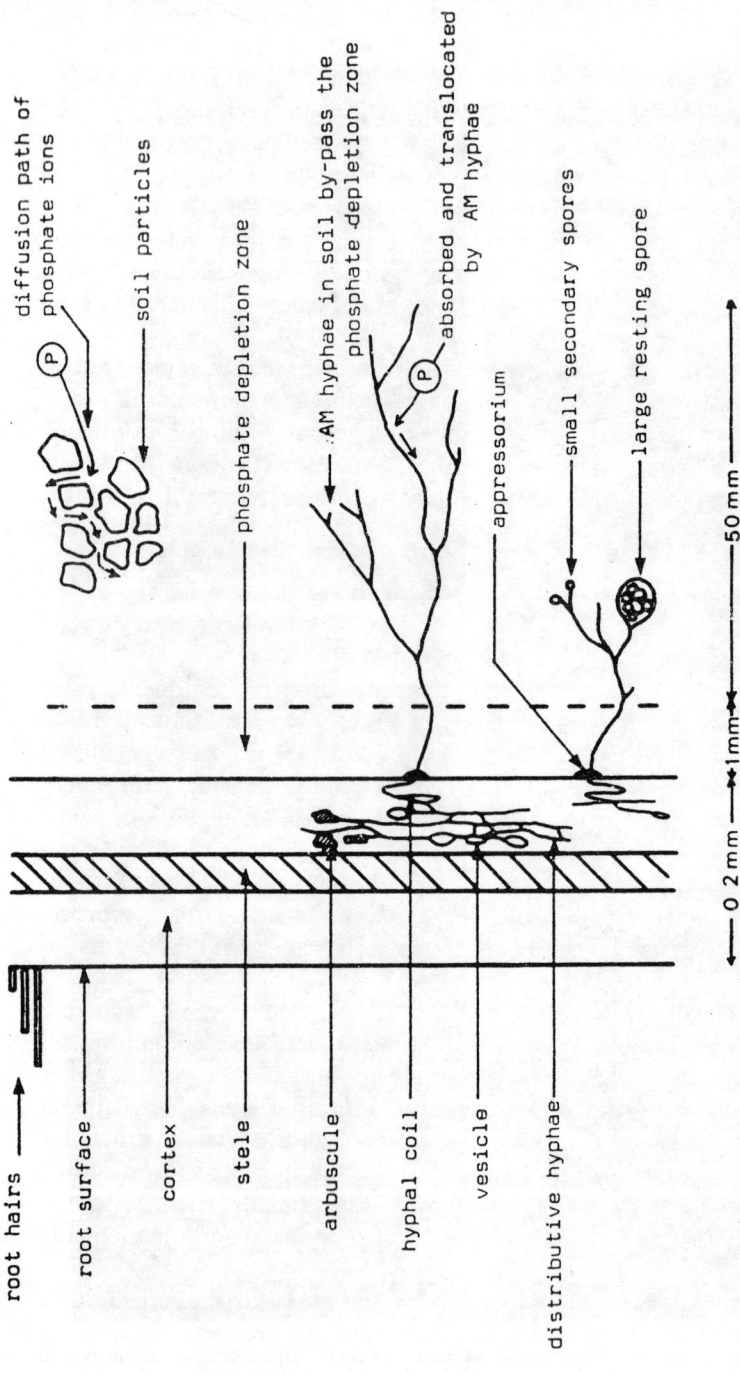

Fig. 96   Diagram (not to scale) illustrating the major features of an arbuscular mycorrhiza and the chief mechanism whereby it is believed to enhance the uptake of phosphate from soil. The slow diffusion path of phosphate ions in soil, resulting from tortuous pathways and reversible adsorption to the soil particles, is short-circuited by direct transfer of phosphate to the root through the fungal hyphae which extend well beyond the root hair zone. P = primarily $H_2PO_4^-$. (From Hayman, 1982).

retrieval of spores depending on their size. Large circular sieves can also be used over glass troughs instead of small sieves in a funnel and the filtrates collected in a serial fashion.

The root pieces and residue from 710 and 250 μm sieves are examined under a dissecting microscope for hypae, spores and sporocarps and stained for light microscopic observations. The residues from 105 and 53 μm sieving must generally show up all the spores of smaller size when examined under a light microscope. The sieving method was originally used in nematode studies and later adopted for studying AM fungal spores and propagules. The method works fairly well in sandy soils but may pose problems with organic matter-rich soils.

The finer roots are washed gently in tap water and simmered in 10 percent KOH at 90°C for 1–2 hours, rinsed in tap water , immersed in 2 per cent HCl and stained with 0.05 per cent trypan blue in lactophenol. This is done by boiling roots in the stain for 3 minutes, draining the excess stain and immersing the stained roots in lactophenol overnight to destain the cortical cells. Later the stained specimens may be examined under a microscope in lactic acid-glycerol medium (1:1) when only the hyphal strands and spores retain the stain. Due to the corrosive nature of the materials used in these studies, the procedures of staining and destaining are done in a fumehood.

To avoid the toxicity of compounds, a modified procedure has been suggested as follows: (1) heat roots in 2.5% KOH for 3 minutes at 121°C or 10–30 minutes at 90°C; (2) rinse roots in water; (3) if necessary, bleach roots in alkaline H_2O_2 for 10–30 minutes followed by rinsing with water; (4) soak roots in 20–50 vol. 1% HCl for 1–24 hrs; (5) stain roots in acidic glycerol/trypan blue for 3 minutes at 121°C or 10–30 minutes at 90°C and (6) destain in acidic glycerol and store roots in the same medium for examination under microscope. The clearing, staining and destaining procedures may be varied by each investigator as the person begins to understand the nature of the specimen handled.

The presence of oval, round or irregularly lobed vesicles occurring between or inside cortical cells, attached to hyphae and containing oil globule is a sign of AM fungal infection. These vesicles act as storage organs. The presence of branched arbuscules is another sign of AM fungal infection. These structures are intended to serve as two way channels for transport of nutrients, more particularly carbohydrates. Structures known as appressoria connect AM fungal ramifications inside roots with the mycelium of the fungus outside the root and serve as absorbing elements from soil to roots.

The morphology or resting spores form the basis for identification of isolates. Spores with straight or angular stalks are grouped together as genus *Glomus* whereas those with bulbous stalk come under the genus *Gigaspora*. Spore types without stalks (sessile) are grouped as genus

Acaulospora. Spores that are arranged regularly on a central core in a sporocarp come under the genus *Sclerocystis*. The size of spores vary from 50 μm (*Glomus microcarpus*), 100–200 μm (*Glomus mosseae*), upto 400 μm (*Gigaspora margarita*) extending to nearly 1mm in few *Gigaspora* isolates. A specialist can only decide the actual identification of AM fungi to the level of species because the classification of these fungi is far from being clear.

Methods of AM Fungal Inoculation

Since it has not been possible to culture AM fungi, alternate methods of plant inoculation have been practiced. Individual spores, soil sievings, infected roots or plants and soil from around the roots can be used for inoculating fresh plants. However, these methods are subject to contamination with unwanted microorganisms and pathogens.

Growing plants infected with specific species of AM fungi in pots is currently one of the recognised practices not only to maintain the fungal germplasm on live roots but also to mass produce inoculum for pot and limited scale field experiments. For this purpose spores that look alike morphologically are isolated by means of glass capillaries from soil sievings and transferred to individual watch glasses. The spores are then surface sterilized by immersing in 2 per cent chloramine T and 200 ppm streptomycin for 15 minutes followed by several washings with sterilized water using sterilized glass pippettes and identified with the help of standard publications (Gerdman and Trappe, 1974). With care, about 20 identical spores are transferred to moist filter paper towellettes and wrapped around roots of aseptically grown seedlings. The seedlings with the wrappings are then individually planted in sterilized sand and soil mixture in small containers. When plants grow to about 2–3 months, a portion of the substratum is subjected to the wet sieving process to determine if that particular species of AM fungus has established well in the root region. Following the capillary method, bulk samples of spores from wet seivings can be obtained and transferred to suitable potting substrates in pots in which the desired host plants are raised. Since specificity between symbionts has not been fairly well established, it is necessary to choose a host plant species that has good root system to generate a mass of hypahe and spores. Species of grasses or millets (pearl millet for instance—*Pennisetum americanum*,) and sorgum (*Sorghum bicolor*) have been used with success in mass multiplying AM fungi. The choice of a suitable substrate to grow the selected host species is also important. Several investigators have used perlite, vermiculite, soilrite or plain sand plus soil mixture with success.

When the host plant species have grown for about 3 months, the roots and soil are examined for purity of culture and then gently macerated in sterile containers. The macarate can be directly used as inoculum by placing it close to the roots (at the rate of 5–10 per cent of the potting mix) to about 2–5 cm below the soil level and watered periodically. Alternatively, seeds can be pelletted, air dried and sown.

There have been improvements in culturing AM fungi on plant roots by a system known as 'aeroponic culture', wherein roots are grown in a fine continuous mist of nutrient solution under aseptic conditions with the shoot system intact in open air to facilitate the production of high quality inoculum without contamination. The roots are checked for colonization by the specific AM fungus, macerated in a waring blender and used as inoculum.

Plant Response to AM Fungal Inoculation

Soil phosphorus is a critical factor in plant response and responses are generally better under low phosphorus levels. Host genotypes and fungal strains seem to influence the response of plants to inoculation. Several investigators have reported response of trees to AM fungal inoculation. The results obtained with *Leucaena leucocephala* are illustrative where it was observed that growth enhancement was correlated with increases in AM fungal colonization of roots and uptake of phosphorus, copper and zinc (Fig. 97). Worldwide field experiments have provided evidence to show that under marginal p-deficient soils lacking in effective AM fungal endophytes, increase in yield of wheat, maize, barley, potatoes, white clover, red clover, lucerne, cowpea and other legumes are possible. Increased uptake of zinc has also been shown in AM fungus inoculated peach, maize, wheat and potato in zinc deficient soils. Other observations implicated in AM associations relate to increased uptake of sulphur and cadmium. Improved water absorption and tolerance of plants to water stress in citrus and avacado seedlings have also been noticed. There are also reports of increased levels of cytokinins and chlorophyll by AM fungus infected plants.

Legume Arbuscular Mycorrhiza Interaction

Rhizobia and arbuscular mycorrhiza often interact synergistically resulting in better root nodulation, nutrient uptake and plant yield. In soils with low P content, this interaction is marked, especially with added phosphate. Such beneficial interactions have been shown in the following legumes: *Stylosanthes guyanensis*, *Centrosema pubescens*, *Medicago sativa*, *Phaseolus* sp., *Glycine max*, *Arachis hypogea*, *Vigna unguiculata*, *Pueraria* sp., *Trifolium repens* and *Trifolium subterraneum*.

The practical utility of this dual effect remains to be explored in spite of the inherent limitation that AM fungi are obligate symbionts and cannot be mass multiplied in pure culture by any known method.

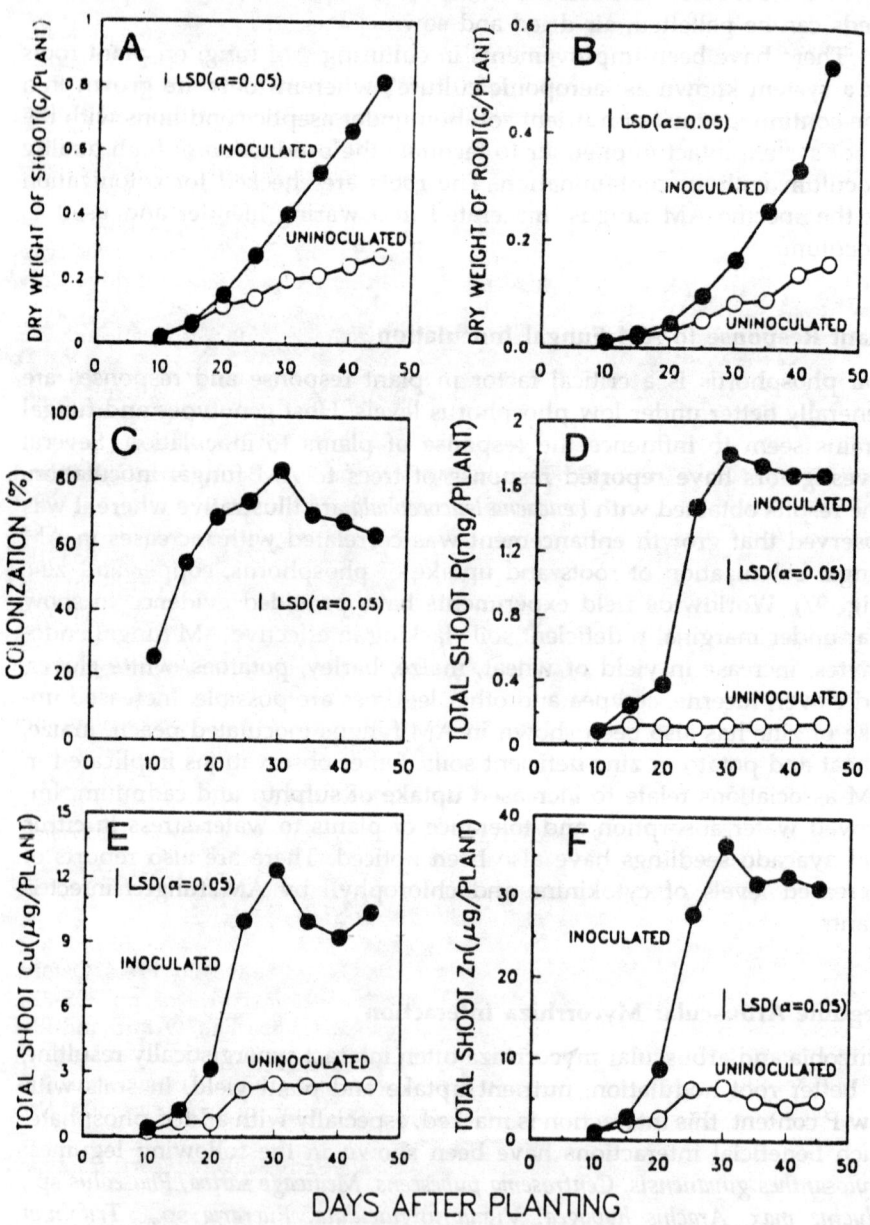

Fig. 97 Influence of soil infestation by *G. aggregatum* on (A) shoot dry weight, (B) root dry weight, (C) root colonization, (D) shoot phosphorus uptake, (E) shoot copper uptake, and (F) shoot zinc uptake of *L. leucocephala*. (Manjunath and Habte, 1988).

Selected References

Bagyaraj, D.J., Manjunath, A. and Patil, R.B. 1979. Interaction between a vesicular-arbuscular mycorrhiza and *Rhizobium* and their effects on soybean in the field. *New Phytologist*, 82, 141–145.

Bjorkman, E. 1960. *Monotropa hypopitys* L.—an epiparasite on tree roots. *Physiologia Pl.*, 13, 308.

Gerdemann, J.W. and Trappe, T.W. 1974. *The Endogonaceae in the Pacific Northwest*, Mycologia Memoir, No. 5, 76 pp.

Harley, J.L. 1969. *The Biology of Mycorrhiza*, 2nd Ed. Leonard Hill, London.

Harley, J.L. 1971a. Associations of microbes and roots. In *Microbes and Biological Productivity*, pp. 309–332. Eds. D.E. Hughes and A.H. Rose, Cambridge Univ. Press, London.

Harley, J.L. 1971b. *Mycorrhiza*. Oxford Biology Readers. Eds. J.J. Head and O.E. Lowenstein, Oxford University Press.

Harley, J.L. and Lewis, D.H. 1969. The physiology of ectotrophic mycorrhiza. *Adv. microb. Physiol.*, 3, 53–81.

Hayman, D.S. 1982. Practical aspects of vesicular-arbsuscular mycorrhiza, pp. 325–373. In *Advances in Agricultural Microbiology*, Ed. N.S. Subba Rao, Oxford and IBH Publishing Co., New Delhi.

Hung, L.L. and Sylvia, D.M. 1988. Production of vesicular arbuscular mycorrhizal fungus inoculum in aeroponic culture. *Appl. and Environmental Microbiol.* 54, 353–357.

Ko, W.H. and Farley, J.D. 1969. Conversion of pentachlorobenzne to pentachloroaniline in soil and the effect of these compounds on soil microorganisms. *Phytopathology*, 59, 64–67.

Koske, R.E. and Gemma, J.N. 1989. A modified procedure for staining roots to detect VA mycorrhizas. *Mycol. Res.*, 92, 486–505.

Krishna, K.R. and Bhagyaraj, D.J. 1984. Growth and nutrient uptake of peanut inoculated with the mycorrhizal fungus *Glomus fasciculatum* compared with non-inoculated ones. *Pl. Soil*, 77, 405–408.

Manjunath, A. and Habte, M. 1988. Development of vesicular-arbuscular mycorrhizal infection and the uptake of immobile nutrients in *Leucaena leucocephala*. *Pl. Soil* 106, 97–103.

Marx, D.H. 1972. Ectomycorrhizae as biological deterrents to pathogenic root infections. *Ann. Rev. Phytopath.*, 10, 429–454.

Melin, E. 1953. *Mycorrhiza. Handbuch der Pflanzen Physiologie*, 11, 606–638.

Mikola, P. Ed. 1981. *Tropical Mycorrhiza Research*. Oxford University Press, Oxford.

Molina, R. and Trappe, J.M. 1982. Applied aspects of ectomycorrhizae, pp. 305–324. In *Advances in Agricultural Microbiology*, Ed. N.S. Subba Rao, Oxford and IBH Publishing Co. New Delhi.

Mosse, B. 1977. Plant growth responses to vesicular-arbuscular mycorrhiza. X Responses of *Stylosanthes* and maize to inoculation in unsterile soils. *New Phytologist*, 78, 277–288.

Mosse, B. 1973. Advances in the study of vesicula:-arbuscular mycorrhiza. *Ann. Rev. Phytopathol.*, 11, 171–196.

Powell, C.L. and Bagyaraj, D.J. Eds. 1984. *VA mycorrhiza*. CRC Press Inc., Boca Raton, Florida.

Schonbeck, F. 1979. Endomycorrhiza in relation to plant diseases, pp. 271–280. In *Soilborne Plant Pathogens*, Eds. B. Schippers and W. Gams. Aademic Press, New York.

Subba Rao, N.S. and C. Rodriguez-Barrueco Eds. 1993 *Symbioses in Nitrogen Fixing Trees*. Oxford and IBH Publishing Co. New Delhi.

Zak, B. 1964. Role of mycorrhizae in root disease. *Ann. Rev. Phytopath.*, 2, 377–392.

16. Biotechnology in Agriculture

The New Green Revolution

The green revolution which gave us plenty of grains to feed millions of people and the revolution in medicine which increased the life span of man are common knowledge even to lay people. All this was possible due to major discoveries and technological innovations in agriculture and medicine. Today, we are witnessing another revolution in biosciences because of some major advances in cell biology and genetics. Many people are inclined to believe that while the battle for green revolution of the type we saw three decades ago was fought in the field, the battle for the new revolution in biosciences known as "Biotechnology" is being fought in modern laboratories. Some argue that the new biotechnology in agriculture is the second green revolution (part II) which will speed up crop improvement by gene manipulation in a Petri dish rather than in an open field.

The meaning or the definition of the word biotechnology has been the subject of hot debate by scientists and technocrats. The definition depends on the extent of expertise a group or an individual has with regard to cell biology. It also depends on the needs of a society or a country one lives in. The use of microorganisms or their products for food, feed, biofertilizers, biopesticides and medicine was known during the last 60 years. The major developments in medicine and industry during those years came with the use of microorganisms for the benefit of mankind through new inventions in microbial technology and fermentation processes but the biotechnology of which we are currently talking about is envisaged by splicing genes and altering genomes by insertion of foreign DNA by genetic recombinant DNA techniques (the so-called genetic engineering). The cell which is being manipulated by genetic recombination may be a microbial cell or a cell from a tissue of a plant, animal or man with the ultimate objective of inserting or cloning useful genes to obviate the use of long and tedious process of conventional breeding often replacing it by tissue culture techniques. David Baltimore, Nobel Laureate, formerly Director, Whitehead Institute and Professor, Department of Biology, Massachusetts Institute of Technology defined Biotechnology as "the application of scientific and engineering principles to the processing of materials by biological agents to provide goods and services". The author defines biotechnology "as the application of science and technology to accelerate or improve nature's processes in producing man's ever increasing needs for a good living."

The path from green revolution to gene revolution or from conventional plant breeding to genetic engineering has been filled with many significant findings. For almost a century plant breeders identified and selected desirable characters and combined them into one individual plant. Since all characters are controlled by genes in chromosomes, plant breeding may be regarded as manipulation of chromosomes. This was done by the sorting and retention of similar chromosomes in the same plant to reach a homozygous state, a method termed pure line selection. Alternatively, different chromosomes can be combined to form a heterozygous state, a method known as hybridization conferring hybrid vigour or heterosis. The next step was the development of genetic variability through spontaneous or artificially induced mutations. Normal plants are diploids but when plants are developed with three or more sets of chromosomes, they become polyploids that tend to be bigger than diploids. Autopolyploid plants have genes similar to their diploid ancestors whereas allopolyploids are combinations of genomes of two different species that differ in characteristics.

The first achievement of hybridization techniques was the development of hybrid maize in 1919 which revolutionized American agriculture. The development of hybrid wheat and rice plants in 1960s filled the bread basket of developing countries, generally known as green revolution, for which Dr. Norman Borlaug was awarded the Noble Peace Prize in 1970.

The discovery that plant cells can develop into entire plants was another land mark in the development of new varieties of crop plants. The term tissue culture was coined to denote the *in vitro* development of plants in test tubes from calluses generated from plant parts. This led to mass production of uniform plants and revolutionized floriculture in the globe. Tissue culturing often leads to progenies which are variable. These progenies are known as somoclonal variations and have been exploited to generate mutations and it has been estimated that *in vitro* tissue cultures can produce ten times more somoclonal variations than that can be induced by chemical mutagens. Fusion of naked genetically compatible protoplasts (inter specific) resulted in successful regeneration of new varieties. Fusions between incompatible protoplasts (inter generic) resulted in abortive cell division and successes in regeneration was never achieved, excepting the instance of crossing between tomato and potatoes forming 'pomatoes' which can only be regarded as a laboratory success not amenable to commercial exploitation.

With the advent of biotechnology, agriculture has reached a science based industrial state. By using recombinant DNA technology, many transgenic life forms have been engineered since 1985. Transgenic plants belonging to both monocotyledonous plants such as maize, millet, wheat, rice and ragi and dicotyledonous plants such as alfalfa, clover, peas, soybean, mothbean, potato, tobacco, cotton, flax, sugarbeet and sunflower have been constructed. New varieties of vegetables and fruits such as cabbage, carrot, cauliflower, celery, cucumber, horseradish, lettuce, rape, grape, muskmelon and strawberry have been developed. The new varieties have incorporated genes capable of resisting one or more of the following: herbicides, insects, stress, frost or virus infections.

Plant biotechnology has opened up the possibility of producing artificial seeds, artificial sweetners (sugar substitutes) and bioplastics. Normal seeds have an embryo surrounded by cotyledons for initial sustenance during germination. By somatic hybridization, plant embryos can be mass multiplied in fermentation tanks and each embryo is then encapsulated in a jelly-like coat that can be called an artificial seed. Some estimates have revealed the possibility of production of 80,000 embryos per day but the cost could well be prohibitive for commercial exploitation. Presently, several companies are engaged in reducing the cost for atleast some crops such as carrots and celery.

The most important sugar substitute is the maize based high fructose corn syrup (HFCS) known as isoglucose in Europe. Some estimates put HFCS production worldwide to 6 million tonnes available in liquid as well as crystal form.

Aspartame is a synthetic chemical thousand times sweeter than sugar. With the advent of this product nearly 38 research institutes and companies around the world are engaged in producing novel chemical sweetners.

Thaumatococcus danielli or commonly known as Katemfe is a plant that grows in humid forests in Western and Central Africa. The berries of this plant contain the protein thaumatin that is 2500 times sweeter than sugar. Tate and lyle, a UK based sugar company had set up plantations of Katemfe in Ghana, Liberia and Malaysia. The frozen berries were processed to obtain purified thaumatin that was sold under the brand name 'Talin'. One drawback of thaumatin is its lingering taste limiting its use in food products. In spite of this, research is underway to understand the gene coding for thaumatin and its transfer to *E. coli* and other plants.

Stevia rebaudiana grows in Paraguay and several countries in South East Asia. The plant is capable of producing proteins several hundred times sweeter than sugar. The product is being marketed in Japan which has also bitter taste. African forests abound in sweet berries such as 'Miraculous berry' (3000 times sweeter than sugar) and Mexico has *Lippia dulcis*, thousand times sweeter than sugar. The search for cheaper substitute to cane sugar is being pursued vigorously and in future years we may have alternate sugar sources.

An interesting example of how a plant can be made to produce novel chemicals such as bioplastics is the transgenic *Arabidopsis* capable of producing granules of polyhydroxybutyrate (PHB), a polyester which is normally obtained from the bacterium *Acaligenes eutrophus*. In fact, PHB is a storage product in many bacteria intended to be used as a source of energy by bacterial cells in times of nutritional stress. This bioplastic material is a delicate product destroyed by pH above 8 and temperatures above 70°C. The product is mixed with polyhydroxyvalerate (PHV) to make it flexible and moulded into any shape, spun as fibre or rendered into a film. It is biodegradable to CO_2 and H_2O with no environmental hazard. The bioplastic is compatible with living tissue and hence can be adapted for medical purposes. It can also be used as a mulch in agriculture.

Transgenic animals and microorganisms are being used for fundamental research, for production of pharmaceuticals such as goats-lactoferrin and for production of biological control agents. Pigs and rabbits are genetically engineered to function as organ donors for human beings and chickens have been exploited for producing foreign proteins in eggs. "Pharm biotechnology" (agricultural production of pharmaceutical products) has to be distinguished from "Farm biotechnology" (productivity related agricultural applications) and very likely agriculturally produced pharmaceuticals will be marketed sooner than agricultural products, despite the fact that the latter could undoubtedly increase global food supply. This has been due to success in pharmaceutically oriented animal experiments as exemplified by the transgenic modification of pigs and sheep for expressing valuable pharmaceutical products in milk where all animals in the offspring appeared healthy contrasting with the transgenic pigs generated to produce leaner meat or more rapid growth whose offspring had adverse effects.

There are about 20 different man made pharmaceuticals involving crops that have been genetically changed to produce a range of prophylactics from cholera vaccines, herpes vaccine and cancer treatments. Potatoes seem to be ideal vehicles for the new generation of vaccines such as vaccine against *E. coli* disorders of the intestine. These are friendly and easier to tolerate than injections.

Phosphorus in seeds is a poor nutrient for monogastric animals such as chickens unless phytase is present to release phosphorus. Feeding chickens with seeds containing the phytase gene from *Asperigillus niger* brought about growth increases in chicken. This biotechnological innovation known as "gene farming" not only improved the quality of chicken feed but also minimised the excretion of phosphate in the environment. Another example is the case of sweet potato which is a staple food in China. The strategy here was to implant twin genes such as viral coat protein gene and *Bacillus thuringiensis* genes into sweet potato to ward off diseases caused by viruses as well as insect pathogens.

Nucleic Acids

It would be helpful to briefly describe some basics of molecular biology before attempting to understand its implications in biotechnology. The classical discovery of the structure of genetic material by Watson and Crick in 1953 revealed the unique suitability of nucleic acids for carrying genetic information and transmitting to subsequent generations. All the information needed for growth and multiplication of most organisms is carried by nucleic acids, especially the double-stranded deoxyribonucleic acid (DNA) or single or doube stranded ribonucleic acid (RNA). RNA differs from DNA in that the single strands have a ribose instead of deoxyribose and uracil in place of thymine. The double stranded DNA occurs in a helical structure with a backbone of alternating phosphate and deoxyribose molecules having a purine or a pyrimidine base linked to the 1-position of each sugar molecule. The two complementary strands of DNA are twined together by hydrogen bonds between the purine and pyrimidine base pairs: adenine-thymine (AT) and guanine and cytosine (CT). An adenine in one strand of DNA occurs directly across a thymine in the other strand. Similarly, a guanine (G) occurring in one strand is bonded to a cytosine (C) across the other strand (Fig. 96). The genetic information is coded by the linear arrangement of bases on the DNA strands. The sequence of nucleotides dictates all the characteristics of an organism and serves as a genetic code. The sequence of nulceotides, read in groups of three (triplet) reflects the sequence of amino acids in the large number of proteins (enzymes) synthesised by a cell. Each triplet is known as a codon and there are 64 possible combinations beginning with four nucleotides (Table 43).

Protein Synthesis

Mitosis (cell division) results in the formation of tissues. Before cell division takes place, the chromosomes in the parent cell have to be duplicated so as to be equally shared by the daughter cells. This replication is carried out within the nucleus of the cell by the action of an enzyme known as DNA polymerase. The synthesis of proteins takes place within the cytoplasm of the cell away from the nucleus. The genetic information in the DNA is conveyed to ribonucleic acid in ribosomes residing in the cytoplasm by a process known as transcription. During transcription, only one of the two strands of DNA becomes translated into RNA by RNA polymerase. The synthesis of RNA always proceeds in a fixed direction beginning at the 5′ end and concluding with 3′ ended nucleotide. The synthesized RNA from DNA of the nucleus in a cell moves into the ribosomes of the cytoplasm carrying with it information needed to synthesize protein in the ribosome. Hence, this RNA has come to be known as messenger RNA (mRNA). The 5′ end of an mRNA molecule attaches

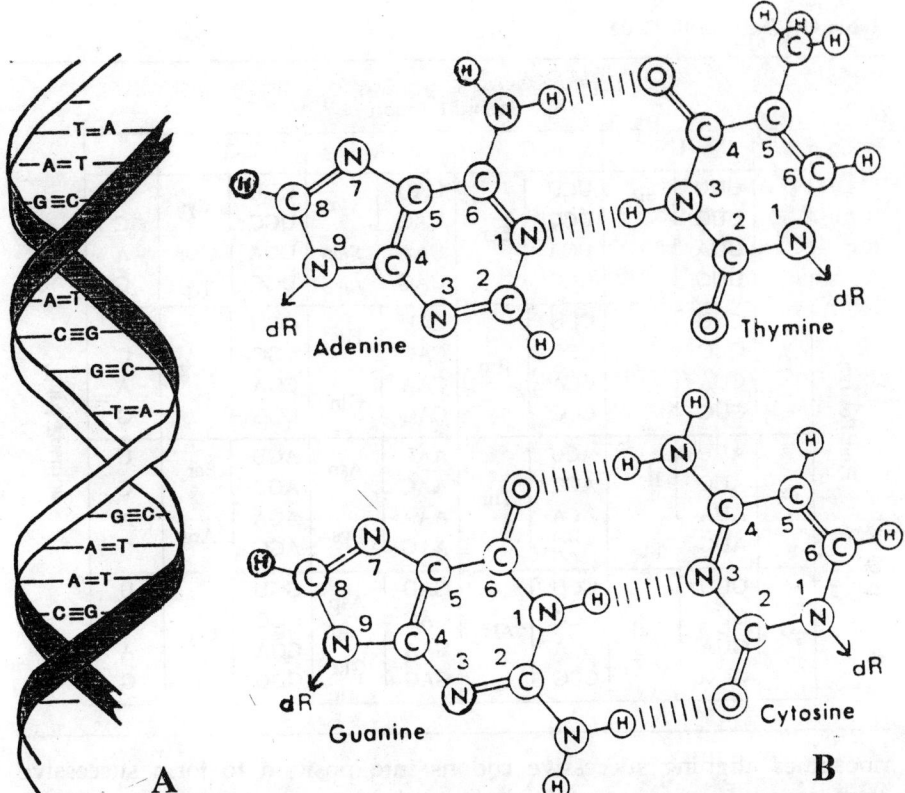

Fig. 98 DNA: A—double helix showing base pairing of nucleotides; B—pairing of adenine with thymine (2 hydrogen bonds) and guanine with cytosine (3 hydrogen bonds); dR means deoxyribose.

to a ribosome. About 4 per cent of total cellular RNA is mRNA. Since only a small segment of mRNA is attached at a given time to a ribosome which is moving across it, a single mRNA molecule can be read at the same time by several ribosomes, occurring as polyribosomes consisting of anywhere from 6 to 50 ribosome units.

Before the process of amino acid polymerization into proteins begins, the 20 different amino acids in a cell are first transformed into energy-rich precursors. These precursors get attached to a small transfer RNA (tRNA) molecule. A group of three nucleotides (triplet) constitute a tRNA and serves as an anticodon that uses base pairing to find three nearby nucleotides which in turn serves as a codon on a mRNA molecule. Specific enzymes known as aminoacyl synthetases now begin to bind or attach aminoacids to specific tRNA molecules (Fig. 99).

Ribosomes are thus minifactories for protein synthesis by a series of codon-anticodon interactions which occur on their surfaces. These interactions take place when mRNA molecules move across the active surface of

Table 43 The Genetic Code

Second Position

First Position		Second Position: U		C		A		G		Third Position
	U	UUU	Phe	UCU	Ser	UAU	Tyr	UGU	Cys	U
		UUC		UCC		UAC		UGC		C
		UUA	Leu	UCA		UAA	Stop	UGA	Stop	A
		UUG		UCG		UAG	Stop	UGG	Trp	G
	C	CUU	Leu	CCU	Pro	CAU	His	CGU	Arg	U
		CUC		CCC		CAC		CGC		C
		CUA		CCA		CAA	Gln	CGA		A
		CUG		CCG		CAG		CGG		G
	A	AUU	Ile	ACU	Thr	AAU	Asn	AGU	Ser	U
		AUC		ACC		AAC		AGC		C
		AUA		ACA		AAA	Lys	AGA	Arg	A
		AUG	Met	ACG		AAG		AGG		G
	G	GUU	Val	GCU	Ala	GAU	Asp	GGU	Gly	U
		GUC		GCC		GAC		GGC		C
		GUA		GCA		GAA	Glu	GGA		A
		GUG		GCG		GAG		GGG		G

ribosomes aligning successive codons into position to form successive aminoacids along polypeptide chains and this process is known as translation. There are 64 potential codons and of these 61 are used to specify amino acids while 3 are made use of to provide signals to terminate the formation of polypeptide chains. Several amino acids are determined by more than one codon. All the codons put together constitute the genetic code and this code is common to all forms of life including microorganisms (Table 43).

Southern and Northern Blot Techniques

Southern blot is the classic technique described by Southern in 1975 for understanding individual genes in a complex mixture of DNA. The procedure involves cutting up high molecular weight genomic DNA into fragments by enzymatic digestion with restriction endonucleases. These bacterial enzymes have been isolated and are available in many laboratories. They have been prepared from several bacteria and are designed as molecular scissors which cut the strands of DNA at specific oligonuleotide recognition sequences (Table 44). The fragments of DNA are then separated on the basis of size by agarose gel electrophoresis. The

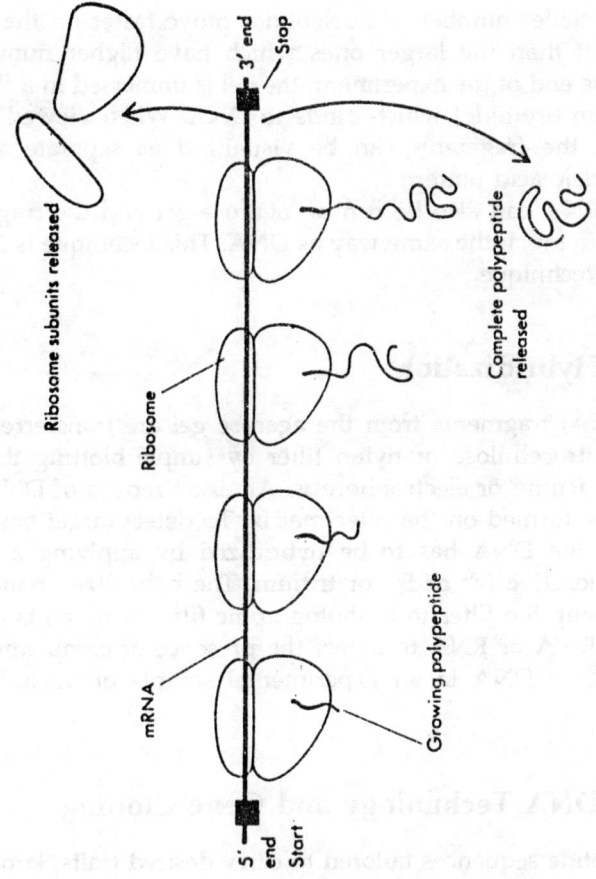

Fig. 99 The mechanism of protein synthesis.

fragments with smaller number of nucleotides move faster on the gel in an electric current than the larger ones which have higher number of nucleotides. At the end of the experiment, the gel is immersed in a fluorescent dye (ethidium bromide) which binds to DNA. When viewed under ultra-violet light, the fragments can be visualized as separate entities reflecting the nucleic acid pattern.

The cellular RNA can also be run on agarose gel and the fragments can be separated in much the same way as DNA. This technique is known as Northern blot technique.

Nucleic Acid Hybridization

The DNA (or RNA) fragments from the agarose gel are transferred to a solid matrix of nitrocellulose or nylon filter by simple blotting through capillary action, vacuum or electrophoresis. An exact replica of DNA pattern on gel is now formed on the filter matrix. To detect target bands by autoradiography, the DNA has to be hybridized by applying a probe labelled with radioactive P^{32} or S^{35} or tritium. The hybridized bands are detected by exposing the filter to a photographic film. A probe is a fragment of labelled DNA or RNA to detect the presence of complementary sequence of RNA or DNA in an experimental sample of nucleic acid. (Fig. 100).

Recombinant DNA Technology and Gene Cloning

Man made nucleotide sequences tailored to carry desired traits, known as molecular probes have been extensively used in molecular biology studies. These nucleotide squences (probes) ought to be correct, pure and available in large amounts. The probe must be labelled in some manner to help detection. These probes are prepared by recombinant DNA techniques using the common enteric bacterium *Escherichia coli*. This organism has been studied so often by molecular biologists that they are now aware of the implications of its genome containing about 4.2 million base pairs (bp). Furthermore, the bacterium has extrachromosomal DNA molecules known as plasmids that can replicate autonomously independent of the nuclear DNA. These plasmids are closed single pieces of supercoiled DNA which can be stably inherited by daughter cells. They have the ability to replicate to high numbers (copies) within each bacterial cell. The restriction endonucleases cleave DNA only to specific oligonucleotide sequences rather asymetrically leaving 'sticky' ends. The sticky ends remain complementary between any two DNA fragments sliced by the same restriction enzyme. This property makes it easy to insert or 'clone' an outsider or foreign gene to the *E. coli* plasmid provided it has been sliced by the same

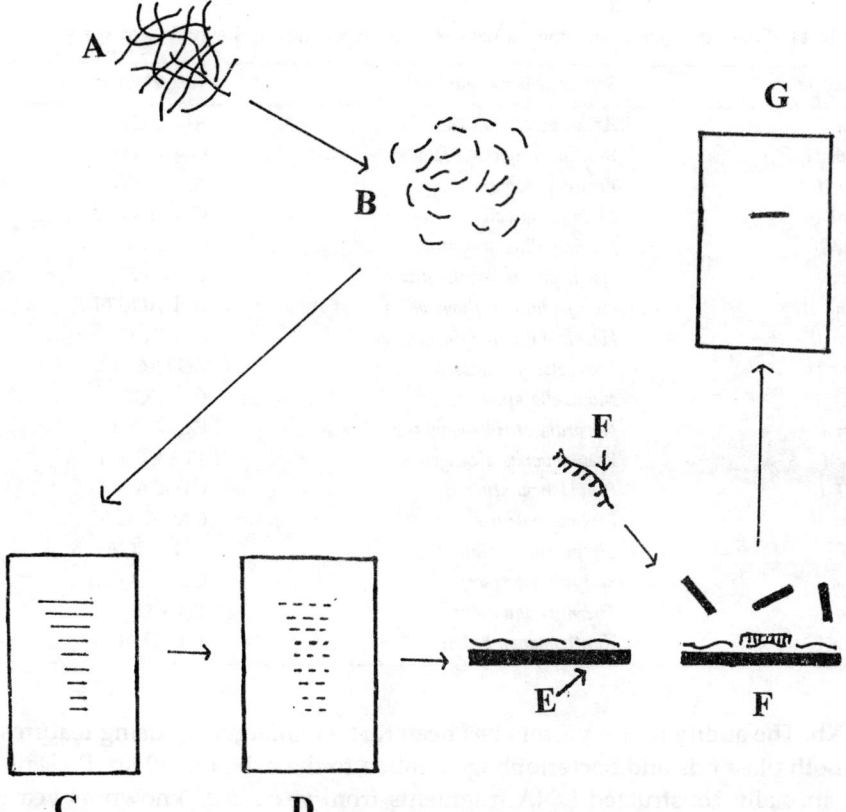

Fig. 100 Separation of DNA fragments by agarose gel electrophoresis and tracing fragments by a fluorescent dye or with a radioactive probe: A-genomic DNA; B—fragments after digestion with restriction enzyme; C—separation by gel electrophoresis and identification with a fluorescent dye such as ethidium bromide in a 'Southern' blotting technique (D); Alternatively, the sequence from the Southern blot is transferred and bound to a solid phase, usually nitrocellulose or nylon membrane (E) followed by the addition of a labelled (F) probe (p^{32} or S^{32} or tritium) with the solid phase (E). The probe hybridizes with the target sequence that is identified by autoradiography (G).

restriction enzyme. The transformed plasmids known as 'vectors' can be amplified to a high copy number in a standard bacterial culture. To retrieve the inserted foreign DNA, the bacterial biomass from the culture medium is separated and lysed. The DNA content is purified and sliced by the same restriction enzyme that was earlier used for original cloning.

Plasmid vectors have been used to transfer DNA from one prokaryotic cell to another, from a prokaryote to an eukaryote and from an eukaryote to a prokarotic cell. The plasmid vectors have the limitations of cloning upto 5000 base pairs (bp) or 5 kilobases (kb). By developing and using bacteriophage lambda chromosomes, foreign DNA have been cloned upto

Table 44 Some restriction enzyme sequences. The arrow indicates actual cleavage site

Enzyme	Source (Microorganism)	Target sequence
Alu I	Arthrobacter luteus	AG ↓ CT
Bam HI	Bacillus amyloliquefaciens	G ↓ GATCC
Bgl II	Bacillus globigii	A ↓ GATCT
EcoRI	Escherichia coli	G ↓ AATTC
HaeIII	Haemophilus aegyptius	GG ↓ CC
HhaI	Haemophilus haemolyticus	GCG ↓ C
Hind III	Haemophilus influenzae	A ↓ AGCTT
Hpa II	Haemophilus parainfluenzae	C ↓ CGG
Kpn I	Klebsiella pneumoniae	GGTAC ↓ C
Msp I	Moraxella sp.	C ↓ CGG
Not I	Nocardia otitidis-caviarum	GC ↓ GGCCGC
Pac I	Pseudomonas alcaligenes	TTAAT ↓ TAA
PsT I	Providencia stuartii	CTGCA ↓ G
Pvu II	Proteus vulgaris	CAG ↓ CTG
Sal I	Streptomyces albus	G ↓ TCGAC
Sma I	Serratia marcescens	CCC ↓ GGG
Taq I	Thermus aquaticus	T ↓ CGA
Xba I	Xanthomonas badrii	T ↓ CTAGA

15 kb. The ability to use vectors has been further enlarged by using features of both plasmids and bacteriophage lambda to the extent of 50 kb. Presently, specially constructed DNA fragments from yeast cells known as yeast artificial chromosomes (YAKs) are available that can be used to clone pieces of DNA upto 1 million bp.

One of the important steps in cloning a foreign gene is to obtain the desired gene in the absolute pure condition. This can be done by a traditional method beginning with the purified protein which the gene produces. Required antibodies are raised which will recognize and precipitate the protein when added to a cell extract from a tissue where the protein is actually synthesized. This results in the precipitation of newly made polypeptides which are being elongated on the polyribosomes. However, the precipitate contains unwanted ribosomes mixed with the mRNA templates required for producing the protein in question. When these mRNA templates are purified from the mixture, a sequence of nucleotides complementary to the gene of interest can be obtained. By using a retroviral ezyme reverse transcriptase, which synthesizes DNA from a RNA template, a cDNA sequence from the mRNA sequence can be obtained. By the addition of DNA polymerase the second strand of DNA can be replicated which results in the formation of a double stranded DNA copy of the gene of interest. This elaborate procedure is known as the polysome precipitation.

In recent years, polysome precipitation method has been replaced by 'DNA libraries'. A cDNA library is made up of all the actively transcribed genes of a tissue inserted into a population of bacterial cells. The bulk of mRNA preparation is reverse transcribed and inserted into plasmids in one lot, with the objective that every possible cDNA sequence will be carried by atleast one bacterial cell in the culture. This cDNA library has to be sorted out for a particular cDNA sequence of interest. The bacterial culture containing the cDNA is spread out on agar plates and filter blot technique is used with a labelled oligonucleotide probe (10–40 bp) to select a colony containing the foreign gene of interest. Presently, many biotechnological companies have developed automated instruments that can rapidly synthesize oligonucleotides of any sequence.

Gene Exchange in Bacteria

Gene exchange in bacteria takes place by processes known as transformation, transduction and conjugation. The addition of foreign DNA to actively growing bacterial culture results in chance entry of foreign DNA into cells by modification of bacterial cell envelope followed by the intake of DNA into the bacterial genome.

Transduction involves bacteriophages or bacterial viruses whose DNA enters the bacterial cell followed by the disintegration of the bacterial chromosome. The phage DNA multiplies in the cells, the cell walls undergo lysis releasing the phages which have multiplied in the mean time. The phages which mediate this type of transduction are known as lytic or virulent phages. On the contrary when a temperate phage (non-virulent type) enters the bacterium, the phage DNA becomes attached to the bacterial chromosome and remains integrated with the bacterial genome for many generations. This process is also known as lysogeny. At times, temperate phages may turn virulent leading to lysis and production of more bacteriophages. The temperate phage, when freed from the cell may carry with it small pieces of DNA which upon delivery to a next host cell can add an additional character to the new cell's capabilities (Fig. 101).

Contact between two cells is required for conjugation, one acting as a donor and the other a recipient. The donor possesses a fertility factor (F^+) and the recipient has no such factor (F^-) but must be viable for successful conjugation.

The A. obacterium Mediated Transfer of Genes

A naturally occurring conjugation phenomenon in *Agrobacterium tumefaciens* (A.t) induced crown-gall disease of plants has been ably ex-

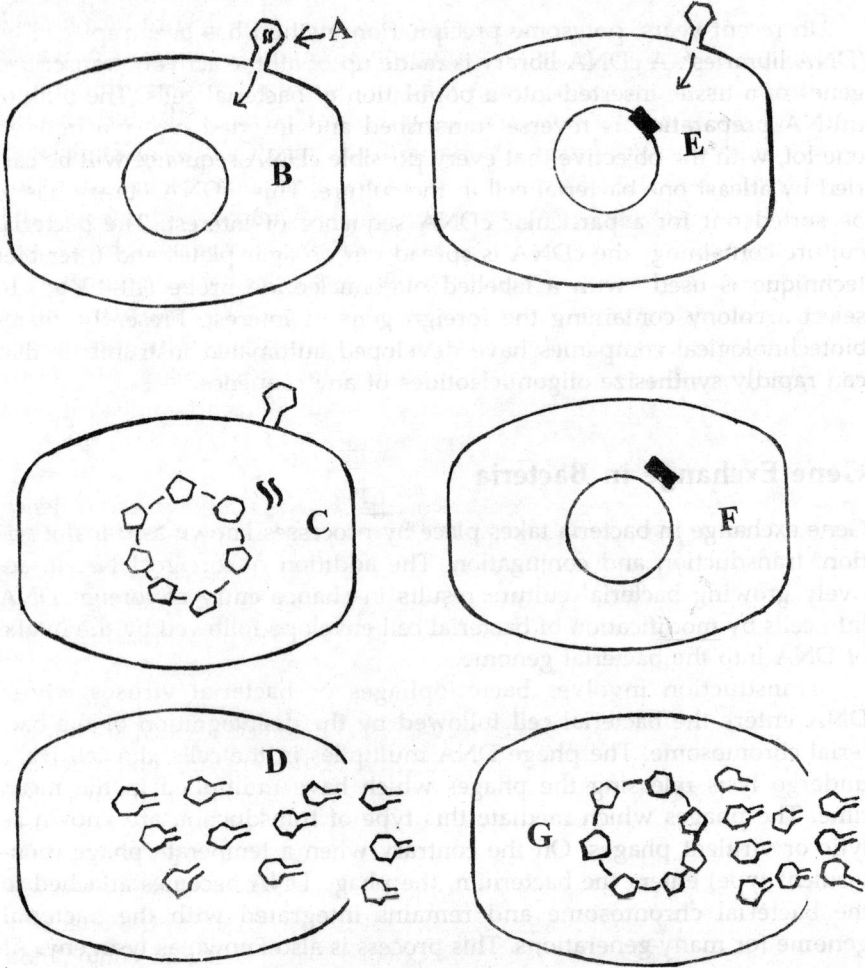

Fig. 101 Bacterial transduction with bacteriophages: A virulent phage (A) attaches to a receptor site on a bacterial cell (B). The phage DNA is injected into the bacterial cell followed by disruption of bacterial DNA and the phage takes control of bacterial cell's functions (C). New bacteriophages are formed and released by cell lysis (D). In a phenomenon called (lysogeny) infection with a temperate phage results in the integration of phage DNA with the bacterial chromosome (E) which is maintained as such for several generations. Occasionally phage DNA gets detached from the bacterial chromosome (F), disrupts the bacterial DNA and takes control of the bacterial cell behaving like a virulent phage causing lysis of the cell with the release of mature bacteriophages (G).

ploited by biotechnologists to genetically engineer foreign genes into dicotyledonous plants rendering them transgenic in characters such as resistance to viral diseases, herbicides or bioinsecticides. A.t is a soil bacterium which infects the crown region of dicotyledonous plants (monocots are resistant) through mechanical wounds to produce crown galls. A.t harbours large extrachromosomal genetic elements known as megaplasmids. Most of the genes required for tumour formation are located on one such 180 kb megaplasmid designated as Ti plasmid, the letters Ti denoting the tumour inducing ability of the plasmid. The Ti plasmid contains a tumour inducing region (T-DNA) which also carries genes for the synthesis of two growth hormones, the IAA and cytokinins and the genes controlling the synthesis of a group of amino acid derivatives known as opines (nopaline and octopine). The expression of bacterial genes controlling growth hormone production is not controlled by the plant but is necessary for the development of crown gall symptoms. The opines synthesized serve as nitrogen and carbon sources for the bacterium. However, the genes controlling the growth hormone production and opine synthesis are not essential for transferring or integration of T-DNA into the host plant cell genome.

The Ti plasmid has also a cluster of about 8 genes known as virulence genes (*vir* genes). This gene cluster of 35 Kb DNA is necessary for the recognition of susceptible cells on the plant surface, to excise T-DNA from the plasmid and transfer the T-DNA region to the host cell. The *Vir* genes are activated only by contact with cell metabolites released by the wounded plant and they are not functional or expressed in pure A.t cultures grown on synthetic media.

There are two border regions to T-DNA (LB and RB) which are known to contain genes involved in the secretion of the enzyme endonuclease that scissors off the T-strand from the Ti plasmid. Thus the genes encoded within the T-strand have all the appropriate signals for efficient transcription and translation in their eukaryotic host.

The process of infection begins with the bacterial surface components of A.t. recognizing the plant surface which is susceptible to the attack by the pathogen, followed by a process analogous to bacterial conjugation whereby a single strand of T-DNA (the T-strand) is transferred to the plant cell probably through pores in the cell wall. Within the host cell, several copies of T-DNA are inserted at single or multiple sites in the host chromosomes and function as typical eukaryotic chromatin.

The earlier procedure to identify transformed cells was to develop plant galls or tumours from such transformed cells. Alternatively, the transformed cells were grown in hormone independent cultures. However, in later experiments, modified T-DNA and Ti plasmids have been used to facilitate rapid experimentation and development of transformed plants. In the modified T-DNA, the genes coding for phytohormones and opines have been excised because, as stated earlier, these genes are not concerned with the transfer and integration of T-strand into the host genome but are essential for the manifestation of crown-gall symptoms. In essence, deleting these genes from T-DNA results in the elimination (disarming) of the

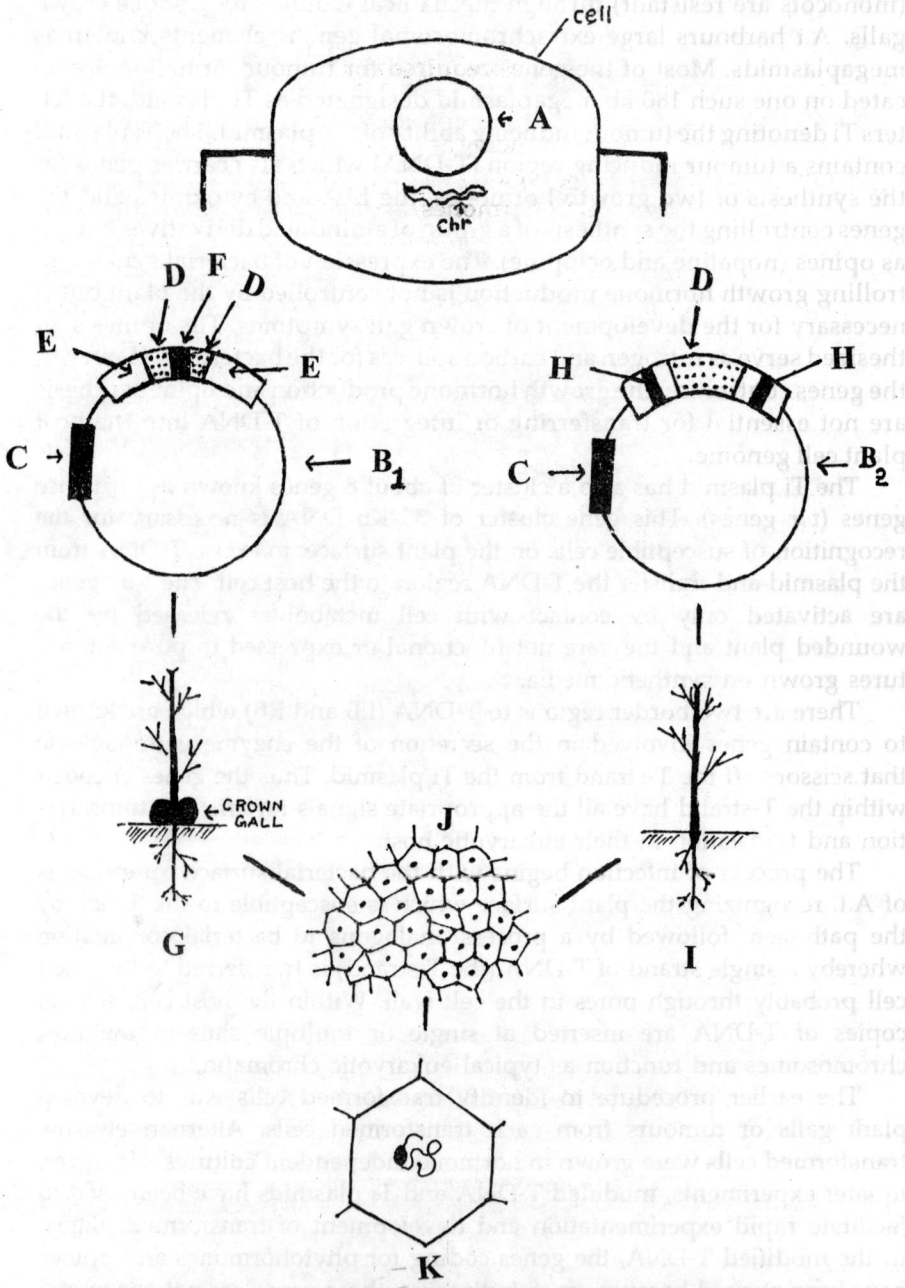

Fig. 102 Diagrammatic representation of *Agrobacterium tumefaciens* (A.t.) mediated cloning of foreign genes into dicotyledonous plants to develop transgenic plants:

(Contd.)

oncogenic (tumourous) phenotype. Foreign DNA can be inserted within the right and left border regions of an appropriate plasmid, thereby enabling the delivery of large multigenic segments of DNA into plants. In this way T-DNA region has been engineered (disarmed) to eliminate the tumour inducing genes (Fig. 102). Selection of transformed cells has been facilitated by engineering a variety of antibiotic markers selectable in plants such as resistance to kanamycin and gentamycin.

Ti plasmids have also been engineered in other *Agrobacterium* spp. and *E. coli* to facilitate efficient plasmid replication and selection. Such manipulated plasmid vectors could be used directly in transformation steps obviating the need to repeatedly infect plants with *A. tumefaciens* cells.

Procedures for *Agrobacterium* Mediated Transformations

Several procedures have been followed till to date: (a) through conventional gall formation on wounded plant stems or leaves that are not amenable to monitoring with disarmed vectors, (b) Co-cultivation of *Agrobacterium* with plant protoplasts, a method that has been successful in plants where regeneration of plants from protoplasts has proved to be successful and (c) tissue transformation with ex-plants (leaf, cotyledon section or somatic embryos) inoculated with *Agrobacterium*, a procedure that has proved faster than protoplast regeneration. When shoots are formed in regenerated plants, the plantlets are transferred to potted soil for acclimatization to natural surroundings.

The introduction of *nif* genes controlling nitrogen fixation in some microorganisms, into higher plants other than legumes is an experimental strategy envisaged by several research groups.

Fig. 102 (*Contd.*)

—Normal non-transformed megaplasmid in a A.t.. cell with bacterial chromosome (chr) underneath; B1—on the left, a megaplasmid enlarged to depict the Vir region (C), the T-DNA (D) and the border regions (E); the T-DNA has genes coding for oncogenic phytohormones and opines synthesis (F). Upon infection, the plant develops crown gall tumour as shown in (G). B2—on the right, a disarmed megaplasmid having the T-DNA region (D) minus genes coding for oncogenic phytohormones and opines synthesis; the plasmid has however the border regions incorporating the required foreign genes (H) with the Vir genes (C) intact. The plant shown as (I) carries only the foreign genes without oncogenes and hence no tumours are present. J—A transformed tissue of either of the two types of plants showing host plant cells with nuclei. K—an enlarged transformed host cell with the chromosome carrying the T-DNA region shown in the diagram as black dot.

Direct DNA-transfer Technologies

Most monocotyledons like cereals and sugarcane are not amenable to *Agrobacterium* technology because they are resistant to infection by the bacterium. Alternative technologies are being developed which include the introduction of foreign DNA into plant protoplasts mediated by polyethylene glycol and poly-L-ornithine, electroporation (use of electric current), calcium phosphate coprecipitation, liposome fusion, microinjection and particle bombardment. Regeneration of plants from transformed protoplasts is labour intensive and subject to the possibility of development of somoclonal variations.

The use of 'microprojectiles' or 'particle bombardment' is a procedure for delivering foreign DNA into plant cells. The technique involves coating small gold or tungsten beads with plasmid DNA and propelling the beads to intact cells, embryos or differentiated tissues using high velocity particle 'guns' or electrical discharges. This technique has been useful in delivering DNA into the nucleus and mitochondria of yeast, nucleus and chloroplasts of *chlamydomonas*, epidermal tissues of *Allium cepa* and suspension cultures of maize. A major limitation of this technology arises from the paucity in the number of particle guns available and the high cost in building them.

Antisense RNA Strategy

The basic idea in antisense strategy is to block the expression of a particular gene product with the help of a transgenic construct containing the gene or part of the gene with the transcript in reverse orientation with respect to the promoter. This procedure results in the formation of a complementary RNA rather than the normal sense RNA. This antisense RNA binds with its homologous sense RNA, preventing translation and/or facilitating degradation or accumulation of a gene product by 90 to 99 per cent, thus creating a phenotypic mutant. There have been many reports of insertion of antisense construct of polygalacturonase enzyme into tomato plants. The resulting transgenic tomato plants had low levels of this enzyme and hence their fruits ripened slowly thereby increasing shelf life. Antisense RNA technology has also been used with success towards minimizing viral diseases.

Frost Control Biotechnology

Frost injury to plants is caused at temperatures less than 0°C. Two kinds of frost injuries have been recognized: those occurring above minus 5°C and those occurring when temperature drops below–5°C. Plants resist frost by restricting freezing to intercellular spaces and adjusting the water potential intracellularly to reach equilibrium with the ice formed in inter-cellular spaces. Frost injury takes place when this equilibrium is upset and the rate of intracellular ice formation exceeds that of intercellular spaces followed by death due to disruption of cell membrane properties. Frost tolerant plants, however, have endogenous ice-tolerance mechanisms.

How does frost sensitive plants overcome frost injury? There appears to be a supercooling mechanism in such plants to avoid ice formation. Ice nucleation or initiation of ice embryos is caused by the orientation of water molecules by organic and inorganic substances. Several plant-associated bacteria are highly active in ice nucleation and may be responsible for frost injury and the biotechnological implications of this activity have been studied with reference to frost injury at temperatures above minus 5°C. Several strains of bacteria such as *Xanthomonas campestris*, *Pseudomonas viridiflava*, *P. fluorescens* and *Erwinia herbicola* inhabit the epidermal crevices and hairs of leaf surfaces and are active in ice nucleation at temperatures above minus 5°C. It should be noted that the efficiency of ice nucleation activity of these bacteria differs not only with the strains of bacteria but also with the plants harbouring them. The population density of ice nucleating bacteria is variable on plant surfaces depending upon the species of plants and the environmental conditions under which they grow. These epiphytic bacteria are known to occur on frost resistant as well as frost susceptible plants and are known to be killed by disinfectants and U.V. light.

The genes conferring ice nucleation characteristic have been partially characterized from *P. syringae*, *P. fluorescens* and *E. herbicola* and are known to be a single contiguous region of approximately 4000 bp. They have been cloned in *Escherichia coli*.

Streptomycin and oxytetracycline as well as copper hydroxide applications to leaf surfaces reduced the incidence of frost injury, despite the fact that dead cells were also known to be active as ice nucleating agents. Likewise, non-ice nucleation-active bacteria can also reduce the population of ice-nucleating ones both in the green house and the field by limiting nutrients to the ice nucleating bacteria. For example, non-ice nucleation-active (Ice⁻) mutants of *P. syringae* reduced the population size of ice nucleation active parental strains of *P. Sryringae* that were co-inoculated on pretreated plants.

Field trials have been conducted to understand the competition behaviour of ice⁻ strains of *P. syringae* by inoculating these strains to potato plants. Ice⁻ strains dominated the leaf surfaces for the first 4–6 weeks after inoculation. The population of Ice⁺ strains on plants colonized by Ice⁻ *P. syringae* strains was significantly decreased in comparison with uninoculated plants. The incidence of frost injury to potato plants inoculated with Ice⁻ strains was significantly lower than uninoculated control plants in natural field frosts in a field experiment in California.

The use of microorganisms for competitive control of frost injury have been found to be effective only when applied to young vegetative plants in the field because such young vegetation may not have been extensively colonized by other epiphytic microflora. Minimal occurrence of extraneous microflora on leaf surfaces which can be achieved by the application of bactericides such as cupric hydroxide can result in micro-habitats most conducive for the functioning of Ice⁻ *P. syringae* in large numbers on the leaf surface so that the Ice⁻ strains can effectively compete for nutrition with frost inducing naturally occurring wild strains of *P. syringae*. Integrated chemical and biological control measures to contain frost injury to plants appear to be a desirable approach. In addition, co-application of copper resistant Ice⁻ strains of *P. syringae* and cupric hydroxide has also been considered as an attractive proposition to control frost injury.

Virus Resistance in Transgenic Plants

There have been many reports about the development of transgenic plants which have become resistant to virus infections. These plants showed resistance to virus infections when they were transformed with sequences related to several gene functions. Some of the areas where successes have been seen related to sequences concerning viral capsid protein, viral movement protein, antisense RNA, antibody-mediated resistance, interferon-related genes and host genes involved in plant protection.

Viruses are biochemical complexes consisting of a RNA or a DNA genome packaged into a protein capsid which may or may not be surrounded by a membrane envelope. The protein coat covered genome is referred to as the nucleocapsid. The proteins on the surface of the capsid and envelope determine the interaction of the virus with the host and elicit the protective immune response against the virus (Fig. 103). Some virus particles also contain enzymes required to facilitate the replication of the virus. The tobacco mosaic virus (TMV) is an example of a virus with helical symmetry whose capsomeres (many protein subunits of the capsid) appear as projections that are assembled on the RNA genome into rods extending to the length of the genome. In other viruses the capsomere arrangement is cubical or icosahedral enclosing its nucleic acid component. Engineering resistance in plants involves either countering the capsid properties or disrupting the virus replicating mechanisms in the host.

Transformation of Sequences Related to Viral Capsid Protein

Coat mediated resistance is the expression of a gene that causes the transformed cell and regenerated transgenic plants to produce the coat protein (CP) of a virus thereby conferring resistance to infection in the transgenic plant. Plants that accumulate large amounts of coat protein escape/mini-

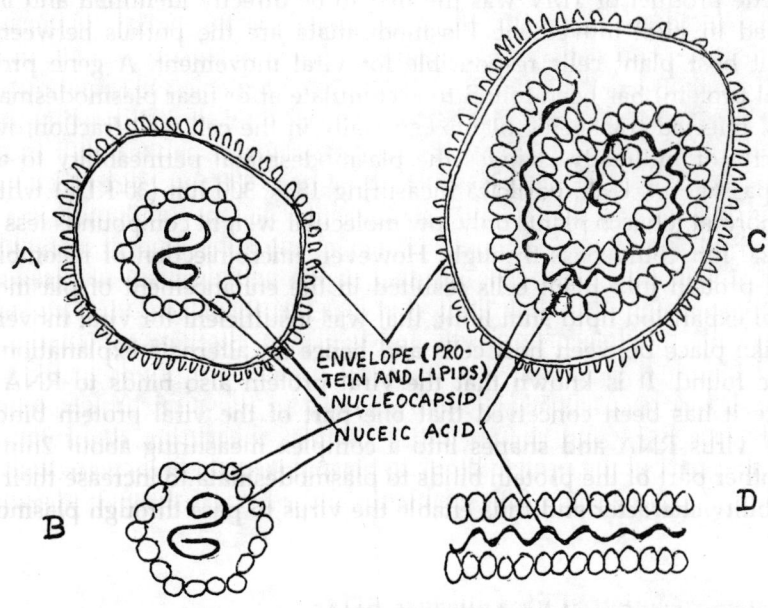

ENVELOPE (PRO-
TEIN AND LIPIDS)
NUCLEOCAPSID
NUCEIC ACID

Fig. 103 Elementary features of typical virus particles: A—Enveloped icosahedral virus; B—Naked icosahedral virus; C—Enveloped helical virus and D—Naked helical virus.

mize virus infection. The CP mediated resistance was first reported in tobacco mosaic virus (TMV) in 1986. Subsequently several instances of resistances for several virus diseases of plants have been reported which include tomato mosaic virus, alfalfa mosaic virus, cucumber mosaic virus, tobacco streak virus, potato virus x and y, tobacco etch virus, tobacco rattle virus, potato leafroll virus, potato virus S. A point which needs to be stressed is that CP mediated resistance does not provide immunity to transgenic plants; for instance TMV resistant plants that expressed CP gene were susceptible to inoculum levels of 10mg/ml while plants that were not transgenic got infected even at inoculum levels of 0.001–0.01 mg/ml. Despite the fact that the mechanism involved in CP mediated resistance has not been clearly understood, some kind of inferences can be made from studies on TMV diseases in tomato and tobacco. Firstly, the transgenic plants are protected due to reduction in the number of infection sites that means fewer number of plants may get infected than controls. Secondly, the transgenic plants are less likely to develop systemic infection and thirdly, such plants may produce less virus particles than the ones not having the resistant gene.

Transformation of Sequences Related to Viral Movement Protein

A gene product of TMV was the first to be directly identified and investigated in viral movement. Plasmodesmata are the portals between adjacent host plant cells responsible for viral movement. A gene product (viral protein) has been found to accumulate at or near plasmodesmata of TMV infected transgenic plants especially in the cell wall fraction, which directly or indirectly changes the plasmodesmatal permeability to allow the passage of TMV particles measuring 18 × 300 nm (30-KDa) while in the normal tobacco plants only low molecular weight compounds less than 1.5–2.0 nm could pass through. However, microinjection of recombinant viral protein into plant cells resulted in the enhancement of plasmodesmatal expansion upto 9nm limit, that was insufficient for viral movement to take place between host cells and hence an alternate explanation had to be found. It is known that the viral protein also binds to RNA and hence it has been conceived that one part of the viral protein binds to TMV virus RNA and shapes into a complex measuring about 2nm and the other part of the protein binds to plasmodesmata to increase their permeability character and thus enable the virus to pass through plasmodesmata.

Resistance Conferred by Antisense RNA

Antisense regulation of gene expression is a natural phenomenon concerning the specific expression of a nucleotide strand which is negative to a certain gene transcript capable of intervening the expression of that gene at different levels. Antisense RNA may bind to the loop initiated by RNA polymerase in the nucleus and interfere in the initiation of transcription. It may function in the cytoplasm by hybridizing with mRNA leading to translation arrest by preventing the binding of ribosomes to mRNA. Antisense regulation of gene expression has been successfully exploited in plants. Some examples are inhibition of flower pigmentation by antisesnse for chalcone synthase, intervention in the expression of ribulose biphosphate carboxylase in tobacco and polygalacturonase in tomato.

Most plant viruses replicate in the cytoplasm and do not go through any nuclear phase and therefore antisense RNA may either intervene in translation or promote mRNA degradation. Antisense mediated resistance has been achieved in transgenic plants against several plant viruses including TMV. This mode of resistance appears to be milder than coat-mediated protection.

Antibody-mediated Resistance

Antibody-mediated protection (akin to hybridoma-derived monoclonal antibodies) from only two plant viruses (TMV and artichoke mottle crinckle virus) are known. The mode of action of antibodies appears to be through neutralization of viral surface proteins in a way that intervenes the initial establishment of infection.

Resistance Due to Interferon Related Genes

Low concentrations of human interferon is transiently antivirally active in plants and several groups have constructed transgenic plants expressing human interferons but the protection offered to viral infections have been rather mixed in the sense that some plants received protection while others did not. Antiviral factors (AVF) have been detected subsequently in virus-infected plants and purification of the factor revealed that a major fraction of AVF was a 22 kDa glycoprotein. AVF gets particularly stimulated in host plants with viral infections which are localized.

Manipulation of Host Genes for Plant Protection

Several stable dominant resistant genes have been introduced into commercial crops from native gene pool in conventional plant breeding. Some examples are the TMV-resistant Tm family genes introduced into cultivated tomato from the native species *Lycopersicum peruvianum* and the N genes of *Nicotiana* spp. conferring resistance to TMV virus. The latter gene has been transferred from *N. glutinosa* to several tobacco cultivars thereby creating transgenic viral resistant tobacco plants. Similarly, the presence of a ribosomal binding antiviral protein (PAP) in *Phytolocca* spp. has been exploited to create transgenic tobacco (*N. benthamiana*) and potato plants and such protection in transgenic plants against some viruses were discernible in plants having PAP less than 5ng/mg. Many of these genes conferring resistance appear to encourage necrosis of tissues at confined loci and offer protection against the spread of the virus while others block viral replication.

Risks and Benefits

The risks involved in the capsid coat protein strategy lies in transpeptidation when a second virus infects an already infected plant. If and when both the viruses mature with both coat proteins, virus particles may emerge which have an outer coat made up of a mixture of both coat proteins conferring the resultant transcapsidated virus, the ability to infect a different range of plants because such mixed infections are commonly known to occur in nature. Other risks cited in literature are the possibility of recombination through switching off of templates, the possibility of heterologous recombination, the possibility of synergistic action of mixed infections and the part played by helper-dependent (RNA) complexes in aphid transmission, all of which have been known to take place in naturally occurring infections.

On the other hand, benefits may accrue in gene combinations. The biotech company Asgrow seeds has genetically engineered a virus resistant squash which reduces the planted acreage by half, yielding as much as it was doing earlier by planting the conventional variety susceptible to virus infections in full acreage using double the quantities of pesticides and herbicides. Papaya (*Carica papaya*) is infected by the ubiquitous ring spot virus, a RNA virus that is transmitted by aphids to the host. Field trials in Hawaii with transgenic papaya plants carrying ring spot virus coat proteins have demonstrated full protection to plants from ring spot virus.

Tolerance to Herbicides

A factor that limits the use of herbicides is the fact that many herbicides not only kill the weeds but also harm the main crop they are intended to protect. Secondly, the residues of herbicides in soil may damage the growth of subsequent crops grown on the same field. Herein lies the objective of developing transgenic plants capable of resisting herbicides applied against target weeds. Many transgenic plants exhibitng herbicide tolerance have been sold in the market and over 45 companies around the world are engaged in developing plants capable of resisting herbicides such as triazines, atrazine, paraquat, sulfonylureas, glyphosate, dalapon and likewise, over 48 research institutes around the world have research projects devoted towards herbicide tolerant plants. In the USA., of the 27.8 million hactares planted with transgenic plants 71.0 per cent is covered with plants (soybean and maize) resistant to "Roundup" herbicide, which in fact is the trade name for glyphosate a widely used herbicide. These transgenic plants carry the gene that neutralizes the herbicide.

Two approaches used in developing resistance to herbicides are (1) genetic engineering so that the plant target enzyme is made less sensitive to the herbicide and (2) metabolic inactivation by which the herbicide molecule is degraded or converted to an inactive form.

Glyphosate is a non-selective herbicide. The enzyme in plants which binds to glyophosphate is 5-enolpyruvyl 3-phosphoshikimate phosphate (EPSP) synthase. Cloning genes into plants which code for a modified EPSP synthase enzyme that has reduced affinity for binding glyphosate has been the strategy for developing transgenic soybeans, cotton and oil seed rape. Similar strategy has been followed for tolerance to sulfonyl ureas, another class of herbicides. In this case, genes that are tolerant to acetolactate synthase (ALS) have been isolated from tobacco and evaluated. The above two examples serve to illustrate the target enzymes that are being made less sensitive to the herbicide.

In another class of herbicides, the bromoxynils, the metabolic inactivation or degradation approach has been used. The enzyme nitrilase is encoded by a gene from the common soil bacterium *Klebsiella* and this enzyme has the capability to degrade the herbicide. Tolerance to this herbicide in cotton is being investigated by cloning the gene responsible for nitrilase production from *Klebsiella* into cotton. A second example is the development of resistance to the herbicide, glufosinate by cloning the *bar* gene for phosphinothricin acetyl transferase from a soil actinomytete into plants.

Gene Protection Technology

This relates to a new biotechnological concept towards controlling plant gene expression, developed and patented by USDA jointly with a private cotton seed company known as Delta and Pine Land Company. If this technology becomes a reality the expression of a particular gene can be blocked at a crucial stage in seed development by programming the plant's DNA. Initially, this technology has been envisaged in plants such as cotton and tobacco by altering genes in the seed with the object of blocking seed germination by killing the embryo. This technology named as 'Terminator Technology' or TT by opponents of biotechnology has been designed to produce seed that go sterile after one sowing. Although the technical aspects of TT are far from being clear, some facts have emerged in broad principles which can be stated as follows:

It is now known that each gene casette has its own promoter close to it to regulate the expression of the gene. This promoter can be manipulated by regulatory proteins to switch on or off the gene under its control, often acting also as a signal for a chemical like an antibiotic to prime the switch. One example of such regulatory genes is the repressor gene. If a repressor binds to a region near the promotor of a gene, it can stop the gene from being expressed into its protein.

With this background, the TT can be understood as a mechanism designed to engineer three casettes of genes into a plant such as cotton. Each casette has its own promoter. One gene produces ribosomal inhibitor protein (RIP), a toxin, which in turn is under the control of a promoter known as late embryogenesis abundant (LEA) that has been so programmed to switch on only during seed development. Between LEA and RIP is a buffer piece of DNA to separate the two genes so that the toxin gene RIP is switched off from action. To switch on the RIP gene, a recombinase gene is introduced that codes for a protein which knocks off the buffer piece of DNA with the result the RIP and LEA come closer for action. This action can be delayed by introducing a third gene that produces a repressor protein which has the ability to block the function of the recombinase gene. This blocking can be undone by spraying tetracycline, an antibiotic; otherwise, the seed is bound to have a live embryo and hence can germinate. On the otherhand if the antibiotic is sprayed before the seed reaches the farmer, the repressor gets set in the off position and hence no repressor protein is produced. The result is that the recombinase gene cuts of the buffer piece of DNA followed by the activation of the toxin RIP gene. The seed produced now has all the protein and other desired qualities of a transgenic seed but does not germinate because the embryo is dead. A simplified scheme to represent TT is shown in Fig. 104.

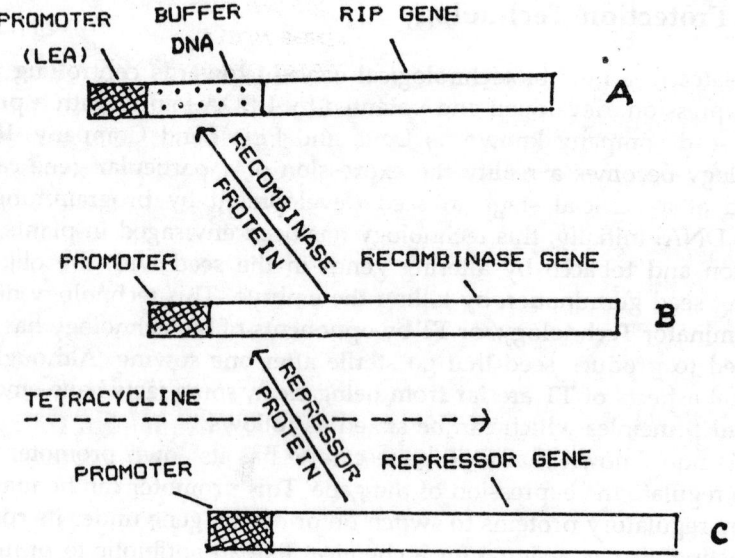

Fig. 104 A simplified scheme to depict terminator technology; A, B, C denote three gene casettes.

The TT has been designed to protect the intellectual property rights of the seed company that produces transgenic plants which have built-in characteristics such as resistance to herbicides, insect predators, viral diseases and so on. The TT bred seed is good for one generation and the seed produced from that generation is destined to be sterile. The cultivator no doubt gets a super produce but has to buy the seed again for resowing. This would mean that he will be in no position to use his own seed from the previous harvest and thus the seed company is assured of its business year after year.

Attempts are being made to protest against the concept of control of gene expression that led to the TT innovation. Monsanto seed company, it appears, has acquired the TT. The import of TT bred seed to India has been banned for fear of the seepage of the self destruction trait into our staple crops such as rice, wheat and so on leading to large scale sterility. However, Monsanto has denied the possession of any seed material generated by TT. Nevertheless, environmentalists as well as scientists have expressed the need to keep vigil to avoid large scale migration of the self destructing gene into other crops.

Ecological Considerations in the Release of Transgenic Plants

Biotechnology research which began in the seventees of this century has now begun to yield fruitful results in industrial pharmaceutical, agricultural and veterinary fields. Currently, there are about 1500 companies operating in the U.S.A alone. The first pharmaceutical product to come out was insulin and the first agricultural biotechnology product was a swine vaccine. Since that time over 2000 transgenic plants and microorganisms have been released in the world depending upon how one defines the word 'release' in each country and the rules pertaining to releasing a product. The commercial use of engineered plants or microorganisms is the largest in China, especially tobacco and tomato. In the U.S.A, as mentioned earlier, a transgenic *Pseudomonas* to which a toxic gene from *Bacillus thuringiensis* has been added is being commercially distributed. Transgenic plant products such as tomatoes and cotton have been approved and tomatoes have been marketed since 1994.

From the current information available with the USDA, it is gathered that the following transgenic crops, with the altred phenotypes indicated in parenthesis, have the approval of USDA for unristricted planting: beet (Glufosinate and glyphosate tolerant); corn (lepidopteran resistant and glyphosate tolerant); cotton (lepidopteran and herbicide tolerant); flax (Sulfonylurea tolerant); melon (delayed ripening); papaya (PRSV resistant); potato (insect resistant); rape (oil profile altered, male sterile and herbicide tolerant); rice (Glufosinate tolerant); soybean (herbicide tolerant, oil profile altered); squash (virus resistant) and tomato (fruit ripening altered and lepidopteran resistant).

The definitions of a microorganism or a recombinant microorganism are not clear. Description and scope covered by the term microorganism rather than its definition per se appears to be more important. The European union has recently decided to use the term 'Biological material' in place of microorganism. Similarly, to decide when a sufficient modification has occurred from the original microorganism to call it a genetically engineered organism (GEO) has also been a debatable issue. Secondly, deciding the appropriate stage when the product could be released is dependent upon field trials for the particular product in one location or many locations (presently size of plots ranging from 0.25 to 10.0 acres) and the attendant assessment of the overall risks involved for the society. The first successful field test of an engineered organism was a bacterium capable of resisting frost injury carried out in 1986. Since that time over 2000 field tests of plants and microorganisms have been conducted in the U.S.A by academicians as well as company managers. Many developed countries have designed protocols for field testing, reviewing and legislation to deal with GEOs. UNIDO has drafted a voluntary code for underdeveloped countries where no such regulations exist.

The United States Department of Agriculture (USDA) Animal and Plant Health Inspection Service (APHIS) regulates the introduction of genetically modified organisms which are or may be derived from plant pests or which are used in veterinary products such as vaccines. APHIS has established a division called Biotechnology, Biologics and Environmental Protection (BBEP) which overseas the regulatory process and provides a circle of protection to safeguard American agriculture. APHIS was the first federal agency to promulgate a set of codified rules governing the introduction of genetically engineered plants and microorganisms.

The reactions of public or the common man to the release of GEO or a transgenic plant are vital; for example, in the U.S.A, India, Germany and Holland transgenic plants have been uprooted and demonstrations held against the introduction of foreign genes into naturally growing plants. Such reactions are mostly based on the lack of knowledge of potential risks involved in planting transgenic plants fuelled by such disasters as Chernobyl nuclear spill over. On the contrary, it was found in a public survey that over 60 per cent of people in the U.S.A. believe that genetically engineered products would usher in the desired change and that hardly 19 per cent of people had really the knowledge needed to perceive the potential risks involved. A recent survey in the U.S.A. revealed that 94 per cent of people felt that the benefits were worth the risks involved.

Will genes from transgenic plants get transferred to other related or unrelated crops or weeds? Will pollinations result in viable progeny? Will the resulting progeny persist, be more aggressive, invasive and persistant than the parent crop? Is it likely that the DNA in transgenic plants transform other forms of life? Is it possible that while insects pollinate transgenic plants (via pollen) or while earthworms graze plant debris or vertebrates consumes them, the gut microflora of these living beings get modified? What happens to DNA from the remains of transgenic plants bound to soil particles?.

The answers to the above questions have not come in a large measure eventhough some indications are available (Table 45) to demonstrate that non-target effects may occur unintentionally. However, these results have come from small scale studies but answers to many other ecological issues arising out of large scale introduction of transgenic plants must come from experience as we learn increasingly from future field trials which are now currently es-

Table 45 Some examples of reported nontarget effects of genetically engineered plants and bacteria (from Watrud and Seidler, 1998 and Seidler *et al.*, 1998)

Plant or bacteria engineered trait	Function normally intended	Unintentional effects
Cotton/*B.t.* Kurstaki	Controls lepidopteran pest	Temporary increase in soil bacteria and fungal numbers and also alterations in species composition of bacteria.
Tobacco-protease inhibitor	Insecticidal tobacco model	Decreased the population of Collembola but increased the population of nematodes.
Tobacco-pathogenesis related proteins.	Disease resistant tobacco model	Decrease in the level of AM infection.
Klebsiella planticola	Model biomass conversion agent for ethanol production	Temporary decrease in AM infection in wheat and phytotoxicity to wheat.
Pseudomonas fluorescens	Degrader of 2, 4-D herbicide	Decreased soil fungal numbers
Streptomyces lividans	Cellulose degrader	Temporary increase in CO_2 evolution upon addition of lignocellulose.
Pseudomonas fluorescens	Biocontrol agent	Plasmid DNA transferred to several genera of indigenous bacteria.
Pseudomonas Sp. B13	3-Chlorobenzoate degradation	Transferred the gene for this character into *Alcaligenes* in a natural aquifer

timated at 400–600 in the U.S.A and hundreds of similar tests in India, Canada, Europe, New Zealand, Australia and Latin America involving about 24 to 36 plant species. The largest number of tests conducted in the U.S.A relate to herbicide tolerance followed by insect and virus resistance tests involving tomatoes, cotton, potatoes, soybeans, tobacco, corn and alfalfa.

Patents

In developing a transgenic plant or a genetically engineered microorganism or for that matter any biotechnological innovation that can cure ailments of man or animal, plenty of money is spent in research and development of a particular product or innovation. To protect these inventions (intellectual property), patents have to be obtained or granted so that the invention is not surreptitiously used by others for profit. Laws for patenting machinery, equipments, devices, new products and improved processes do exist in many countries including India but many countries have not seriously considered patenting live organisms, plant resources, varieties, genetically engineered microorganisms and other innovations coming out of biotechnology or genetic engineering research. In the U.S.A, genetically engineered microorganisms (GEOs) are being patented beginning with the classical oil eating microbe developed by the India born Anand Chakravarthy of the General Electric Company and the same is true of transgenic plants, many of which are still undergoing tests. India is beginning to wake up to the situation following the alleged piracy of haldi (*Curcuma longa*) and basmati name for new rice varieties developed by the U.S.A. companies. The necessity to extend our patent laws to biotechnological innovations is urgent since India is a signatory to the General Agreement on Tariffs and Trade (GATT) and the World Trade Organisation (WTO). It has been stipulated that all signatory countries are required to enact national legislation to be effective from 2005 so as to grant intellectual property rights in biology including plant, animal and microbial innovations that are derived by altering genetically heritable characteristics. To accomplish this, we need training centres, manpower to handle applications and to screen the veracity of claims and above all legislative procedures that are quick to handle disputes. Presently, microorganisms and those created artificially are patentable in several countries such as India, Austria, Brazil, Bulgaria, Denmark, Canada, Germany, Hungary, Israel, The Netherlands, Philippines, Spain, Sweden, Switzerland, Thailand and U.K. However, the discovery should be associated with an industrial, medical or farm application for it to be patented. In other words, the criteria of novelty, inventiveness and originality will have to be satisfied for grant of a patent.

Tissue Culture

Meristems and root apices of plants can be axenically cultured on special tissue culture media to generate a mass of undifferentiated cells known as 'callus' and from tiny bits of this callus material, numerous calluses could be generated. Individual cells from a macerated callus can often be regenerated into new calluses by growing them further on special media. From these callus cultures, new plants can be raised initially by transferring plantlets into small pots and then into natural soil once they have adjusted to the surroundings. This technique, known as early as 1930, has now reached a stage of commercial application by generating clones of plants which are uniform in certain traits such as freedom from seed-borne diseases, viruses, frost damage, salt tolerance and many other attributes which cannot be achieved by plant breeding methods. There are various types of tissue cultures which are frequently used—callus cultures, cell suspension culture, organ culture, meristem tip culture and protoplast cultures. In the case of protoplasts, the cell walls are removed by lysozyme or suitable cell-wall dissolving enzymes and they are cultured in a suitable medium, a technique which facilitates manipulation of cell units without the interference of cell walls (Table 45).

Table 46 Various plant tissue culture techniques and their corresponding applications in crop improvement

Techniques	Applications
Anther culture	Haploid plants
Embryo culture	Hybrids between distant and closely related species which are sexually incompatible
Protoplast culture	Somatic hybrids between sexually incompatible distant and closely related species
Somoclonal variations	Variants of agricultural importance
Meristem culture	Rapid micropropagation and virus free plants of elite clones
Somatic embryogenesis, encapsulation of embryo	For plant transformation and production of synthetic seeds
Cell or suspension culture	Production of phytochemicals
In vitro cryopreservation	Long-term preservation of plant genetic resources
Recombination DNA techniques or gene transfer	Genetically engineered plants or transgenics

Some examples of tissue culture application in agriculture are as follows: Cassava (*Manihot utilissima*) is normally propagated by pushing a large section of a mature plant's stem (stake) into soil. These stakes are bundled and moved from place to place or country to country which present quarantine problems because disease causing germs may move with cassava stakes. The Centro International de Agricultura Tropical (CIAT) and the International Institute of Tropical Agriculture (IITA) bred new varieties of cassava which possess resistance to diseases and insects and developed disease-free lines through meristem cultures for shipment under aseptic conditions to other African countries. CIAT has now an *in vitro* cassava germplasm with 700 meristem-cultured accessions in the bank. Similarly, haploid plants were raised from anthers (anther culturing) and homozygous plants produced in one generation, a process that requires five or six generations by conventional plant breeding methods.

The International Rice Research Institute (IRRI) obtained salt tolerant variants of Taichung 65 variety by tissue culture techniques with 20% higher yield than the parent which was most suitable to salt stressed conditions. IRRI also developed strains of Taichung 65 variety which overcame aluminium toxicity. In Asia due to low temperatures at high altitudes where rice is cultivated, yields are invariably low. By anther culture technique, IRRI has attempted to develop cold tolerant strains of rice. Disease-free potato plantlets and tubers can be developed by tissue culture technology. The International Crop Research Institute for the Semi-Arid Tropics (ICRISAT) is using meristem cultures to produce disease-free groundnut germplasm. The Indian Agriculutral Research Institute (IARI) has been able to overcome the vexing problem of male sterility in papaya (*Carica papaya*) by tissue culture techniques.

Azolla is a water fern which can be successfully harnessed as a nitrogen input in rice cultivation because the system fixes nitrogen through an alga *Anabaena azollae* inhabiting its fronds. Protoplast fusion and hybrid cell generation techniques could be used to cross an *Azolla* that has low yields but tolerates high temperature with an *Azolla* that has high yields but likes cold weather. If this can be achieved, *Azolla* strains could be used which can generate 400 kg N/ha in high temperature tropical fields. In the area of biological nitrogen-fixation, the possibilities of exploiting tissue and cell culture techniques are still open. Nitrogen-fixing bacteria and blue-green algae can be forced into isolated protoplasts or calluses and plantlets regenerated. The plants developed from such callus cultures with nitrogen-fixing bacteria could possibly develop into nitrogen fixing plants. Chloroplasts can be made to take in nitrogen-fixing blue-green algae. One of the primary bottlenecks in these attempts is the physiological barriers between the protoplasts of an eukaryote and a prokaryote. Experiments have been done to transfer *nif* genes from a simple prokaryote (*Klebsiella pneumoniae*) to a simple eukaryote (*Saccharomyces cerevisiae*). The results so far obtained indicate that while the *nif* operon has been moved from the bacterium into the yeast cell, the expression of the desired trait, namely, nitrogen fixation or nitrogenase activity has not been achieved, thereby suggesting that a better understanding of the physiological steps needed for expression of nitrogenase activity by the recombinant yeast cells is a prerequisite to achieve the desired goal to transfer *nif* to higher plant species.

In India, tissue culture techniques have been satisfactorily used (Figs. 105, 106) to rapidly multiply elite cultivars of sugarcane, turmeric, ginger, rubber, mustard, cardamom, citrus, pineapple, pomegranate, almond, banana, apple, *Disoscorea*, *Bougainvillea*, teak, bamboo, sandal, eucalyptus, rosewood and pine. Tissue culture propagation ensures preservation of extinct species and freedom from diseases.

Vaccines

Viral vaccines consist of either virulent virus particles that have been in-activated, or of live particles of virus strains that have been weakened or attenuated, so that they no longer cause disease but immunize against the disease causing virulent strains. The disadvantage of such conventionally made vaccines is that there is a small chance, that one or more virus particles have survived inactivation and vaccination with such a vaccine could therefore lead to isolated cases of disease. Such accidents have indeed happened more than once, for example in cowherds vaccinated against foot-

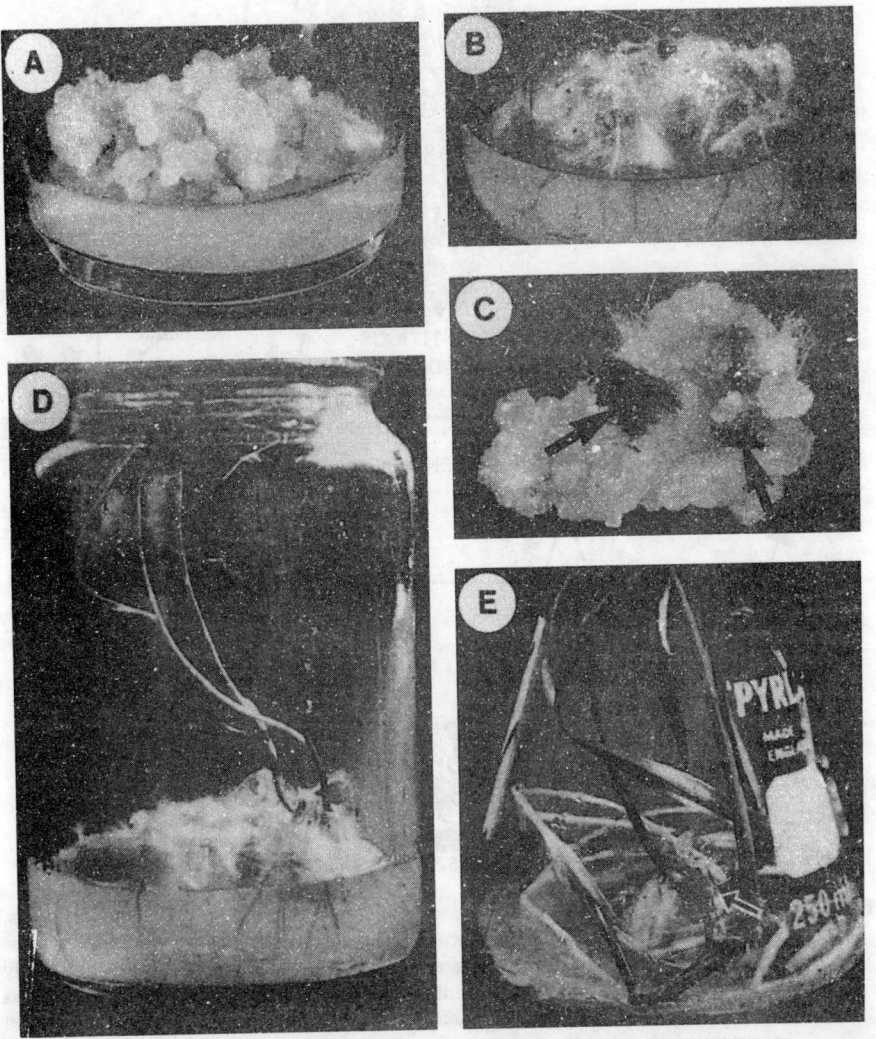

Fig. 105 Plant regeneration from root callus of wheat (*Triticum aestivum*): (A) Callus; (B) Callus showing profuse rooting (C) Callus with many shoot buds (arrow marked); (D) Same as (C) after another three weeks. Note that the shoot buds have grown into plants; (E) The plant formed in cultures has developed a floral spike (arrow marked). (Courtesy: Dr. Sant S. Bojwani, Department of Botany, University of Delhi).

(hoof-) and mouth-disease virus. Moreover, the viruses used to make vaccines are grown in animal cells and therefore the vaccines are very often contaminated with cellular material that can cause adverse immunological

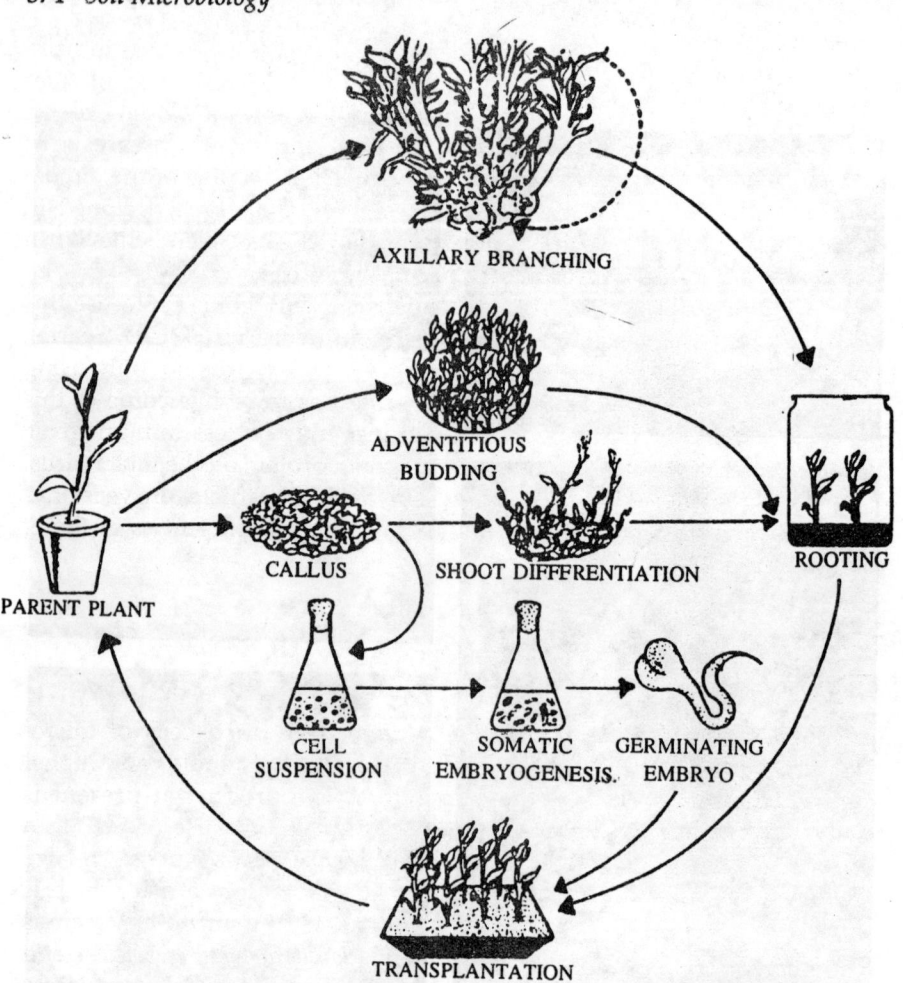

AXILLARY BRANCHING

ADVENTITIOUS
BUDDING

CALLUS   SHOOT DIFFERENTIATION   ROOTING

PARENT PLANT

CELL   SOMATIC   GERMINATING
SUSPENSION   EMBRYOGENESIS.   EMBRYO

TRANSPLANTATION

Fig. 106 Diagram of steps involved in micropropagation (aseptic multiplication) of plants. Shoot multiplication is achieved through forced axillary branching, adventitious budding from pieces of plant parts or callused tissue. The shoots are individually rooted in a medium containing a suitable auxin. When satisfactory root system has developed, the plantlets are transferred to a well drained potting mix. Plant regeneration from callus may occur through shoot differentiation or somatic embryogenesis. In the latter case special treatment for rooting is not required because the embryos have pre-formed root primordia. (Courtesy Dr. Sant S. Bhojwont, Department of Botany, University of Delhi.)

reactions in certain people. Since the proteins forming the coat of virus are the major antigens that induce antibody formation, it should be possible to use only the protein part and not the whole virus (which contains the genetic material, capable of causing the disease), to make a vaccine. This is where recombinant DNA techniques could be used by producing viral proteins from genetically engineered yeast and *E. coli* which in principle should be less expensive and might lead to the production of safer vaccines. In this regard much work has been done on the cloning of the major antigenic viral protein (VPI) of the foot- and mouth-disease virus (FMDV) of cattle. The complementary DNA (cDNA) copies of the single-stranded genomic RNA (MW 2.6×10^8) of FMDV have been cloned in plasmid vectors of *E. coli* and synthesis of the VPI has been achieved in quantity of 1 to 5 million molecules per bacterial cell.

In addition to FMDV, there is also intensive work going on to use this technology to make vaccines for flu causing influenza virus (G.D. Searle), hepatitis B virus (Biogen/Merck) and Herpes simplex virus not only using *E. coli* but also potato tubers. One hurdle that has to be overcome in this approach is that the purified protein vaccines may be less immunogenic than the macromolecular assemblies of the same protein on the intact virus. Recent work on the expression of surface antigen particles in yeast has been promising in overcoming the above mentioned deficiencies in the *E. coli* expression of the antigenic protein.

Single Cell Protein

Commonly referred to as SCP, these products are dried cells of micro-organisms such as algae, actinomycetes, bacteria, yeasts, molds and higher fungi which are grown in large fermentors. The product at present is largely used as animal feed. One of the earliest known use of SCP as a natural source of food comes from Africa: *Spirulina*, a blue-green alga growing in the lake Chad region of Africa develops into a mat which is scooped out periodically by the natives and dried in sun to be eaten as food. During the world war, bakers yeast (*Saccharomyces cerevisiae*) and Torula yeast (*Candida utilis*) were grown on a large scale using molasses and sulphite waste liquor from pulp and paper industry as media to produce a protein supplement for animals and human beings.

There have been several candidates for SCP production from time to time and the substrate used for the cultivation of these microorganisms are also variable (Table 47). The most advanced process for SCP production on a methanol substrate was developed by the then Imperial Chemical Industries of U.K. by growing *Methylophilus* (*Pseudomonas*) *methylotrophus* in large fermentors. There are several limitations in the use of SCP for human consumption and they are due to (1) high nucleic acid contents of many microorganisms that would result in kidney stone formation or gout (2) poor digestibility, gastrointestinal problems and skin reactions and (3) the possible presence of toxic or carcinogenic compounds from residues of substrates. However, dried food-grade yeasts and their autolysates have been used for many years. Dried mycelia of mushrooms, by virtue of their

Table 47 Some examples of Single Cell Protein (SCP) from microorganisms grown on different substrates

Organism	Substrate	Crude protein (%)
Algae (Photosynthetic)	CO_2, Sunlight	55
Scenedesmus acutus	CO_2, HCO_3, CO_2, sunlight	62
Spirulina maxima		
Chlorella sp.	CO_2, cane syrup, molasses	–
Bacteria (Photosynthetic)		
Rhodopseudomonas capsulata	Industrial wastes, sunlight	61
Bacteria (nonphotosynthetic)		
Cellulomonas sp.	Bagasse	87
Methylococcus capsulatus	Methane	–
Methylophilus methylotrophus	Methanol	72
M. clara	Methanol	–
Yeasts		
Candida sp.	n-Alkenes	65
Candida utilis (Torula)	Ethanol, Sulphite waste liquor	50–55
Kluyveromyces fragilis	Cheese whey (lactose)	45–54
Saccharomyces cerevisiae	Molasses	53
Fungi		
Fusarium graminearum	Glucose	–
Cephalosporium eichhorniae	Cassava starch	48–50
Chaetomium cellulolyticum	Agriculture and forestry wastes	45
Paecilomyces varioti	Sulphite waste liquor	55
Penicillium cyclopium	Cheese whey (lactose)	47
Scytalidium acidophthlum	Acid, hydrolyzed waste paper	44–47

pleasant flavour can be used in soups, sauces or gravy formulations. The high capital costs and the need for sterility controls render SCP expensive in developing countries where food shortages are common. As animal feed, SCP grown on agricultural residues should certainly find a place in the future economy of developing nations. It is however noteworthy that Dabur in India is manufacturing and selling *spirulina* for human consumption.

Biopesticides

Commercial production of biopesticides is restricted to few bacteria, fungi and viruses (Table 48). Artificially formulated media with carbon and nitrogen sources serve as substrates for some bacteria and fungi whereas few bacteria and viruses need live insect larvae grown aseptically in large fermentors as substrate because of the obligatory nature of these insect predators. For instance, *Bacillus popilliae* is produced on a commercial scale in the form of a dust on larvae of Japanese beetles. On the other hand, *B. thuringiensis* and *B. moritai* are produced by conventional media fermentation techniques by various companies. In fact, *B. thuringiensis* in-

Table 48 List of some microorganisms used commercially in microbial insecticide preparations

Microbial source	Commercial name	Country of use	Disease/insect controlled
Bacteria			
Bacillus popilliae Dutvy	Doom	USA	Lawn grass insects
B. thuringiensis Berliner	Thuricide	USA France USSR	Silkworm (*Bombyx mori* L.), mosquito larvae (*Anopheles, Culex, Aedes*), cabbage worm, Tobacco cutworm
Viruses			
Heliothis nuclear polyhedrosis virus (NPV)	Elcar	USA	Tobacco budworm (*Heliothis virescens* F.) Cotton bollworm (*H. zea* Boddie)
Lymantria dispar NPV	Gypcheck	USA	Gypsy moth (*Lymantria dispar* L.)
Orgyva pseudotsugata NPV	Biocontrol-1	USA	Douglas-fir tussock moth (*Orgyva pseudotsugata* McDunnough)
Fungi			
Beauveria bassiana (Bals.) Vuill.	Boverin	USSR	Codling moth (*Carpocapsa pomonella* L.) and colorado potato beetle (*Leptinotarsa decemlineata* Say.)
Hirsutella thompsoni Fisher	–	USA	Citrus rust mite (*Phyllocoptruta oleivora* Ashmead)
Protozoa			
Nosema locustae Canning	–	USA	Grasshoppers
Nematodes			
Romanomermis culicivorax Coman	Skeeter, Doom	USA	Mosquito

secticide has come to stay as an alternative to chemical pesticide in many countries.

Bacillus thuringiensis

Bacillus thuringiensis (B.t) is a Gram positive sporulating bacterium that produces a crystalline inclusion known as parasporal crystal in sporulating cells. The parasporal crystal contains crystal proteins known as 'endotoxins' that are toxic to certain insect pests. Commercial formulations differ but they essentially contain a mixture of B.t spores and crystals which are sprayed on plants. Susceptible insects feed on sprayed plant foliage resulting in gut paralysis, stoppage of feeding and death within 3–5 days due to the endotoxins ingested by the insect.

There are more than 30 sub-species of B.t distinguished by differences in immunological properties of parasporal crystals. A stain of B.t

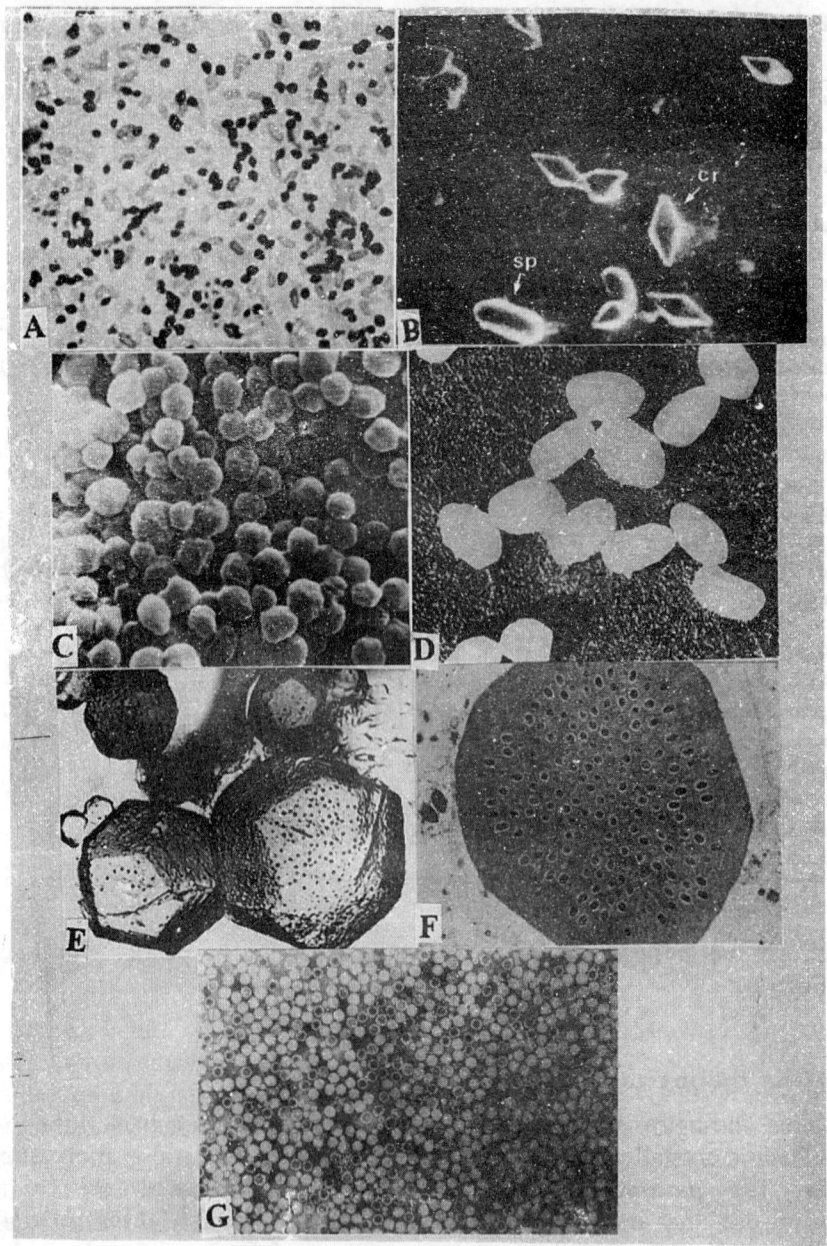

Fig. 107 Some examples of bacterial and viral insecticides: A—*Bacillus thuringiensis* var. *Kurstaki*, a stained preparation showing crystals as darkly stained inclusions in the bacterial cell; B—Scanning electron micrograph of the specimen of *B.t..* var *Kurstaki* showing spore (sp) and Crystal (cr) by arrows (Courtesy V. Sekar); C—Nucleopolyhedrosis virus (Courtesy C.M. Ignoffo); D—Granulosis virus (Courtesy C.M. Ignoffo); E—cytoplasmic polyhedrosis virus (Courtesy, J. Adams); F—Entomopox virus (Courtesy, M. Bergoin); G—Non-inclusion virus (Courtesy, E. Vago). The figures C to G from Microbial Processes, National Academy of Sciences, USA publication.

developed by the U.S.D.A., called HD-1 var. *krustaki* and its derivatives have been used since 1970 for producing commercial preparations by fermentation. The preparations have been used to control lepidopteran caterpillars such as cabbage loopers, cababage worms, spruce and tobacco budworms and gypsy moth larvae. Another stain of B.t var *Israelensis* has been in usage by several manufacturers for preparation of products that control larvae of mosquito and blackfly. Some companies have also developed B.t products against coleoptera beetles. Some examples of companies actively engaged in B.t. biopesticide products are Sandoz crop protection Corporation, Abbot laboratories and Mycogen Corporaticn. Each company has its own trade mark for their products such as Certain Vectobac AS, Tenkar, Skeetal, Dipel, Bactospeine, Thuricide, Javelin and M-one.

Different strains of B.t have different crystal morphologies and a single strain can contain more than one type of crystal. The strain B.t. var. *Kurstaki* produces a small cuboidal crystal (Fig. 107), which may be attached to a large bypyramidal crystal, each containing different type of protein possessing different immunological and insecticidal properties. The bipyramidal crystal contains a protein of 130 to 140 KDa size and is designated as P1 type while the cuboidal protein contain a protein of about 65 KDa designated as P2 type. P1 protein appears to be active against lepidopteran insects whereas P2 protein seems to be active against certain dipterans, especially mosquitoes and lepidopterans as well. The P1 protein in bipyramidal crystal is a protoxin that is at first inactive but upon prolonged incubation gets activated by proteases of the midgut of insects converting it into 65–70 KDa protein that is presumed to be actively toxic.

Toxicity varies depending upon the three types of lepidopterans classified on the basis of susceptibility to B.t endotoxin—type 1 insects are killed by the endotoxin alone, type II killed by endotoxin but toxicity enhanced by spores and type III killed only when both spores and endotoxin are present.

Plasmids control endotoxin production. The Cry or deltaendotoxin crystal protein genes from B.t code for the highly active, linear polypeptide insecticidal proteins. These proteins bind to receptors on midgut epithelial cells and cause disrupted metabolism, resulting in cessation of insect feeding, paralysis and death in 24 hours. The crystal producing (Cry⁺) B.t. strains can serve as plasmid donors to B.t. strains cured of Cry⁺ plasmid or to other bacterial species such as *Bacillus cereus* grown in broth cultures where matings are allowed to take place. The genes that control P1 protein production have been cloned to *E. coli* and *Bacillus subtilis* host vector systems. The B.t. var. *Isreaelensis* toxic protein is controlled by a 75-MDa transmissible plasmid. Similarly, the genes coding for insecticidal protein from *B.t. var tenebrionis* active against coleopteran insects such as colorado potato beetle have been cloned and characterised.

The development of B.t. strain with a broad spectrum of activity has been the aim of several manufacturing companies. A second current strategy has been to introduce B.t toxin gene into the genome of plants or plant-associated microorganisms. Research workers at Agrigenetics and Plant Genetic Systems and Monsanto have successfully engineered B.t toxin genes into tobacco, cotton and tomato plants and into root colonizing bacteria of the genus *Pseudomonas*. As a matter of fact, Monsanto is field testing transgenic 'Bollgard' variety of cotton resistant to bollworm in several parts of the world including India. The plant endophyte *Clavibacter xyli* subsp *cynodontis* was the target for expression of the endotoxin protein of B.t var. *Kurstaki*. This modified endophyte was later transferred into corn by Crop Genetic International. However, *C.xyli* expression of toxin was very low in rapidly growing parts of the plant and it is not clear if that level would be sufficient to effectively control the insect pest.

The stability of the toxin genes in plants of future generations is an important consideration while engineering endotoxin traits into transgenic plants and this problem becomes pronounced if the strategy is to incorporate gene actions for multiple insect targets. Currently, many of the transgenic plant species resistant to insect pests have been considered for commercial purposes and field tests have been permitted in the USA. and many other countries.

Mass production of B.t insecticides is carried out in liquid broth fermentation tanks inoculated with a selected strain. The broth is then concentrated and formulated into an aqueous oil-based flowable product. Two considerations have been uppermost in the minds of scientists involved in B.t improvement—1) improvement of the quantum of protoxin from the present 20 to 30 per cent of dry weight of bacterial cells to a higher level with single component protein capable of killing diverse insect pests and (2) incorporation of desirable insect feeding stimulants in the final product for quick consumption by the target insect to avoid photoinactivation of the endotoxin.

In the U.S.A, a transgenic *Pseudomonas* to which a toxic gene from B.t has been added is being commercially distributed under the brand name 'Dagger G' to control damping off of cotton.

Insect Viruses

Insect viruses (Fig. 107) fall into five major groups—1) nucleopolyhedrosis viruses (NPV), cytoplasmic polyhedrosis viruses (CPV), granulosis viruses (GV), entomopox viruses (EPV) and non-inclusion viruses (NIV)

Insect viruses, because of their specificity and obligate nature of parasitism on insects can only be mass produced on live insects. The process of production of the bollworm-budworm nuclear polyhedrosis virus (NPV), *Baculovirus heliothis* consists of raising bollworm larvae on a semi-synthetic diet containing water-based mixture of casein, sucrose, wheat germ, yeast, wesson salts and growth factors. Alphacel is used as a filler and agar-agar is used to solidify the diet. Chemicals like formalin, sorbic acid, methyl parahydroxybenzoate, aureomycin are often used to inhibit the growth of contaminating bacteria, yeasts and fungi. The diet is dispensed while hot from a large tank by a filling machine into plastic trays and sealed. These are inoculated by caterpillars from an insectary and incubated under controlled conditions to produce a mass of larvae. A known volume of virus is sprayed on the diet which replicates on the caterpillars to the extent of 5000–10,000 times in 5–7 days. The infected caterpillars are suctioned and then macerated with water, filtered, centrifuged, precipitated and spray-dried. After quality control testing, the preparation is packaged and sold under brand names. In the U.S.A., there are several companies producing this virus commercially. Some of these companies also produce NPV for the following insect hosts: *Lymantria, Mamestra, Neodiprion, Orgyia, Pieris, Prodenia, Spondoptera* and *Trichoplusia*. NPVs are most promising because of their safety, virulence and stability. The CPVF, EPV, and NIV are not so promising and they are still being investigated.

The first field trial of a genetically improved nuclear polyhedrosis virus of the alpha looper, *Autographa californica* (AaNPV) that expresses an insect selective toxin gene (AaHIT) derived from the venom of the scorpion *Androctous australis* was done in 1994. Earlier results of laboratory experiments with the cabbage looper *Trichoplusia ni* showed a 25 per cent reduction in time of death compared to wild type virus without alteration in pathogenicity or host range. The results of another field experiment demonstrated that the modified baculovirus was very active, resulting in decreased crop damage and also reduction in the secondary cycle of infection compared to the wild type virus.

Entemopathogenic Fungi

Fungi which produce spores or toxins are also mass produced in culture medium or on living hosts but the former method is more popular. Among the different entomopathogenic fungi, the successful ones are *Aschersonia aleyrodis, Beuvaria bassiana, Metarrhizium anisopliae, Entomophthora thaxteriana* and *Nomuraea rileyi*. North America produces commercial products under the trade name 'Biotrol FBB' with *B. bassiana* and 'Biotrol FMA' with *M. anisopliae* to control scale insects, whitefly, plant hoppers, aphids etc.

Biofertilizers

Biofertilizers are carrier based preparations containing beneficial microorganisms in a viable state intended for seed or soil application and designed to improve soil fertility and help plant growth by increasing the number and biological activity of desired microorganisms in the root environment. The term biofertilzers is of recent origin and is in vogue only in third world countries especially in India. Biofertilziers are ecofriendly and cannot at any rate replace chemical fertilizers that are indispensable for getting maximum yield of crops. These products can at best minimize the use of chemical fertilzers not exceeding 40 kg N./ha under ideal agronomic and pest-free conditions.

Nobbe and Hiltner in 1895 produced the first laboratory grown rhizobia for nearly 17 different legumes in the U.S.A. under the brand name 'Nitragin'. By 1920 there were several research institutions and private manufacturers in the business but today hardly few have stayed in the market and they produce inoculants mainly for soybean cultivation. In Australia, rhizobial inoculants are produced and used mainly for pasture legumes. In India, however, biofertilizers are produced and used for a variety of crops by incorporating non-symbiotic nitrogen fixing bacteria like *Azotobacter* and *Azospirillum*. The crops include, sugarcane, cotton and millets. Rice being a staple diet is grown extensively and under hot and humid conditions, blue-green algae (*Nostoc, Aulosira, Anabaena* etc.) containing biofertilizer is being advocated while *Azolla* organic manure is advocated in cooler areas with moderate temperature.

There is need for new and improved strains of bacteria of both N_2 fixing and p-solubilizing bacteria (*Pseudomonas striata, Bacillus megatherium* etc.) which can compete with native strains. These improvements in strain quality can come through natural selection or by genetic engineering methods. Another important factor is the shelf life of the product. Quality control measures are also important in the interest of farmers.

Bioconversion

To obtain liquid fuels from sources other than fossil fuel, the procurement or production of fermentable sugar from nature's biomass is the first step. Biomass includes renewable resources such as residues of crops, sugarcane bagasse, forestry residues, etc., which are estimated to yield 40 billion gallons of ethanol/year. The major constituents of biomass are hemicellulose, cellulose and lignin (3:4:3). Hemicellulose hyrolysis gives xylose, a sugar not so easily fermented to ethanol. Glucose is the major end product of cellulose hydrolysis. Lignin protects cellulose and is difficult to degrade enzymatically.

By mechanical methods cellulosic residues are shredded to fine pieces followed by acid or alkali hydrolysis. The hydrolysate is neutralized to yield a substrate to be acted upon by cellulolytic microorganisms to provide a syrup very poor in glucose content. *Trichoderma* spp. have been known to be good producers of cellulases but some of the commercially available cellulases are not very potent.

Trichoderma reesei produces three classes of enzymes: endoglucanases, exo-cellobiohydrolases and B-glucosidases. It is generally accepted that the first two classes of enzymes act cooperatively and synergistically in deploymerizing cellulose to glucose and oligosaccharides, which are then converted by B-glucosidase to glucose. Cellulose biosynthesis in *Trichoderma* is under multiple regulatory control of induction and repression. Several attempts have been made to obtain mutants of *Trichoderma* and later selecting them for high cellulase production by Cetus Corporation, Berkeley, California and U.S. Army Natick R and D Laboratories, Natick, Massachusetts, U.S.A.

The use of ethanol as a fuel has come to the forefront because admixture of ethnol with gasoline (2:8) has been practised in Brazil and used as an automobile fuel. Brazil has surplus of sugarcane production and can afford to spare sugarcane juice for ethanol production by conventional *Saccharomyces cerevisiae* fermentation whereas other developing countries cannot afford to do so. A question has often been asked whether biomass could be used for production of ethanol. To accomplish this, research is required to evaluate the genetic transfer and regulation of genes for glycolysis and hexose monophosphate shunt pathway. Most of the structural genes involved in the reactions between glucose and pyruvate have been identified and mapped.

Attempts at cloning of glycolysis genes in *E. coli* have been made and genes for most of the glycolysis enzymes of yeasts have been cloned. However, the solution to scaling up problems in alcohol production clearly lie in devising a prototype microorganism which overcomes product (ethanol) inhibition and tolerates high temperature. The ethanol sensitivity could also be overcome by continuous dialysis of the culture or by some other continuous process involving immobilization of the yeast and separation, perhaps by magnetic means.

Another factor in the utilization of biomass is the expensive physical process of shredding the materials and the chemical hydrolysis process which could only be overcome by ethanol production directly by using cellulose and yeast concurrently. Attempts in this direction have to be combined with efforts to concentrate the glucose syrup by inexpensive means so as to enrich the substrate for yeast fermentation.

Futurology

Is biotechnology the answer to feed the ever increasing human population in the next millennieum? Given the present acreage planted to staple crops and the increasing cost of agricultural inputs, coupled with the diminishing buying power of the rural masses, the prospect of meeting the food needs of developing countries by conventional agriculture appears rather bleak. Apart from staple crops such as wheat and rice, the production of pulse grains (including oil yielding grains), the traditional source of protein, has never shown any sizeable change during the last several years. Research achievements in breeding high yielding and early maturing varieties of pulses and coarse grains (millets) have not been as spectacular as those of rice or wheat. The application of chemical fertilizers has been generally practiced in the cultivation of rice, wheat, sugarcane, cotton or other cash crops whereas most pulses and millets are grown on marginal rain-fed lands with very little agronomic practices or inputs by way of fertilizers and pesticides. Elite cultivars of pulses and millets, eventhough developed by research institutes, are rarely available for large scale planting by the average small-scale land holder. Intensive farming is no doubt practiced in wheat cultivation but this is rarely seen in pulse or millet cultivation. Biotechnology research in developing countries is rudimentary and not expansive because of the cost involved. With this scenario, transgenic crop technology, so well entrenched with private research enterprise in the USA and Europe, has to obviously cater to the needs of developing countries. These large multinational companies who have patented their inventions are only eager to sell the technology to the developing world. Granting that a novel genotype meets with the rigorous standards of any country that is willing to buy such an innovation, the products from such an invention obviously become expensive. Therefore, it is of utmost importance to put all efforts in developing transgenic plants from our own research endeavors.

Supplementing traditional sources of food with unconventional sources of food has always been an attractive alternative to food shortages. One of the unconventional source is the single cell protein (SCP) from bacteria and fungi grown on inexpensive media containing hydrolysates of biomass. The biotechnology involved in converting lignocelluloses from biomass has unfortunately not reached a stage of perfection to yield growth media rich in monosaccharides. Secondly, SCP has always remained as a source of animal feed and more rigorous tests including acceptability tests have to be done before advocating SCP as a supplement to traditional food.

Another strategy is to minimize chemical inputs in agriculture or to abolish them completely by making plants self-sufficient with regard to their nutrition and protection from pests and diseases. Significant headway in this direction has been made in recent years by developing transgenic plants resistant to diseases, pests and herbicides. However, success has not been achieved in *nif* gene transfer to crop plants and therefore, fertilizer nitrogen and phosphorus will continue to be vital for obtaining sustained crop yields. What has been a significant achievement in recent years is the success in nodulating

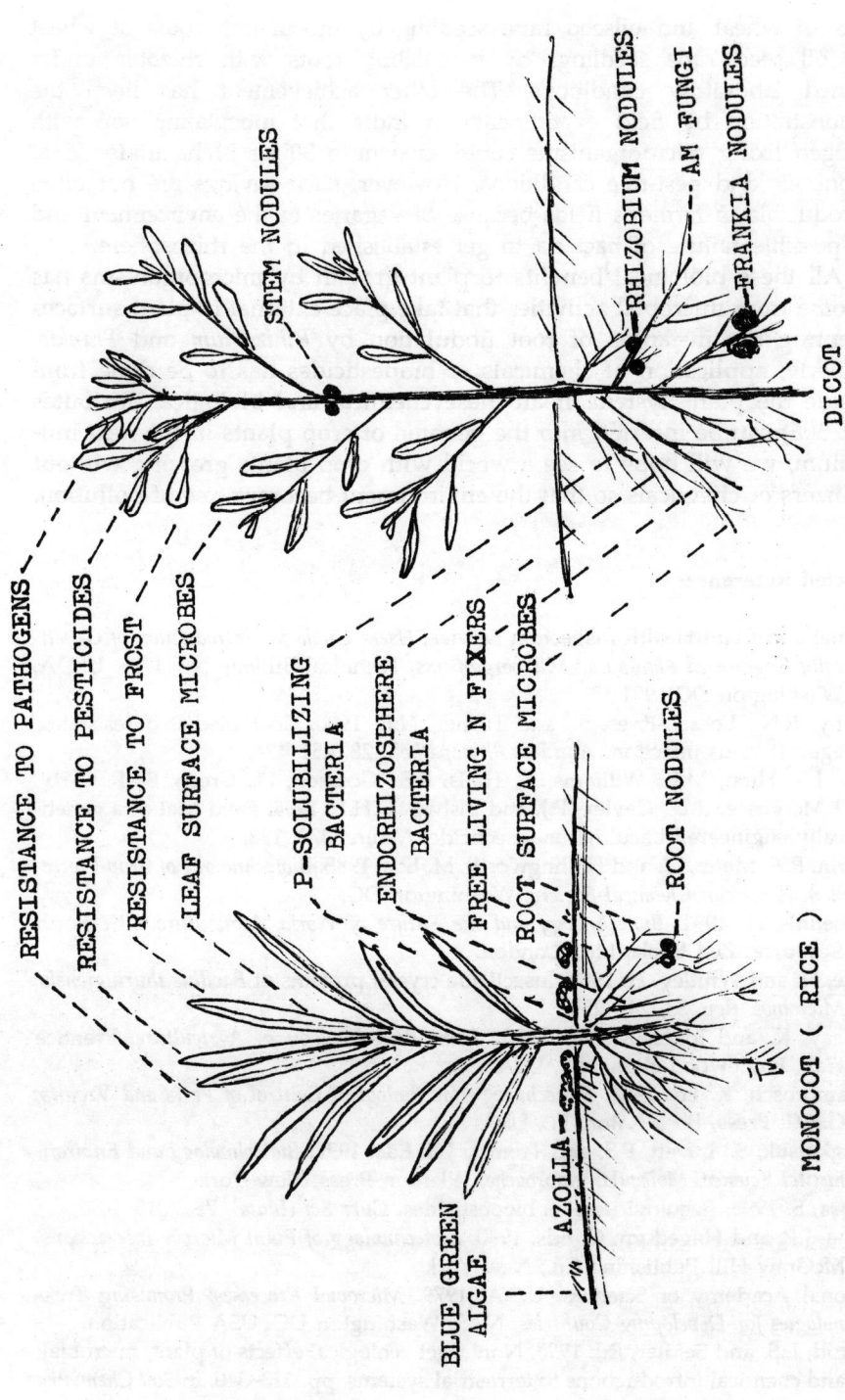

Fig. 108. Conceptual transgenic plants of the next millennium. On the left is a rice plant, a representative of the monocotyledonous plants and on the right is a typical dicotyledonous plant (Agricultural biotechnology marvels)

roots of wheat and oilseed rape seedling by inoculating roots of wheat and oil seed rape seedlings by inoculating roots with rhizobia under defined laboratory conditions. The other achievement has been the demonstration by field experiments in India that inoculating soil with nitrogen fixing microorganisms could save upto 30 kg N/ha under ideal agronomic and pest-free conditions. However, such savings are not often reproducible in farmer's fields because of vagaries in the environment and the possible failure of bacteria to get established in the rhizosphere.

All these biological benefits to plant growth by microorganisms has to come from microbial activities that take place external to plant surfaces excepting the instances of root nodulation by *Rihizobium* and *Frankia*. Similarly, application of chemicals or biopesticides has to be done from outside the plant system. If all these chemical and biological attributes (Fig. 108) can be inserted into the genome of crop plants in the next millennium, we will hope to see a world with crop plants growing without fertilizers or chemicals so that the environment becomes free of pollution.

Selected References

Animal and Plant Health Inspection Service, *Users Guide for Introduction of Genetically Engineered Plants and Microorganisms*, Technical Bulletin No. 1783. USDA, Washington DC 1991.

Beachy, R.N., Loesch-Fries, S., and Turner, N.E. 1990. Coat mediated resistance against virus infection. *Ann.Rev.Phytopathol.*, 28, 451–474.

Cory, J.S., Hirst, M.L., Williams, T., Hails, R.S., Goulson, D., Green, B.M., Carty, T.M., Posse, R.D., Cayley, P.J. and Bishop, D.H.L. 1994. Field trial of a genetically engineered baculovirus insecticide. *Nature*, 370, 138.

Hedrin, P.A. Menn, J.J. and Hollingworth, M. Eds. 1988. *Biotechnology of Crop Protection. American Chemical Society*, Washington DC.

Hobbelink, H. 1991. *Biotechnoogy and the Future of World Agriculture*. The Forth Resource, Zed Books Ltd., London.

Hofte, H. and Whitley, H. 1989. Insecticide crystal proteins of *Bacillus thuringiensis*. *Microbiol. Rev.* 33, 242–255.

Lindsay, K. and Jones, M.G.K. 1990., *Plant Biotechnology in Agriculture*. Prentice Hall, Englewood cliffs, N.J., USA.

Maramorosch, K. Ed. 1991. *Biotechnoogy for Biological Control of Pests and Vectors*, C.R.C. Press, Boca Raton, FL, USA.

Mongkolsuk, S., Lovett, P.S. and Trempy J.E. Eds. 1992. *Biotechnology and Environmental Science, Molecular Approaches*. Plenum Press, New York.

Mishra, S. 1998. Baculoviruses as biopesticides. *Curr Sci (India)*, 75, 1015–1022.

Nakas, J.P. and Hagedorn, C. Eds. 1990. *Biotechnology of Plant Microbe Interactions*. McGraw-Hill Publishing Co., New York.

National Academy of Sciences. U.S.A. 1979. *Microbial Processes: Promising Technologies for Developing Countries*, NAS, Washington DC. USA Publication.

Watrud, L.S. and Seidler, R.J. 1998. Nontarget ecological effects of plant, microbial, and chemical introductions to terrestrial systems. pp. 313–340. In *Soil Chemistry and Ecosystem Health* Ed. P.M. Huang, Soil Science Society of America Inc. Madison, Wis., USA.

Seidler, R.J., Watrud, L.S. and George, S.E. 1998. Assessing risks to ecosystems and human health from genetically modified organisms. pp. 110–146. In *Handbook of Environmental Risk Assessment and Management.* Ed. P. Calow. Blackwell Science Ltd. London, England.

Sekar, V. and Carlton, B.C. 1985. Molecular cloning of the delta-endotoxin gene of *Bacillus thuringiensis* var *israelensis. Gene,* 33, 151–158.

Appendix

Appendix on Media

Unless otherwise stated the media are all sterilized by autoclaving at 15 lbs pressure for 15 minutes. The quantity of agar used to solidify the different media depends on its quality and may vary from 15 to 20 g/litre.

I. Media for Isolating Soil Microorganisms

Bacteria

Nutrient Agar

Beef extract	3.0 g
Peptone	5.0 g
Agar	15.0 g
Distilled water	1000 ml

Heat until agar and peptone dissolve. Adjust pH to 6.6 to 7.0, using bromothymol blue as an indicator.

Soil-extract Agar

Glucose	1.0 g
K_2HPO_4	0.5 g
Agar	15.0 g
Soil extract (Stock*)	100 ml
Tap water	900 ml

*1000 g of sieved garden soil is mixed with 1000 ml of tap water and steamed in an autoclave for 30 minutes. A small amount of $CaCO_3$ is added and the whole is filtered through a double filter paper. Dissolve the agar in 900 ml of water by steaming it for an hour or more. Add 100 ml of the stock soil extract solution. Add glucose just prior to tubing. Reaction should be pH 6.8.

Asparagine Mannitol Agar (Thornton, 1922)

K_2HPO_4	1.0 g
KNO_3	0.5 g
$MgSO_4 \cdot 7H_2O$	0.2 g
$CaCl_2 \cdot 6H_2O$	0.1 g

NaCl	0.1 g
FeCl₃·6H₂O	trace
Asparagine	0.5 g
Mannitol	1.0 g
Agar	15.0 g
Distilled water	1000 ml

After the agar and salts have been dissolved add the mannitol and adjust the pH to 7.4.

Plate Count Agar

Tryptone	5.0 g
Yeast extract	2.5 g
D-Glucose	1.0 g
Agar	15.0 g
Distilled water	1000 ml
pH	7.0

Reinforced Clostridial Medium

Yeastrel yeast extract	3.0 g
Peptone (Evans)	10.0 g
Lab-Lemco meat extract	10.0 g
D-Glucose	5.0 g
Sodium acetate	5.0 g
Cysteine	0.5 g
Soluble starch	1.0 g
Agar	0.5 g
Distilled water	1000 ml

pH after autoclaving should be 7.1 to 7.2. The medium is used for the growth of anaerobes and for use as a diluent in the enumeration of anaerobes.

Fungi

Czapek-Dox Agar (Thom and Raper, 1945)

NaNO₃	3.00
K₂HPO₄	1.00 g
MgSO₄·7H₂O	0.50 g
KCl	0.50 g
FeSO₄·7H₂O	0.01 g
Sucrose	30.00 g
Agar	15.00 g
Distilled water	1000 ml

One gram of yeast extract per litre may be added. Original Czapek's sucrose nitrate agar contains only 2 g NaNO₃ per litre.

Rose-bengal Agar (Martin, 1950)

Glucose	10.00 g
Peptone	5.00 g
KH₂PO₄	1.00 g
MgSO₄·7H₂O	0.05 g
Streptomycin	30.00 mg
Agar	15.00 g
Rose-bengal	0.035 g
Distilled water	1000 ml

The antibiotic is sterilized separately and added aseptically to the sterilized medium. Aureomycin (35 to 2000 µg) may be substituted for streptomycin. The medium is specially recommended for the isolation of fungi in the presence of large numbers of bacteria.

Potato Dextrose Agar

Potatoes, peeled and diced	200 g
D-Glucose	20 g
Agar	15 g
Distilled water	1 litre

Boil 200 g of peeled diced potatoes for 1 hour in a litre of water. Filter and make up the volumes to one litre. Add glucose and agar and steam until agar is dissolved.

Richard's Synthetic Medium

KNO₃	10.00 g
KH₂PO₄	5.00 g
MgSO₄·7H₂O	2.50 g
FeCl₃·6H₂O	0.02 g
Sucrose	50.00 g
Agar	15.00 g
Distilled water	1000 ml

Actinomycetes

Kenknight and Munaier's Medium

Dextrose	1.00 g
KH₂PO₄	0.10 g

NaNO₃	0.10 g
KCl	0.10 g
MgSO₄·7H₂O	0.10 g
Agar	15.00 g
Distilled water	1000 ml

Asparagine mannitol agar and soil extract agar can also be used for the isolation of actinomycetes.

Yeasts

Malt Extract Agar

Malt extract	30.0 g
Mycological peptone	5.0 g
Agar	20.0 g
Distilled water	1000 ml
pH	5.4

In order to inhibit bacterial growth 10% sterile lactic acid solution can be added to the molten medium just before pouring the plates so as to bring down the pH to 3.5.

Glucose Yeast Extract Peptone Agar

Glucose	20.0 g
Bacto peptone	10.0 g
Bacto yeast extract	5.0 g
Agar	20.0 g
Distilled water	1000 ml
pH is not adjusted	

Blue-Green Algae

Pringsheim's Medium (Pringsheim, 1964)

KNO₃	0.02%
MgSO₄·7H₂O	0.001%
(NH₄)₂HPO₄	0.002%
CaCl₂·6H₂O	0.0005%
FeCl₃	0.00005%

Chu's Medium No. 10 (Chu, 1972)

Ca(NO₃)₂	0.004%
MgSO₄·7H₂O	0.0025%

K_2HPO_4	0.0005% to 0.001%
Na_2CO_3	0.002%
Na_2SiO_3	0.0025%
$FeCL_3$	0.0008%

Soil Algae

Modified Bristol's Medium (Allen, 1949)

KH_2PO_4	0.50 g
$NaNO_3$	0.50 g
$MgSO_4 \cdot 7H_2O$	0.15 g
$CaCl_2 \cdot 6H_2O$	0.05 g
NaCl	0.05 g
$FeCl_3 \cdot 6H_2O$	0.01 g
Tap water	1000 ml

Protozoa (Singh, 1946)

Holozoic protozoa are isolated using pure cultures of bacteria known to be edible by protozoa as the substrate. Silica gel plates are usually employed where glass rings are embedded in to contain different strains of bacteria so as to prevent the spreading of one to another. 0.05 CC of the inoculum is added at the centre of the bacterial circle.

II. Media for Nitrogen Fixing Organisms

Azotobacter

Ashby's Mannitol Agar

Mannitol	20.0 g
K_2HPO_4	0.2 g
$MgSO_4 \cdot 7H_2O$	0.2 g
NaCL	0.2 g
K_2SO_4	0.1 g
$CaCO_3$	5.0 g
Agar	15.0 g
Distilled water	1000 ml

Jensen's Medium

Sucrose	20.0 g
K_2HPO_4	1.0 g
$MgSO_4 \cdot 7H_2O$	0.5 g

NaCl	0.5 g
FeSO₄	0.1 g
Na₂MoO₄	0.005 g
CaCO₃	2.0 g
Agar	15.0 g
Distilled water	1000 ml

Burk's Nitrogen-free Medium

Burk's salt	1.3 g
Fe-Mo Mixture	1.0 ml
Sucrose	20.0 g
Distilled water	1000 ml

Burk's Salt

MgSO₄	20 g
K₂HPO₄	80 g
KH₂PO₄	20 g
CaSO₄	13 g

Fe-Mo-Mixture

FeCl₃	1.45 g
Na₂MoO₄	0.253 g
Distilled water	1000 ml

Beijerinckia (Becking, 1956)

Sucrose	20.0 g
KH₂PO₄	0.8 g
K₂HPO₄	0.2 g
MgSO₄·7H₂O	0.5 g
FeCl₃	0.1 g
Na₂MoO₄	0.005 g
Agar	15.0 g
Distilled water	1000 ml
pH	6.5

Derxia (Campelo and Dobereiner, 1969)

Starch	20.00 g
K₂HPO₄	0.05 g
KH₂PO₄	0.15 g
MgSO₄·7H₂O	0.20 g
CaCl₂	0.02 g
FeCl₃	0.01 g
Na₂MoO₄·2H₂O	0.002 g

Bromothymol blue	0.5 ml
(0.5% in absolute alcohol)	
NaHCO₃	1.0 g
Unrefined agar	20.0 g
Distilled water	1000 ml

Clostridium

Nitrogen-free media used for *Azotobacter* in which mannitol is replaced by glucose can be used. Immediately before inoculation dissolved oxygen is removed by steaming for 30 minutes and then allowing to cool. Plates or tubes are incubated in an anaerobic jar for 7 days at 30°C. Enrichment in the liquid medium followed by plating out is necessary to obtain pure culture.

Rhizobium

Yeast Extract Mannitol Agar

Mannitol	10.0 g
K_2HPO_4	0.5 g
$MgSO_4·7H_2O$	0.2 g
NaCl	1.0 g
Yeast extract	0.1 g
Agar	20.0 g
Distilled water	1000 ml

Bergersen's Synthetic Medium

Mannitol	10.00 g
$Na_2HPO_4·12H_2O$	0.45 g
Sodium glutamate	1.10 g
$MgSO_4·7H_2O$	0.10 g
$FeCl_3·6H_2O$	0.02 g
$CaCl_2·6H_2O$	0.04 g
Thiamine†	100 µg
Biotin	200 µg
Distilled water	1000 ml
pH	6.8–7.0

Azospirillum

Semisolid medium (Bulow and Dobereiner, 1975)

K_2HPO_4	0.1 g
KH_2PO_4	0.4 g
$MgSO_4$	0.2 g
NaCl	0.1 g

†Vitamin solutions are sterilized by filtration through sintered glass funnel and added to the medium after sterilization.

CaCl$_2$	0.02 g
FeCl$_3$	0.01 g
Na$_2$MoO$_4$	0.002 g
Sod. malate	5.0 g
Bromothymcol blue (0.05% Ethanol)	5.0 ml
Agar	1.75 g
Distilled water	1.0 litre
pH	6.8

Okon's medium

K$_2$HPO$_4$	6.0 g*
KH$_2$PO$_4$	4.0 g
MgSO$_4$	0.2 g
NaCl	0.1 g
CaCl$_2$	0.02 g
NH$_4$Cl	1.0 g
DL malic acid	5.0 g
NaOH	3.0 g
Yeast extract (Difco)	0.1 g
FeCl$_3$	10.0 mg
Na$_2$MoO$_4$	2.0 mg
MnSO$_4$	2.1 mg
H$_3$BO$_3$	2.0 mg
Cu(NO$_3$)$_2$	0.04 mg
ZnSO$_4$	0.24 mg
Bromothymol blue (0.5% in ethanol)	2 ml
Distilled water	1 litre
Agar	1.5% to 1.8%
pH	6.8

III. Special Media for Studying Rhizobia and Root Nodulation

Congo-red Medium

Congo-red (2.5 ml/l of 1% solution) is incorporated into yeast extract mannitol medium. Congo-red solution is sterilized separately and added to the sterilized medium. Colonies of rhizobia stand out as white, translucent, glistening and elevated, with entire margins.

Glucose Peptone Agar

Glucose	5.0 g

*Mixed in 100 ml water and then autoclaved separately followed by mixing with cold medium later.

Peptone	10.0 g
Agar	15.0 g
Bromo cresol purple	
(1% alcoholic solution)	10.0 ml
Distilled water	1000 ml

Rhizobium grows poorly in this medium and causes little change of pH.

Hofer's Alkaline Medium (Hofer, 1935)

The reaction of yeast extract mannitol medium is raised to pH 11.0 with 1N NaOH. 1 ml of 1.6% thymol blue per litre is added. In contrast to *Rhizobium* the allied genus *Agrobacterium* can grow in this medium, a characteristic feature which serves as a useful criterion to separate the two bacteria on agar plates during routine isolation of root nodule bacteria.

Lactose Medium (Bernaerts and De Ley, 1963)

Lactose is substituted for mannitol in the yeast extract mannitol medium. Sterillization is done by steaming for 30 minutes on two successive days. The medium is then left at room temperature (30 ± 2°C) for 3–4 days to check and confirm its sterility.

Raggio's Medium for Excised Roots (Modified by Bunting and Horrocks, 1964)

Inorganic Medium

$CaCO_3$	3000.00 mg
$CaCl \cdot 6H_2O$	446.00 mg
KCl	165.00 mg
KH_2PO_4	200.00 mg
$MgSO_4 \cdot 7H_2O$	700.00 mg
Na_2SO_4	200.00 mg
KI	0.75 mg
$FeCl_3 \cdot 6H_2O$	2.50 mg
H_3BO_3	1.50 mg
$Na_2MoO_4 \cdot 2H_2O$	0.25 mg
$MnSO_4 \cdot 4H_2O$	6.64 mg
$ZnSO_4 \cdot 7H_2O$	2.67 mg
$CuSO_4 \cdot 5H_2O$	0.07 mg
Distilled water	1000 ml

Organic Medium

| Nicotinic acid | 10 mg |
| Pyridoxine | 2 mg |

Thiamine	2 mg
Glycine	60 mg
Inositol	1000 mg
KNO₃	80.0 mg
Ca(NO₃)₂·4H₂O	300.0 mg
Sucrose	10.0 mg
Agar	10.0 g
Distilled water	1000 ml

Organic medium is used in the small vials into which the cut end of the hypocotyl is pushed in. Inorganic medium is used to moisten the sand (or used as such) into which the suspended roots grow.

Jensen's Seedling Agar (Jensen, 1942)

CaHPO₄	1.0 g
K₂HPO₄	0.2 g
MgSO₄·7H₂O	0.2 g
NaCl	0.2 g
FeCl₃	0.1 g
Distilled water	1000 ml
pH	7.0

Agar (8–15 g/l), according to use as deep or slope.

Suggested Medium for Growing Plants for Collection of Root Exudates

(i) NH₄NO₃	3.2%
(ii) CaH₄(PO₄)₂·H₂O	1.0%
(iii) K₂SO₄	2.0%
(iv) MgSO₄	0.8%
(v) FeCl₃	0.1 g in 250 ml of distilled water

10 ml each of (i) to (iv) and 1 ml of (v) are taken and diluted to 1000 ml.

Four Procedures for the Isolation of *Frankia* from Root Nodules of Non-legumes such as *Comptonia peregrina*, *Alnus crispa* and *A. rugosa* (as summarized by Wheeler, 1984)

1) Nodule homogenates centrifuged in sucrose density gradient → this separates vesicle clusters from contaminating microorganisms → plate out separated fractions on nutrient agar.

2) Surface sterilize nodules → homogenization → microdissection of the homogenate to separate vesicle clusters → pass through changes of sterile water → transfer to liquid media.

3) Grow plants axenically → crush nodules → pour into nutrient agar.

4) Sterilize outermost side of nodules with 3 per cent osmium tetroxide →fragment nodules in liquid medium → subculture fragments. Great caution has to be used in this method because of the highly toxic osmic tetroxide.

Media Used in the Isolation of Frankia:

Composition of BAP Medium (Murry et al., 1984)

Component	Concentration (g per litre)
K_2HPO_4	0.591
KH_2PO_4	0.952
NH_4Cl	0.267
$MgSO_4 \cdot 7H_2O$	0.095
$CaCl_2 \cdot 2H_2O$	0.010
FeNa EDTA	0.010
Na Propionate*	0.480
Trace elements solution**	1 ml
Vitamin solution***	1 ml

pH adjusted to 6.7 (propionate) or 6.3 (pyruvate).
*Na Propionate may be replaced by Na Pyruvate (1.1 g per litre). They are filter sterilized.
** Trace elements solution (g per litre) H_3Bo_3: 2.86; $MnCl_2 \cdot 4H_2O$: 2.27; $ZnSO_4 \cdot 7H_2O$; 0.22; $CuSO_4 \cdot 5H_2O$: 0.08; $Na_2MoO_4 \cdot 2H_2O$: 0.02: $CuSO_4 \cdot 7H_2O$: 0.001.
*** Vitamin solution (mg per litre) thiamin HCl: 10; nicotinic acid: 50; pyridoxin HCl: 50; biotin: 225; folic acid: 10; Ca pantothenate: 10: riboflavin: 10. Phosphate was added after autoclaving.

Composition of M6B medium (Modified from Baker and Torrey, 1980)

Component	Concentration (g per litre)
Yeast extract (Difco)	5
Dextrose	10
Casamino Acids (Difco)	5
KH_2PO_4	1
$MgSO_4 \cdot 7H_2O$	0.1
$CaCl_2 \cdot 2H_2O$	0.01
$CaCl_2$	0.001
FeNa EDTA	0.01
Trace elements solution*	1 ml
Vitamin solution**	1 ml

pH adjusted to 6.5.
*Trace elements solution: see previous Table.
**Vitamin solution: see previous Table.

Composition of Qmod medium (Lalonde and Calvert, 1979)

K₂HPO₄	300 mg/l	
NaH₂PO₄	200 mg/l	
MgSO₄·7H₂O	200 mg/l	
KCl	200 mg/l	Lipid supplement:
Yeast extract	500 mg/l	Dissolve 500 mg of
Bacto-peptone (Difco)	5 g	1-a-lecithin in 50 ml
Glucose	10 g	absolute ethanol and
Ferric citrate (citric acid		add 50 ml distilled
and ferric citrate, 1% soln.)	1 ml	water. Add this at the
Tace element soln.	1 ml	rate of 0.5–50 mg.
(as in previous tables)		
Adjust to pH	6.7–7.0 mg.	
(as in previous tables)	6.7–7.0	
Autoclave	20 min.	
Agar, if used	15 g	

Benson's medium (Benson, 1980) contains (g/litre): K₂HPO₄, 3.0; KH₂PO₄, 2.0; MgSO₄. 7H₂O, 0.2; NaCl, 0.3; ferric sodium-ethylene diaminetetracetic acid (EDTA), 0.16; 1 ml of trace elements and 1 ml vitamin mixture. Calcium carbonate is added to liquid cultures at 0.05 mg/ml. Carbon sources are added at 3 g/litre.

IV. Media for Cellulose/Lignin Decomposing Microorganisms

Omeliansky's Medium

(NH₄)₂SO₄	1.0 g
K₂HPO₄	1.0 g
MgSO₄·7H₂O	0.5 g
CaCO₃	2.0 g
NaCl	trace
Distilled water	1000 ml

NH₄Cl, (NH₄)₂HPO₄, KNO₃, or NaNO₃ may also be used. When nitrates are used CaCO₃ may be omitted. Add 2–3 strips of filter paper to each tube, part of the paper remaining above the surface of the medium. Cellulose decomposing organisms appear within 15–20 days.

Czapek's-Dox medium where sucrose is replaced by filter paper can also be used for isolating cellulose decomposers.

Medium for Lignolytic Fungi (Garren, 1938)

K₂HPO₄	1.0 g

KCl	0.5 g
MgSO$_4$·7H$_2$O	0.5 g
FeSO$_4$	0.1 g
Lignin	2.5%
Agar	20.0 g
Distilled water	1000 ml

By substituting cellulose for lignin, the same medium can be used for celluloytic fungi.

V. Media for Studying Ammonification, Nitrification and Denitrification

Ammonifying Bacteria

KH$_2$PO$_4$	3.00 g
KCl	0.20 g
MgSO$_4$·7H$_2$O	0.20 g
NaCl	0.20 g
CaSO$_4$	0.10 g
FeSO$_4$	0.01 g
Peptone	10.00 g
Distilled water	1000 ml

Bacteria are inoculated into tubes containing 5 ml of the above medium. After 10 days of incubation production of ammonia is tested by means of Nessler's reagent.

Nitrifying Bacteria

(NH$_4$)$_2$SO$_4$	1.0 g
K$_2$HPO$_4$	1.0 g
NaCl	2.0 g
MgSO$_4$·7H$_2$O	0.5 g
FeSO$_4$	trace
Distilled water	1000 ml

Excess of sterilized CaCO$_3$ is added to the medium after sterilization.

Test

Reagent 1:8 g sulfanilic acid in 1 litre of 0.5 N acetic acid. Reagent 2:5 g naphthalamine in 1 litre of 0.5 N acetic acid. Add to the test tubes in which bacteria are growing, 1.2 ml of reagent 1 followed by 1 ml of reagent 2. Development of pink colour indicates the presence of nitrites. (This medium is in addition to the medium described in the chapter on Soil Microorganisms.)

Denitrifying Bacteria

Asparagine-nitrate-citrate Solution

(a)	KNO$_3$	1.0 g
	Asparagine	1.0 g
	Distilled water	250 ml
(b)	Neutral sodium citrate	8.5 g
	KH$_2$PO$_4$	1.0 g
	MgSO$_4$·7H$_2$O	1.0 g
	CaCl$_2$·6H$_2$O	0.2 g
	FeCl$_3$·6H$_2$O	trace
	Distilled water	250 ml

Solutions (a) and (b) are mixed, made up to 1 litre and 15 g agar is added.

VI. Sulphur Bacteria

Asparagine Sodium Lactate Gelatin

Asparagine	1.0 g
Sodium lactate	5.0 g
K$_2$HPO$_4$	0.5 g
MgSO$_4$·7H$_2$O	1.0 g
FeSO$_4$(NH$_4$)$_2$·6H$_2$O	trace
Gelatin	120 to 150 g
Distilled water	1000 ml

Sterilize in an autoclave at 10 lbs pressure for 15 minutes. Cool in ice water.

Sulphite Polymyxin Sulphadiazine Agar (Angelotti et al., 1962)

Tryptone	15.0 g
Yeast extract	10.0 g
Ferric citrate scales	0.5 g
Agar	15.0 g
Distilled water	1 litre
pH	7.0

Dissolve the ingredients by steaming and adjust the pH to 7.0. Distribute in 100 ml in screw capped bottles. Sterilize by autoclaving at 15 lbs pressure for 15 minutes. Sterilize the following solutions by filtration separately and add to 100 ml of medium melted and cooled to 45°C.

(a) 5 ml of freshly prepared 10% solution of sodium sulphate.
(b) 1.0 ml of 0.1% solution of polymxin B.
(c) 1.0 ml of 1.2% solution of sodium sulphadiazine.

Mix well and pour the plates. The medium is used for the enumeration of sulphite reducing clostridia.

Thiosulphate Agar

$Na_2S_2O_3 \cdot 5H_2O$	5.0 g
K_2HPO_4	0.1 g
$NaHCO_3$	0.2 g
NH_4Cl	0.1 g
Agar	20.00 g
Tap water	1000 ml

For certain organisms it is advisable to add an excess of $CaCO_3$

Sulphur-Phosphate Medium

$(NH_4)_2SO_4$	0.20 g
KH_2PO_4	3.00 g
$MgSO_4 \cdot 7H_2O$	0.50 g
$CaCl_2 \cdot 6H_2O$	0.25 g
$FeSO_4 \cdot 7H_2O$	trace
Sulphur powdered	10.00 g
Distilled water	1000 ml

Sulphur (1 g) is weighed out into individual 250 ml Erlenmeyer flasks and 100 cc of the liquid medium added. The reaction of the medium is about pH 4.0. The flasks are sterilized in flowing steam for 3 consecutive days.

VII. Phosphate Solubilizing Organisms

Katznelson and Bose (1956)

Soil extract	100 ml
Glucose	1.0 g
Agar	2.0 g

Sterilize in 100 ml lots, cool and add 5 ml of 10% K_2HPO_4 and 10 ml of 10% $CaCl_2$ to each flask. Adjust pH to 7.0 with sterile N/10 sodium hydroxide. Pour the plates immediately and allow to solidify. The bacteria are inoculated on to the agar and incubated at room temperature for 7 days. A clearing develops around the colonies of those capable of solubilizing phosphate.

Pikovskaya's Medium (Modified by Sundara Rao and Sinha, 1963)

Glucose	10.0 g
Tricalcium phosphate	5.0 g
$(NH_4)_2SO_4$	0.5 g
KCl	0.2 g

MgSO₄·7H₂O	0.1 g
MnSO₄	trace
FeSO₄	trace
Yeast extract	0.5 g
Agar	15.0 g
Distilled water	1000 ml

Clearing around the bacterial growth indicates PO_4 solubilization.

VIII. Media for Iron and Manganese Bacteria

Iron Bacteria

(NH₄)₂SO₄	0.5 g
NaNO₃	0.5 g
MgSO₄·7H₂O	0.5 g
K₂HPO₄	0.5 g
CaCl₂	0.2 g
Ferric ammonium citrate	10.0 g
Agar	15.0 g
Distilled water	1000 ml

Manganese Bacteria (Bromfield, 1956)

KH₂PO₄	0.005 g
MgSO₄·7H₂O	0.002 g
(NH₄)₂SO₄	0.010 g
Ca₃(PO₄)₂	0.010 g
Difco yeast extract	0.005 g
MnSO₄·4H₂O	0.005 g
Distilled water	100 ml

pH adjusted to 6.0. Medium is autoclaved at 10 lbs for 15 minutes. To solidify the medium add 2% Difco agar.

Selected References

Allen, O.N. 1949. *Experiments in Soil Bacteriology*. 1st Edition. Burgess Publishing Co., Minneapolis, Minn.

Angelotti, R., Hall, H.E., Foter, M.J. and Lewis K.H. 1962. Quantitation of *Clostridium perfringens* in foods. *Appl. Microbiol.*, 10, 93–199.

Becking, J.H. 1959. Nitrogen fixing bacteria of the genus *Beijerinckia* in South African Soils. *Pl. Soil*, 11, 193–206.

Bernaerts, M.J. and De Ley, J. 1963. A biochemical test for crown gall bacteria. *Nature*, Lond., 197, 406–407.

Bromfield, S.M. 1956. Oxidation of manganese by soil microorganisms. *Austr. J. Biol. Sci.*, 9, 238–252.

Bunting A.H. and Horrocks, J. 1964. An improvement in the Raggio technique for obtaining nodules on excised roots of *Phaseolus vulgaris* L. in culture. *Ann. Bot.,* 28, 229–237.

Campelo, A.B. and Dobereiner, J. 1969. Soil Biology. *Int. News Bull.,* 11, 40–44.

Chu, S.P. 1942. The influence of the mineral composition of the medium on the growth of planktonic algae. 1. Methods and culture media. *J. Ecol.,* 30, 284–325.

Diem, H.G. and Dommergues, Y.R. 1988. Isolation, characterization and cultivation of *Frankia*. In *Biological Nitrogen Fixation—Recent Developments,* pp. 227–254, Ed. N.S. Subba Rao, Oxford and IBH Publishing Company, New Delhi.

Fred, E.B. and Waksman, S.A. 1928. *Laboratory Manual of General Microbiology.* McGraw Hill Book Co., Inc., New York and London.

Garren. K.H. 1938. Studies on *Polyporus abietinus.* II. Utilization of cellulose and lignin by the fungus. *Phytopathology,* 28, 875–878.

Harrigan, W.F. and McCana, M.E. 1966. *Laboratory Methods in Microbiology.* Academic Press, London and New York.

Hofer, A.W. 1935. Methods for distinguishing between legume bacteria and their most common contaminant. *J. Amer. Soc. Argon,* 27, 228–230.

Jensen, H.L. 1942. Nitrogen fixation in leguminous plants. II. Is symbiotic nitrogen fixation influenced by *Azotobacter? Proc. Linn. Soc., N.S.W.,* 57, 205–212.

Johnson, L.F., Curl, E.A., Bond, J.H. and Fribourg, H.A. 1959. *Methods for Studying Soil Microflora—Plant Disease Relationships.* Burgess Publishing Co., Minneapolis, Minn.

Katznelson, H. and Bose, B. 1969. Metabolic activity and phosphate dissolving capability of bacterial isolates from wheat roots, rhizosphere and non-rhizosphere soil. *Can. J. Microbiol.,* 5, 79–85.

Martin, J.P. 1950. Use of acid, rose Bengal and streptomycin in the plate method for estimating soil fungi. *Soil Sci.,* 69, 215.

Okon, Y., Albrecht, S.L. and Burris, R.H. 1977. Methods for growing *Spirillum lipoferum* and for counting it in pure culture and in association with plants. *J. Appl. Environ. Microbiol.,* 33, 85–88.

Pringsheim, E.G. 1964. *Pure Cultures of Algae: Their Preparation and Maintenance.* Hafner Publishing Co., New York and London.

Singh, B.N. 1946. Silica jelly as a substrate for counting holozoic protozoa. *Nature,* Lond., 157, 302.

Sundara Rao, W.V.B. and Sinha, M.K. 1963. Phosphate dissolving organisms in the soil and rhizosphere. *Ind. J. agric. Sci.,* 33, 272–278.

Thom, C. and Raper, K.B. 1945. *A Manual of the Aspergilli.* Williams and Wilkins Co., Baltimore, U.S.A.

Thornton, H.G. 1922. On the development of a standardized agar medium for counting soil bacteria with special regard to the repression of spreading colonies. *Ann. appl. Biol.,* 2, 241–274.

Tuite, J. 1969. *Plant Pathological Methods: Fungi and Bacteria.* Burgess Publishing Co., Minneapolis, Minn.

Vincent, J.M. 1970. *A Manual for the Practical Study of the Root Nodule Bacteria.* IBP Handbook, No. 15, Blackwell Scientific Publications, Oxford and Edinburg.

Wheeler, C.T. 1984. Frankia and its symbiosis in non-legume (actionorhizal) root nodules, pp. 173–195. In *Current Developments in Biological Nitrogen Fixation.* Ed. N.S. Subba Rao, Oxford and IBH Publishing Co., New Delhi.